Advances in Architectural Geometry 2014

Philippe Block • Jan Knippers • Niloy J. Mitra •
Wenping Wang
Editors

Advances in Architectural Geometry 2014

Springer

Editors

Philippe Block
ETH Zurich
Inst. for Technology in Architecture
Zurich
Switzerland

Jan Knippers
University of Stuttgart
Inst. of Building Structures and
Structural Design
Stuttgart
Germany

Niloy J. Mitra
University College London
Dept. of Computer Science
London
United Kingdom

Wenping Wang
The University of Hong Kong
Dept. of Computer Science
Hong Kong
Hong Kong SAR

ISBN 978-3-319-11417-0 ISBN 978-3-319-11418-7 (eBook)
DOI 10.1007/978-3-319-11418-7
Springer Cham Heidelberg New York Dordrecht London

Library of Congress Control Number: 2014957535

Mathematics Subject Classification (2010): 00A67, 97M80

Cover illustration: "Constructing Complex Geometries: A Case Study on the Cité des Civilisations du Vin in Bordeaux, France" by B. Soquier, R. Mizzi, D. Dureisseix and JB. Valette with kind permission by XTU architects 2013, Paris.

Printed on acid-free paper

Springer is part of Springer Science+Business Media (www.springer.com)

Preface

Architectural geometry lies at the core of the architectural design process, playing important roles from the initial form finding to the final construction phase. Its multi-disciplinary nature provides a world stage for dialogues between architecture practitioners, engineers, geometers, and computer scientists, as testified by the multitude of innovative solutions and new challenges arising from this fruitful exchange.

We are pleased to bring the fourth edition of the *Advances in Architectural Geometry 2014* (AAG'14) to London, arguably the epicentre of architectural geometry with an immense density of advanced and prolific architecture, engineering practices, design firms, and universities that push development in this exciting field.

The *Advances in Architectural Geometry 2014* is the fourth version of the conference series, which was founded under the guidance of Helmut Pottmann and colleagues in 2008. The wealth of knowledge and experience to be shared in the program at AAG'14 reflects the success of the initial purpose and wide interests across the communities of architecture design, structural engineering, computational geometry and applied geometry.

We invited widely recognized and highly distinguished experts in the field as keynote speakers. Cristiano Ceccato is responsible for the geometrical and technological development of a wide series of internationally recognized iconographic buildings of Zaha Hadid Architects and contributed to the development of rule based design processes and parametric form finding in digital design processes; Daniel Bossia, Director at AKT II and Head of the specialist team at P.art®, has both structural engineering and architecture degrees, and is an expert in computational and structural design with extensive knowledge in programming, form finding and non-linear analysis; Behrokh Khoshnevis, Professor at University of Southern California, is widely regarded as a pioneer in contour crafting that represents culmination of knowledge across robotics, motion planning, and civil engineering leading to smart robots that can possibly automate the construction of single buildings or colonies of buildings in the very near future; and finally, Mark Pauly, Professor of Computer Science at the Ecole Polytechnique Federal de Lausanne, brings us back to core studies in geometry and computational approaches to form finding.

The papers co-chairs with the help of a scientific committee have supervised and curated the scientific program of AAG'14. From a pool of 65 submissions, 24 papers have been accepted and included in the present proceedings published by Springer. In addition, 12 workshop proposals and 20+ posters have been accepted. We would like to thank all the authors, reviewers, keynote speakers, and workshop lecturers, and the attendants who have made tremendous efforts to contribute to the exciting program of AAG'14. We would especially like to thank the University College London (UCL) and the sponsors who have made this AAG financially viable. We also thank Dr. Loretta Choi for her excellent administrative support during the paper submission process, and Stephen Merchant, Dawn Bailey, and Moos Heuting from the UCL for their untiring support during the months of preparation leading to the actual conference.

Stuttgart, Germany — Jan Knippers
Zurich, Switzerland — Philippe Block
Hong Kong, Hong Kong SAR — Wenping Wang
Vienna, Austria — Alexander Schiftner
London, UK — Niloy J. Mitra

Contents

Simulation of Aggregate Structures in Architecture: Distinct-Element Modeling of Synthetic Non-convex Granulates

Karola Dierichs and Achim Menges

Abstract Aggregate Architectures are full-scale structures made from large numbers of non-convex, geometrically interlocking designed granules. They form a novel class of material systems which are in many ways directly opposed to conventional architectural assembly systems. Whereas in an assembly structure both local parts and global formation can be clearly defined, aggregates can only be observed in their behavior as a granular mass. Thus one of the core challenges in working with granulates is the development of appropriate tools of observation. Both experiments and simulations are applied and need to be used in combination with each other. In this context the paper will present the most recent development of Distinct-Element Modeling (DEM) Simulations for Aggregate Structures. Previous results have been presented in terms of the geometric principle used to compute the individual grain and its contacts. The new results shown here lay their focus on developing accurate models simulating the construction process as well as specified load cases. Initially the overall field of Aggregate Architecture will be introduced. Consequently a brief description of Distinct-Element Modeling in general and for non-convex granulates in specific will be given. The exact modeling approach for a large excavated dome structure will be introduced both in its concepts and detailed parametric settings. The results of this simulation will be discussed and areas of further developments indicated.

1 Introduction: Aggregate Architecture

Aggregate Architectures are architectural material systems consisting of large numbers of elements in loose contact, which are at the same time performing typically architectural tasks, such as defining spatial organizations, showing structural stability or modulating light transmission and thermal insulation (Dierichs and Menges 2012) (Fig. 1).

K. Dierichs (✉) • A. Menges
Institute for Computational Design, University of Stuttgart, Stuttgart, Germany
e-mail: karola.dierichs@icd.uni-stuttgart.de; achim.menges@icd.uni-stuttgart.de

P. Block et al. (eds.), *Advances in Architectural Geometry 2014*,
DOI 10.1007/978-3-319-11418-7_1

Fig. 1 Aggregate vault 2013 and 3D laser-scan of the structure (K. Dierichs, ICD, A. Scheider, IIGS)

These type of systems are termed aggregates or granulates in the field of soft matter physics (Neddermann 2005, p. 1, Jaeger et al. 1996). They are mainly known from natural systems like sand or snow (Bagnold 2005; Nicot 2004). In an architectural context they are only rarely deployed in their loose form but most commonly known as an addition in a binding concrete matrix (Hensel and Menges 2008). Yet an aggregate's capacity to fully re-cycle, form multiple stable states and be rapidly configured, makes loose granulates a very relevant field of investigation in an architectural context. Especially if the individual grain is artificially made, it can be geometrically defined and thus functionally graded in its architectural performance, mostly and in the first place with regards to structural behavior but also integrating for example light transmission or heat transfer.

Previous applications of loose granulates in an architectural context range from Building Physics (Hausladen et al. 2006), Earth Architecture (Houben and Guillaud 1994) and Geo-Engineering (Hensel and Menges 2006d; Hensel et al. 2010; Trummer 2008) to Form-finding (Gaß and Otto 1990; Hensel and Menges 2006a, b, c, d). Especially in the latter field, the notion of an unbound granulate as a structural-architectural system in its own right has been introduced. Frei Otto at the Institute for Lightweight Structures at the University of Stuttgart conducted initial tests with sand (Gaß and Otto 1990). These experiments were later taken on and further developed at the Architectural Association in Diploma Unit 4 and the Emergent Technologies and Design Program (Hensel and Menges 2006a, b, c, d). Especially the introduction of Designed Granulates presented a significant widening of the potential architectural scope of these aggregate systems (Hensel and Menges 2006b, c; Tsubaki 2011). Parallel developments can in recent years be observed in the field of Granular Physics (Athanassiadis et al. 2014; Miskin and Jaeger 2013).

One of the core challenges in designing with granular systems is the development of appropriate tools for observation and quantitative measurements. Whereas in a conventional architectural assembly system the exact geometry of both local components and global form can be exactly defined by the architect, this paradigm needs to change when one works with aggregates. Granulates require a mode of designing, where the evolving formation is merely observed and consequently interacted with (Hensel and Menges 2008; Dierichs et al. 2012). Both material experiments and mathematical simulations need to be applied for that purpose and ideally combined into a joint model using methods such as laser-scanning and hull-surfaces as a basis for comparison (Dierichs and Menges 2010; Scheider 2014).

In this context, the focus of this paper is to present recent developments in the numerical simulation of non-convex synthetic granulates. The new model shown here allows for analyzing formation and behavior under load-cases of a granular vault structure formed through an excavation process using a clump-based Distinct-Element Modeling (DEM) Simulation.

Initially the DEM method will be introduced both in the wider context of particle simulations and specifically with regards to the physical-mathematical principles it is implementing. Consequently the excavation model, its geometric basics and parametric settings will be described and the results of the simulations will be discussed. Conclusively an outlook on further developments will be given both with regards to the simulations themselves and their incorporation into the wider field of Aggregate Architectures.

2 Distinct-Element Modeling (DEM) of Non-convex Granulates

Various mathematical-numerical approaches to modeling the behaviour of granular substances have been developed (Pöschel and Schwager 2005). The two main categories are Molecular Dynamics (MD) with its sub-category the Distinct-Element Method (DEM) and Event-Driven Molecular Dynamics (ED) (Pöschel and Schwager 2005, pp. 13–30 and pp. 135–136; Luding 1994). Other approaches are Rigid-Body Dynamics (RBD), which is especially wide-spread in gaming applications, or the Distinct- Simulation Monte Carlo (DSMC) (Pöschel and Schwager 2005, pp. 191–192). Depending on the application, these approaches need to be compared to each other with regards to their suitability to solve the problem in question. ED for example uses event-based time-stepping and mainly hard collisions, whereas DEM is based on a clock-based time-stepping and allows for both hard and soft collisions (Pöschel and Schwager 2005, pp. 18–22 and pp. 135–136). DEM is specifically suited for the observation of dynamic micro-mechanical behavior, DSMC for that of the probable behavior of a granulate as a whole at rest (Lanier and Radjaï 2009; Duran 2000, pp. 202–206).

For this specific application, where large amounts of synthetic non-convex granulates need to be observed, a DEM modeling approach has been chosen, since at one point in time there is a multitude of contacts and the observation of micro-mechanical behavior is relevant (Fig. 2).

The main challenge is the fact that the particles are geometrically non-convex polyhedral, whereas the DEM simulation is initially based on the computation of spheres, which allow for a comparatively easy collision detection model based on

Fig. 2 Simulation of non-convex granules which are composed of convex polyhedral elements (K. Dierichs, ICD, F. Fleissner, ITM)

the spheres' centers and radii. A non-convex body by virtue of its geometry does not allow for this relatively simple model of boundary detection. Incorporating the accurate modeling of non-convex geometries and their collisions in a DEM model is thus one of the core aims of this project.

2.1 DEM

The DEM model has first been introduced by Cundall and Strack and uses the force and torques as well as the Euler's angles of the individual elements to compute the Newtonian equations of motion (Cundall 1971; Cundall and Strack 1979; Pöschel and Schwager 2005, p. 13). The class of algorithms uses the following basic sequence of approximating the differential equations according to Pöschel and Schwager The model is initialized with the particle coordinates (1), consequently the coordinates of the particles at time $t + \Delta t$ are predicted (2), then forces are computed by defining the interacting pairs and their contact forces (3), coordinates are corrected based on the previous results (4), data are extracted (5) and finally the program is terminated (6) (Pöschel and Schwager 2005, p. 28).

2.2 Clump Modeling of Non-convex Granulates

As introduced earlier, the simulation of non-convex polyhedral particles is a challenge within DEM models due to the complexity of collision detection. Two basic

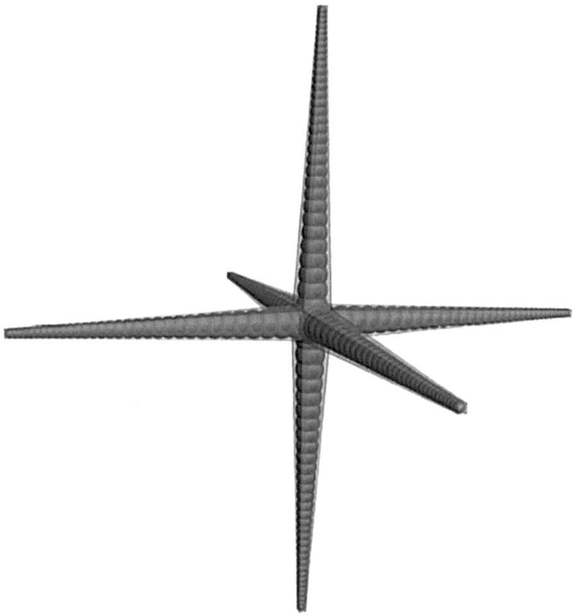

Fig. 3 Clump model of a six-armed, non-convex granulate showing the pebble distribution with 32 pebbles per arm (M. Purvance, ITASCA)

geometric modeling approaches for this problem exist and have been benchmarked (Dierichs and Menges 2013). The first one uses a so-called 'clump'-model, which means that the non-convex shape is composed of individual spheres, also called 'pebbles', which are rigidly bound to each other (Fig. 3).

The second one uses convex polyhedral or 'blocks' to compose the non-convex granule. Benchmarking has shown that for particle numbers below 1,000 the block-model is faster, yet above that critical number the speed of computation increases in the clump-model and it is actually faster than the block-based one (Dierichs and Menges 2013).

Since any of the aggregate systems which are used in this context has at least 1,000 particles, the clump-based simulation method has been chosen. The basic particle is a six-armed cumulated cube measuring 10 cm in diameter. For the purpose of the simulations presented here it has been modeled using 20 pebbles per arm, yet pebble-amount needs to be reduced if particle numbers further increase. The pebbles are references to a surface model of the polyhedral hull, such that the actual surface is used to compute the moments of inertia.

3 Methods: The Hopper- and Excavation-Model

The physical construction experiment has been conducted at a one-to-one scale: Fifty-thousand synthetically produced particles are poured into a tank with a bottom-lid measuring 2.25 m by 2.25 m by 1.35 m which is placed on a sub-construction measuring 0.75 m in height. The bottom is consequently removed, parts flow out and un-loaded material is excavated to leave a dome made of only 10 % of the entire material. The redundant parts are entirely re-usable in this process.

In order to develop a simulation of this process a hopper- and excavation-model has been set-up at a smaller scale using 1,284 particles with the bounding box measuring 600 by 600 by 850 mm. It uses a clump generate command to distribute the particles in the box. This implies that particles are generated individually and consequently their settling is computed which allows for exact computation of contacts and forces.

Material calibration values have been gathered by experiment and on that basis set to a friction coefficient of 0.58 and normal viscous damping of 0.25, the need for deformation modeling has been excluded using high-speed photography showing that energy is dissipated within the granulate (Fig. 4).

The clump template allows for setting different pebble numbers to model the arms, thus for higher numbers less pebbles can be chosen to accelerate the model at the expense of accuracy. After simulation of the excavation process, large and small planar loads are applied to the simulated dome.

The adjustable parameters of the model are thus the material friction, damping, particle surface geometry, pebble number per clump, clump number per model, the cycling, the excavation radius, as well as the size of the planar load.

Fig. 4 High-speed analysis at 2,000 fps of a six-armed granule falling onto a pile (K. Dierichs, ICD, S. Poppinga and M. Thiele, PBMG Freiburg)

4 Results

4.1 Simulation of Hopper and Excavation

The entire sequence is split in four parts: simulation of the granular mass in the bounding box, removal of the bounding box and bottom lid, excavation of material from below and loading of the aggregate (Fig. 5).

The model is set to 510,000 fixed time-step cycles and n cycles for close packing in total. The clump generation and settling is computed using an equilibrium parameter to achieve dense packing in n numbers of cycles. The hopper and excavation are both computed for 5,000 cycles each and the planar loading is cycled for 500,000 times. With an average clump number of 1,284, pebbles are set to only 20 per arm in order to accelerate computing time. The overall model run takes an average of 120 min clock time to compute.

An analysis of the clump surface velocity shows that the granular mass comes to rest at the bottom layers first in stage 1. After removal of the lid, the structure

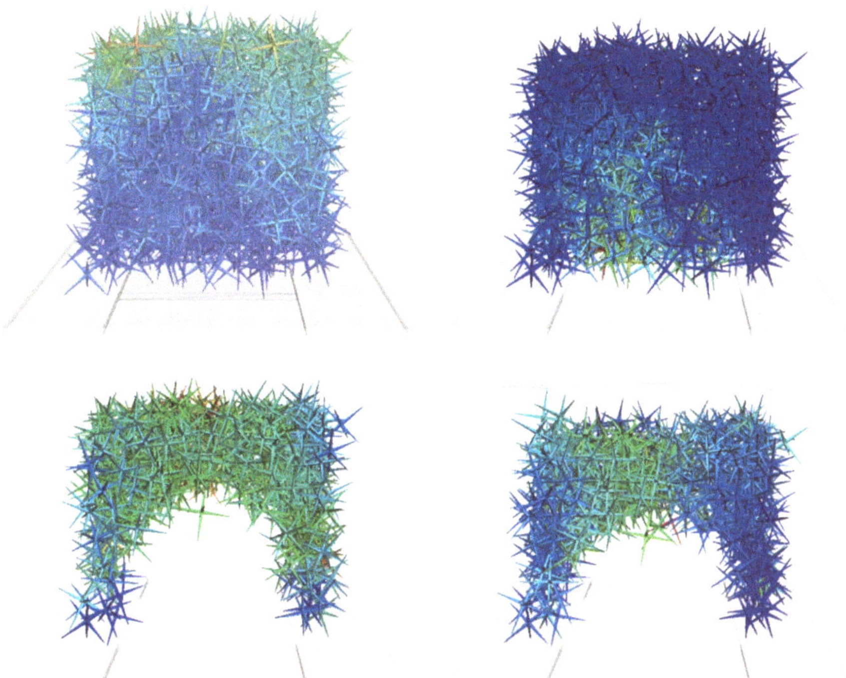

Fig. 5 Hopper- and excavation-sequence in a surface velocity plot shown in a filtered section of the entire granulate formation with the scalar ranging from *blue* indicating low speeds to *red* indicating high ones. *Top left*: simulation of the clump distribution, *top right*: removal of lid and bounding box, *bottom left*: excavation of 200 mm, *bottom right*: loading with a large planar load

is excited at the bottom of the hopper, which proliferates into the top middle layers.

After excavation the motion moves almost symmetrically throughout the entire vault-structure yet the dome does not collapse. Planar loading compresses the configuration with surface speeds remaining higher over the excavated core whereas the corner regions of the aggregate are remaining at relative rest (Fig. 5).

4.2 Load Cases and Excavation Radii

Both excavation radii and dimensions of planar loads can be varied. A set of simulations has been run that uses an excavation radius of 100 and 200 mm in combination with planar loads of 50 by 50 mm and 500 by 500 mm each weighing 30,000 g (Fig. 6). The load is simulated using a servo-control mechanism.

Excavation radius and consequent thickness of the remaining aggregate vault affect the load bearing behavior under both a large and a small planar load, with

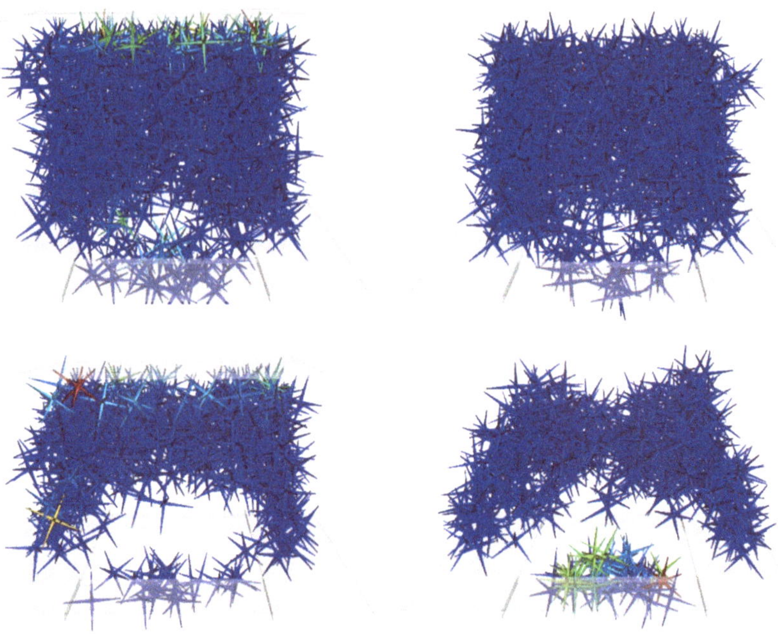

Fig. 6 Load cases and excavation radii in a surface velocity plot shown in a filtered section of the entire simulated granulate formation with the scalar ranging from blue indicating low speeds to *red* indicating high ones. *Top left*: large planar load with 100 mm excavation radius, *Bottom left*: large planar load with 200 mm excavation radius, *Top right*: small planar load with 100 mm excavation radius, *Bottom right*: small planar load with 200 mm excavation radius

thinner structures deforming stronger under the same pressure, which is an expected result.

Load dimensions lead to different effects as well. A large planar load tends to compress yet not deconstruct or deform the current formation, whereas a small planar load strongly deforms or destroys the aggregate vault. Whether or not the common notion of destruction is appropriate within the design theoretical discourse of aggregate structures as reconfigurable systems is of course to be discussed.

4.3 *Statistical Simulation of Contact Forces*

Contact force can give one indication of how loads are transferred in a designed granulate on a micro-mechanical level. A statistical set of 20 simulations has been run in order to understand the behavior of the six-armed granule under compression in the excavation sequence described in Sect. 4.1 (Fig. 7).

Contact forces have been measured in the centre of the supportive columns using a vector-definition to select the clump measured. Most contact force diagrams show a peak load in the early stages of cycling and consequently plateau with minor oscillations. Only a small fraction of the probes deviate from that pattern (Figs. 8 and 9).

This type of statistical testing of micro-mechanical behavior is especially relevant with regards to analyzing how a variation in particle geometry affects the structural performance of a granulate made from designed particles.

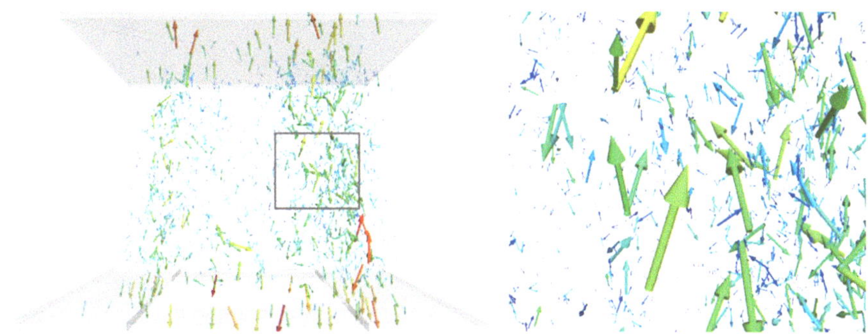

Fig. 7 Global view and detail view of contact forces in a granular probe with a large planar load after 50,000 cycles

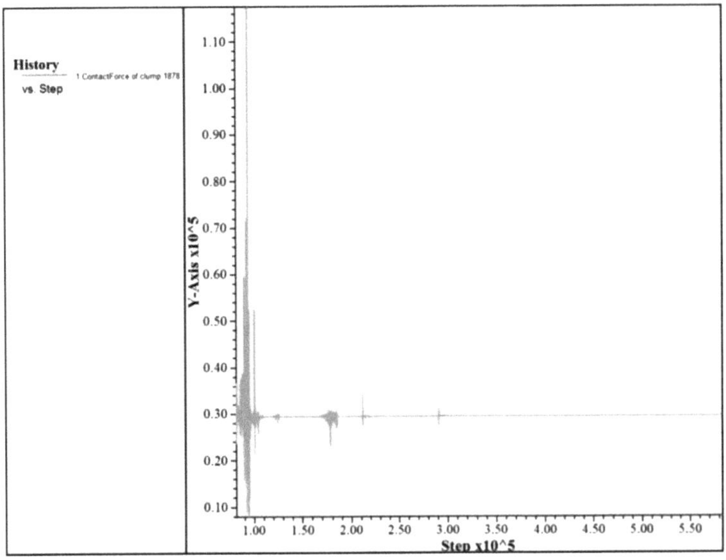

Fig. 8 Typical load response of a single particle, this behavior can be observed in 14 out of 20 probes

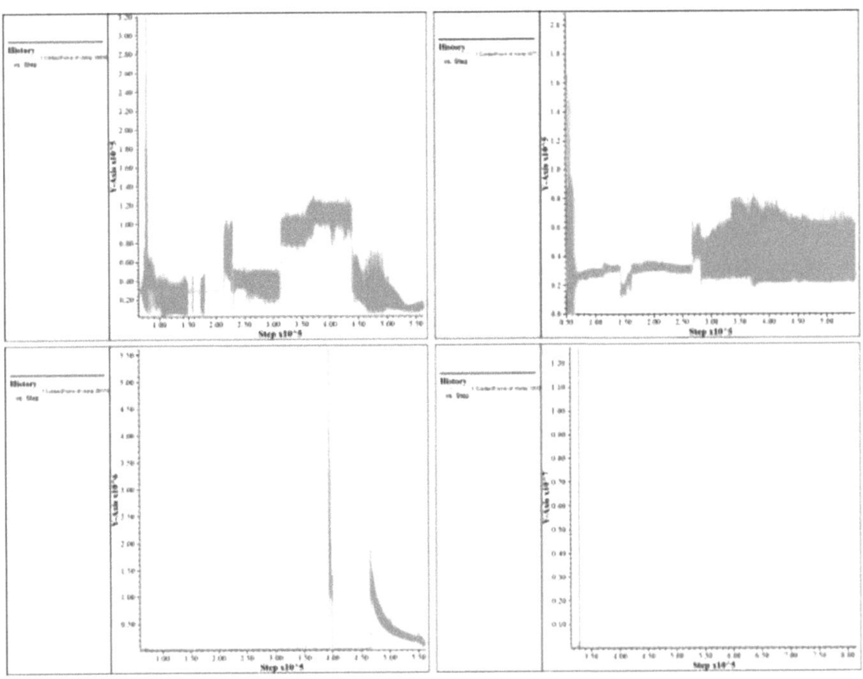

Fig. 9 Samples of four load responses of individual clumps that deviate from the typical pattern

5 Discussion

The model developed allows for the numerical analysis of both the overall and the micromechanical behavior in a designed granulate during construction and under load conditions. It is especially suited to analyze variants in particle morphologies, pouring conditions and load cases.

Challenges are the further calibration of the simulation model with experiments in order to render results more reliable especially when extremely large numbers of particles are simulated and computational time increases. Due to the fact that granulates allow for merely probable predictions of their behavior these comparisons need to be based on statistical measures.

6 Outlook

Future developments of this research project include the numerical comparison with physical experiments using hull-models over point clouds generated from both the simulations and the 3D scans. These will be used for the above-mentioned calibration and for deformation modeling of the Aggregate Structures.

Another branch of research will be the simulation of extremely large systems of up to 100,000 particles and the development of suitable geometrical modeling strategies.

Acknowledgements The authors would like to thank Sacha Emam, Lothar te Kamp, Matt Purvance and Reza Taghavi of ITASCA Consulting Inc. for their great support in the simulation model development as well as the ITASCA Education Partnership Program for providing the software. They would also like to thank Florian Fleissner, Institute of Engineering and Computational Mechanics at the University of Stuttgart for initial simulations in Pasimodo as well as Anette Scheider and Professor Volker Schwieger at the Institute of Engineering Geodesy, University of Stuttgart for their ongoing collaboration in 3D-scanning of the Aggregate Structures. They would like to express their gratitude to Simon Poppinga and Marc Thiele of the Plant Biomechanics Group, University of Freiburg for their support with the high-speed photography of granules in free-fall.

References

Athanassiadis, A., Miskin, M., Kaplan, P., Rodenberg, N., Lee, S., Merritt, J., Brown, E., Amend, J., Lipson, H., Jaeger, H.: Particle shape effects on the stress response of granular packings. Soft Matter **10**, 48–59 (2014)

Bagnold, R.: The Physics of Blown Sand and Desert Dunes, 1st edn. 1954, 2nd ed. Dover Publications, Mineola (2005)

Cundall, P.A.: A computer model for simulating progressive large-scale movements in blocky rock systems. In: Proceedings of the Symposium of the International Society of Rock Mechanics, Nancy, 1971. 1: paper no. II-8

Cundall, P.A., Strack, O.D.L.: A discrete numerical model for granular assemblies. Géotechnique **29**(1), 47–65 (1979)
Dierichs, K., Menges, A.: Aggregate structures: material and machine computation of designed granular substances. In: Menges, A. (ed.) Architectural Design, Special Issue: Material computation-Higher integration in morphogenetic design, pp. 74–81. Wiley, London (2012)
Dierichs, K., Menges, A.: Material computation in architectural aggregate systems. In: Formation, Proceedings of the 30th Conference of the Association for Computer Aided Design (ACADIA), pp. 372–378. New York (2010)
Dierichs, K., Schwinn, T., Menges, A.: Robotic pouring of aggregate structures – responsive motion planning strategies for online robot control of granular pouring processes with synthetic macro-scale particles. In: Proceedings of the Robots in Architecture Conference 2012, pp. 196–205. Springer, Vienna (2012)
Dierichs, K., Menges, A.: Aggregate architecture: simulation models for synthetic non-convex granulates. In: Beesley P., Khan O., Stacey M. (eds.) Proceedings of the 33rd Annual Conference of the Association for Computer Aided Design in Architecture (ACADIA) – Adaptive Architecture, pp. 301–310. Waterloo/Buffalo/Nottingham (2013)
Duran, J.: Sands, Powders and Grains: An Introduction to the Physics of Granular Materials. Springer, New York (2000)
GAß, S., OTTO, F.: Experimente/Experiments, Form-Kraft-Masse 5/Form-Force-Mass 5. Mitteilungen des Instituts für leichte Flächentragwerke (IL) Universität Stuttgart Nr. 25/Information of the Institute for Lightweight Structures (IL) University of Stuttgart Nr. 25. Karl Krämer Verlag, Stuttgart (1990)
Hausladen, G., de Saldanha, M., Liedl, P.: ClimaSkin: Konzepte für Gebäudehüllen, die mit weniger Energie mehr leisten. Callway, München (2006)
Hensel, M., Menges, A.: Hani Fallaha – adaptive pneumatic shelters 2003–2004. In: Hensel, M., Menges, A. (eds.) Morpho-ecologies, pp. 232–241. AA Publications, London (2006a)
Hensel, M., Menges, A.: Eiichi Matsuda – aggregates 01 2003–2004. In: Hensel, M., Menges, A. (eds.) Morpho-ecologies, pp. 262–271. AA Publications, London (2006b)
Hensel, M., Menges, A.: Anne Hawkins and Catie Newell – aggregates 02 2004. In: Hensel, M., Menges, A. (eds.) Morpho-ecologies, pp. 274–283. AA Publications, London (2006c)
Hensel, M., Menges, A.: Gen Takahashi – aggregates 03 2005–2006. In: Hensel, M., Menges, A. (eds.) Morpho-ecologies, pp. 286–295. AA Publications, London (2006d)
Hensel, M., Menges, A.: Aggregates. In: Hensel, M., Menges, A. (eds.) Architectural Design, Special issue: Versatility and Vicissitude, Performance in Morpho-ecological Design, pp. 80–87. Wiley, London (2008)
Hensel, M., Menges, A., Weinstock, M.: Emergent Technologies and Design: Towards a Biological Paradigm for Architecture. Routledge, Abingdon (2010)
Houben, H., Guillaud, H.: Earth Construction: A Comprehensive Guide. Intermediate Technology Publications, London (1994)
Jaeger, H., Nagel, S., Behringer, R.: Granular solids, liquids, and gases. Rev. Mod. Phys. **68**(4), 1259–1273 (1996)
Lanier, J., Radjaï, F.: Experimental and numerical analysis of local variables in granular materials. In: Micromechanics of Granular Materials, pp. 1–49. STE Ltd./Wiley, London/Hoboken (2009)
Luding, S.: Modelle und Simulationen granularer Materialien (models and simulations of granular materials). Ph.D. Dissertation, University of Freiburg, Freiburg, 1996/519 (1994)
Miskin, M., Jaeger, H.: Adapting granular materials through artificial evolution. Nat. Mater. **12**, 326–331 (2013)
Nedderman, R.M.: Statics and Kinematics of Granular Materials. Cambridge University Press, Cambridge (2005 [1992])
Nicot, F.: Constitutive modelling of snow as a cohesive-granular material. Granul Matter **6**(1), 47–50 (2004)
Pöschel, T., Schwager, T.: Computational Granular Dynamics: Models and Algorithms. Springer, Berlin/Heidelberg (2005)

Scheider, A.: Detecting und modeling fine structures from TLS data. In: Proceedings on International Workshop: Integration of Point- and Area-wise Geodetic Monitoring for Structures and Natural Objects. Russian Federation, Novosibirsk (2014)

Trummer, P.: Engineering ecologies. In: Hensel, M., Menges, A. (eds.) Architectural Design, Special issue: Versatility and Vicissitude. Performance in Morpho-ecological Design, pp. 96–101. Wiley, London (2008)

Tsubaki, K.: Tumbling units – tectonics of indeterminate extension. In: Gail, P., Meredith, M. (eds.) Matter: Material Processes in Architectural Production, pp. 187–203. Routledge Press/Taylor & Francis, London (2011)

Post-tensioned Discrete Concrete Elements Developed for Free-form Construction

Ole Egholm Pedersen, Niels Martin Larsen, and Dave Pigram

Abstract This paper presents a method for the construction of non-uniform precast concrete shell structures from unique parts. A novel method of discontinuous post-tensioning is introduced which allows tension to be taken through the connections. This increases the formal possibilities of the system beyond compression-only funicular forms. It also allows live loading to be catered for without significant thickening of the structure. Most significantly, the post-tensioning system allows for an almost total elimination of falsework a significant impediment to the realistion of complex shell structures. This advancement reduces resources needed for assembly, and increases the overall robustness of the system.

Also presented are a series of refinements that add to the geometric precision of individual concrete elements cast in lasercut PET moulds. These include the development of a casting rig used to fix casting moulds in the right position during casting and algorithms to produce mold details to mediate between the unique components and the repeated rig forms.

The last design experiment is the most elaborate, demonstrating a pavillion structure which incorporates all areas of development. The successful execution of this design experiment allows us to conclude that the building system in its current state of development is ready to be tested in a scale larger than the typical research pavilion.

O.E. Pedersen (✉) • N.M. Larsen
Aarhus School of Architecture [AAA], Norreport 20, 8000 Aarhus C, Denmark
e-mail: oleegholm.pedersen@aarch.dk; niels.martin.larsen@aarch.dk

D. Pigram
Graduate School of Architecture, Preservation and Planning [GSAPP], Columbia University, New York, NY, USA

University of Canberra, Canberra, Australia

University of Technology Sydney, Sydney, Australia

Supermanoeuvre, 42/56-60 Foster Street, Surry Hills, Sydney, Australia
e-mail: dave@supermanoeuvre.com

P. Block et al. (eds.), *Advances in Architectural Geometry 2014*,
DOI 10.1007/978-3-319-11418-7_2

1 Introduction

The framework provided by industrialisation and mass-production yielded architecture of repeated, standardised elements. The dominance of this framework is waning as the logistical impediments to a return to individualisation in our built environment are overcome. Within this context, this paper describes a method for the design and construction of complex gridshell structures in precast concrete comprising unique parts (Fig. 1). It is the ambition of the research to take advantage of concrete's amorphous qualities, and to challenge the prevailing design of concrete elements as standardised, flat, rectilinear panels while addressing key feasibility issues (Larsen 2012).

The process of defining the geometry of the components and extracting fabrication information is carried out through use of bespoke algorithms, i.e. custom written Python scripts in McNeel's Rhinoceros 3D environment. Each component's unique geometry is calculated, unrolled into flat shapes, and all parts are nested into digital drawings representing cutting templates. The mould material utilised is amorphous Polyethylene Terephthalate [aPET] plastic, normally used for packaging and soda bottles. The aPET plastic sheets are then laser cut and folded to produce the unique casting templates.

The overarching goal has been to try to establish a robust building system that may become applicable to larger architectural structures. A series of design experiments are examined, and the state of the system is demonstrated in a pavilion realised in Sydney in March 2014.

Fig. 1 Form-found structure comprising unique precast concrete elements (Design Experiment 3)

1.1 Contributions

The advances presented in this paper include:

- The design of a discontinuous post-tensioning system allowing for tension to be taken across joints both during and after installation.
- The elaboration of numerous techniques to increase part accuracy during casting.

1.2 Related Work

This research extends a method published in 2012 by the authors for constructing free-form structures from post-tensioned discrete precast concrete elements as demonstrated in the Pre-Vault pavilion constructed in 2011 (Larsen et al. 2012a). That method demonstrated that digital form finding techniques, computational file-to-fabrication workflows and innovative sustainable concrete casting techniques can be integrated to construct funicular structures using prefabricated elements in a practical, affordable and materially efficient manner. That method was tested with compression-only structures, shaped with custom written dynamic relaxation software (Day 1965), based on an interactive near real-time particle-spring simulation similar to CADenary (Kilian and Ochsendorf 2005). Similar ways of using digital form-finding with digital production techniques in order to construct shell and funicular structures have been carried out by the Block Research Group (Block et al. 2010). Some of these case studies have focused on constructing thin tile shells through use of cardboard falsework (Block et al. 2011), as did Larsen et al. in 2012a (Fig. 2). The elimination of this falsework is a key contribution here. Other related work by the Block Research Group includes the creation of thin shells using flexible fabric formwork and alternate method for addressing the viability of freeform concrete structures, they have also published a detailed history of the use of flexible formwork which is relevant (Block and Veenendaal 2014). Finally, the most obvious and significant related work towards the realisation of free-form concrete shells comes in the many built examples by Hans Isler (1960). These were created onsite with insitu-concrete. Issues with the control of cracking and reliance on rigid formwork are avoided with the precast method here described.

Fig. 2 Pre-Vault Pavilion constructed in 2011 as demonstrative of the previously published method for producing a concrete gridshell (Larsen et al. 2012a)

2 Towards Increased Buildability

Shell and vault structures can be extremely materially efficient. Shells due to the incorporation of membrane action (Block and Veenendaal 2014) and vaults through the total avoidance of bending in compression-only form. However, the realisation of such structures is typically heavily reliant on falsework that is custom fabricated to fit unique geometries. The production of this falsework is costly in terms of material, time and labour (physical and intellectual), and represents a significant impediment to the feasibility of architectural shells.

The reliance on falsework is due to the structure not achieving its ultimate structural performance until it is complete, and to the geometry of the shell relying on the falsework (or guidework) for its definition. In Larsen et al.'s earlier published case study this challenge was met in the typical manner: by designing and building an extensive falsework to support the elements during construction (Fig. 3). However, the typical problem of geometric definition was solved via the use of unique parts which each embed the structure's final form in their own.

This method avoids the need for falsework by employing a novel post-tensioning system that enables the transfer of tensile forces between members during construction.

Fig. 3 In earlier experiments extensive falsework was needed for construction

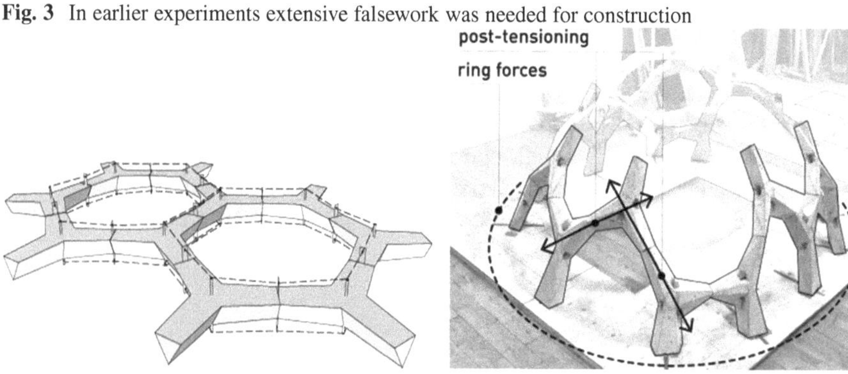

Fig. 4 *Left*: external post-tensioning. *Dashed lines* show the position of tension wires above and below concrete components. *Right*: internal post-tensioning. *Arrows* show the position of threaded tension rods

3 Discontinuous Post-tensioning

Two strategies for adding post-tensioning across the joints between components were tested. The first is an external post-tensioning system where cables are connected to the top and bottom of the assembly, forming loops around the hexagonal holes in the structure (Fig. 4 Left). The second is an internal system of discontinuous straight threaded rods that each connect between only two components (Fig. 4 Right). Both strategies deal with two conclusions drawn from Larsen et al.'s previous work:

- The concrete elements have a small cross section, and this makes it difficult to fit a detail in the center of the Y-shaped component where up to three tendons cross.
- Builders should have easy access to placing and tensioning the tendons during construction in order to maintain speed in the assembly process.

Post-tensioning has been previously proposed as a strategy to replace traditional reinforcement and to counteract live loads upon completion of concrete shell structures (Dallinger and Kollegger 2008). However their method uses continuous tendons, not optimal with a system of individual elements which need to be mounted and tied-in a few elements at time, and therefore does not alter the need for falsework.

3.1 External Post-tensioning

The external post-tensioning concept was to connect all joints and component centres with steel cables running both above and below the construction. By attaching the cables outside the concrete the cross section remains intact and the detail is easy to access during assembly. The assembly procedure was planned so six concrete parts could be assembled with tightened wire before lifting them in place. As Fig. 5 (Left) shows, it was possible to use this approach to assemble a full-scale prototype assembly. Two ends of a large structure were subsequently made without falsework, though considerable propping was necessary. However, due to the weight of the sections and difficulties in tightening the wires (especially to tighten them equally top and bottom) it was not possible to continue to mount

Fig. 5 *Left*: full-scale prototype assembly of external post-tension system. *Right*: a large-scale construction attempt is stopped due to shortcomings in post-tensioning the system

sections of six preassembled parts. It was concluded that the external tensioning system here described was not a viable solution for avoiding the use of falsework.

3.2 *Internal Post-tensioning*

In order to address the shortcomings of the external tensioning scheme, a discontinuous post-tensioning concept strategy was developed. It deals in particular with the problems of assembly described above. The concept is that elements are tied together with their closest neighbours, allowing for the addition of one element at a time and avoiding the need to lift six elements into place at once. A threaded rod forms the tendon spanning centre to centre of two adjacent elements. It is secured with a nut and washer assembly. The tensile member (threaded rod) permits each element to temporarily cantilever in space. Furthermore, live loads are counteracted by the threaded rods that serve as reinforcement once the structure is complete.

To test the principle of distributed post-tensioning, a series of physical prototypes were constructed. Of main interest was the strength of the joint, and whether it would enable the construction process to be carried out with no – or at least reduced use of falsework. It was also important to test the buildability implications of mounting and connecting the elements with straight rods passing through each.

Early attempts were made to cast the elements with an insert in the centre of each component where three tensile members could meet. This proved disadvantageous due to clearance issues as well as the accompanying reduction of mass in the centre of the element, which would reduce its compressive strength (Fig. 6, Left). The problem was addressed by allowing each tensile member to continue past the centre of the element and exit at the wrist (Fig. 6, Right).

In order to investigate the structural potential of the principle, simple fracture tests were carried out (Fig. 7, Left). The experiments indicated that there is a torque capacity of the structure during construction. In the tests, a 4 mm threaded rod was used, and the height dimension (thickness) of the concrete was 80 mm. The tests showed that it is the tension rod that breaks before the concrete is crushed, which

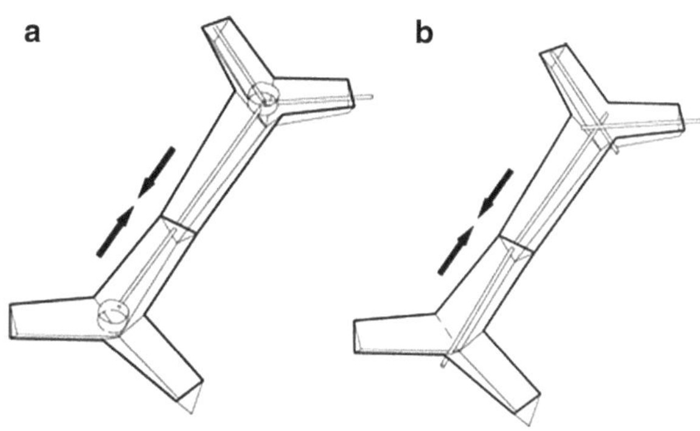

Fig. 6 *Left*: early studies of detail in the centre of each element (**a**). *Right*: final scheme (**b**). Ending tendons outside the element prevents weakening of the element in the centre

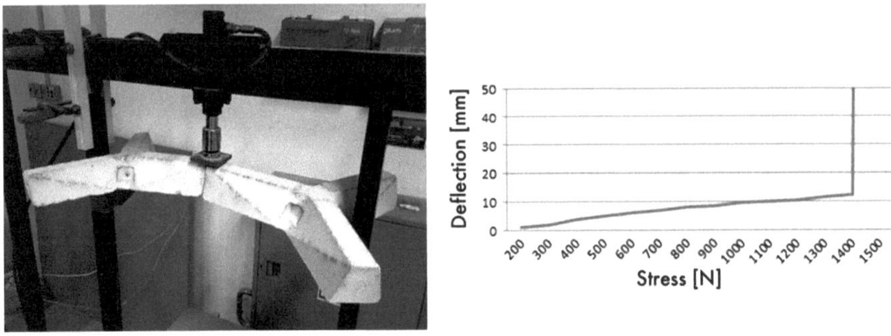

Fig. 7 *Left*: fracture test of post-tensioned concrete elements. *Right*: diagram showing 12 mm deflection before the fracture at 1,400 N

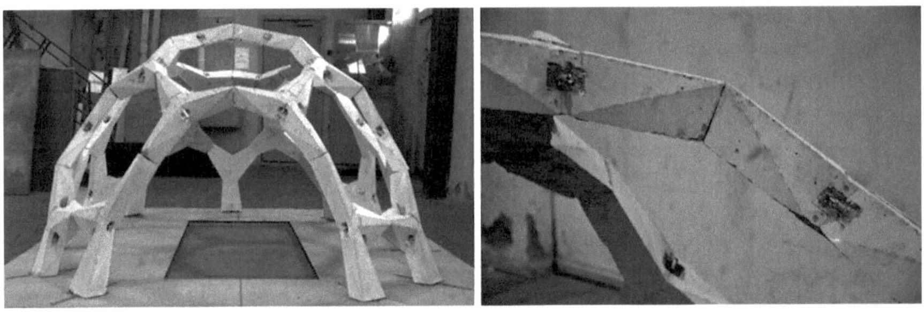

Fig. 8 *Left*: testing of distributed post-tensioning principle. *Right*: improved concrete element precision

can be seen as an advantage because the break is not sudden, but happens after gradual deflection (Fig. 7, Right). To avoid the brittle fracture, the tension rod could be embedded in the concrete. In the case of funicular structures, this may not be relevant, since the structural behaviour changes when the construction is completed, and reinforcement therefore only has to transfer secondary loads in the finished construction.

The first larger prototype assembly was made of 24 concrete components (Fig. 8, Left). It demonstrated the viability of discontinuous post-tensioning because of the strength of the joint. During construction only a few supports were needed, meaning that falsework as such could be avoided. After completion the structure was remarkably stable and precise (Fig. 8, Right) compared to previous prototypes and the Pre-Vault pavilion from 2011 (Larsen et al. 2012a).

4 Increasing Precision in Complex aPET Casting Moulds

The mould production technique used in this method is developed by O. E. Pedersen as an alternative to related means for casting geometrically complex concrete elements such as CNC milling of EPS and fabric formwork (West 2009). The technique addresses issues with precision (fabric formwork) and slow manufacturing speeds (CNC milling). Laser cutting and folding of flat sheets yields complex yet geometrically precise casting moulds in recyclable aPET plastic (Pedersen 2013). The creation of geometry involves scripting a set of parametric relationships in 3D, from which each discrete element and a corresponding unfolded version is generated (Larsen et al. 2012b). The moulds are constructed from flat laser cut sheets by means of folding; following scored and dashed lines (Fig. 9).

Despite the deliberate overall planarity and relatively shallow profile of the components, the 1–2 mm thick APET tends to deform due to the hydrostatic pressure of the concrete. The problems caused by this deformation (in order of importance) are: imprecision in the angles between arms; twisting of individual arms along the longitudinal axis, and; bulging of wrist points, which in turn drag the top endpoints inwards towards the centre (Fig. 10, Left).

A casting rig was developed as a means for preventing arm rotation and twist. Parametrically added flaps position the mould within an outer frame, fixing all three arms along the vertical and longitudinal axis (Fig. 10, Centre). The algorithm for generating the geometry of the parts and the casting templates is based on previous research by D. Pigram and N. M. Larsen (Larsen et al. 2012b). This algorithm has been further developed to be able to analyse and select from a series of possible

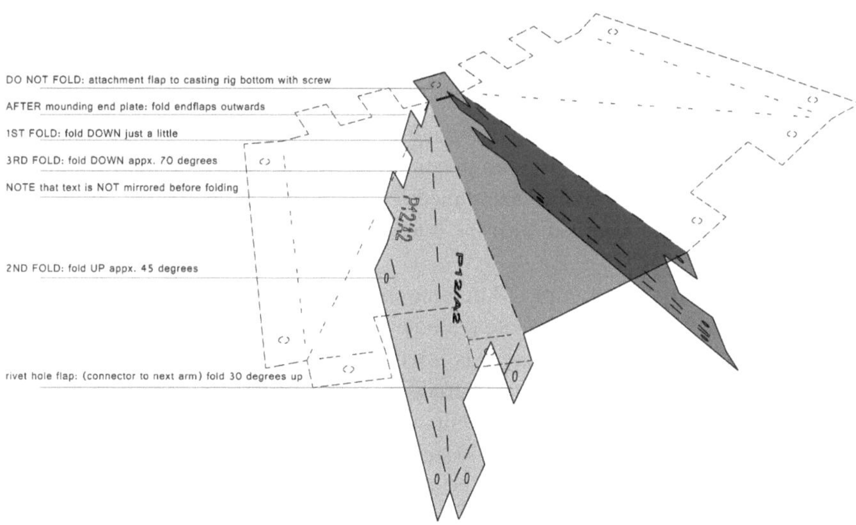

Fig. 9 Flaps and triangulations to add stability to the casting moulds

Fig. 10 *Left*: deformations prevented by casting rig and lid. *Centre*: APET mould placed in rig. *Right*: an algorithm seeks for the optimal positioning in the casting frame

orientations of the component in the casting rig (Fig. 10, Right). Additional flaps connecting the mould to the casting frame are defined and added, including the positioning of fixing points (Fig. 10, Centre). A lid connecting all three wrist points eliminates displacement around the centre. These modifications reduce tolerances to +/− 1 mm as opposed to +/− 6 mm measured in earlier experiments.

5 Case Study: Utzon 40 Pavilion

A case-study pavilion entitled *Utzon:40* was constructed in Sydney to coincide with the Utzon Symposium at the Sydney Opera House in March 2014. The structure was designed with 62 discrete concrete elements and 130 timber pieces with concrete meeting the ground. To counteract the structure's thrust forces a network of 20 mm wide and 2 mm thick aluminium bars was connected to the bottom of each base. This network ensured that horizontal forces were distributed to the other bases and thereby obtained within the system. The aluminium network also served as a setout mechanism to guide the positioning of the base elements during assembly. Figure 11 shows how the base components for the Utzon/40 Pavilion (described below) changed geometry in order to meet the floor. Behind is seen an aluminium tension floor, which is joined with the concrete structure in a recess in each base element.

The construction was completed in three steps. First, the aluminium tension floor was assembled. Second, the concrete elements we added. Along the opening arcs in the pavilions sides the concrete was lifted to form trusses above the openings to ensure structural integrity during construction. It was necessary to be able to adjust the position of previously mounted components while mounting new ones.

It was the completion of this second step that delivered the most conclusive proof of this methods success in eliminating the need for falsework. Figure 12 clearly shows the incomplete shell structure with multiple courses of precast concrete cantilevering without support. It should be noted that temporary propping was needed at various stages, especially to complete the regions above openings.

Fig. 11 Geometry change in base elements

Fig. 12 Post-tensioned concrete elements standing freely without falsework before timber is mounted

The third and final step was to complete the pavilion by mounting the timber components distributing forces across the top of the structure (Fig. 13). It is important to recognise that timber does not form part of this method. Its existence in this case study pavilion relates to the context within which the pavilion was

Fig. 13 Completed Utzon: 40 pavilion

produced. However, as the timber elements were created on a high-end gantry 5-axis CNC router, they serve as a useful guide to the precision of the concrete components and a test of this method's ability to interface with other systems. In both aspects the precast concrete system performed admirably, interfacing without issue with cumulative assembly tolerances in the order of +/− 5 mm.

The case study pavilion also revealed an area where the method could be improved. The corners of concrete elements had a tendency to be crushed due to their sharp edges and small dimensions. A chamfer of around 5 mm needs to be added to the component design.

6 Conclusion

As formulated in the introduction, an important goal for this research is to utilise digital form-generation and digital fabrication techniques to take advantage of concrete's amorphous qualities and develop a novel construction system. The recent developments of the system were mainly focused on three specific properties of the system:

- Development of a post-tensioning system
- Increasing precision in complex aPET casting moulds
- Development of complete digital workflow

Concerning the problem of post-tensioning, the solution based on distributed post-tensioning threaded rods form tendons spanning centre to centre of adjacent components permits each element to temporarily cantilever in space, allowing assembly to occur row-by-row. This fulfilled the aim of almost total elimination of falsework and instead required only simple propping with reusable elements. Furthermore, live loads are counteracted by the threaded rods serving as reinforcement once the structure is complete.

Precision of the casting principle has been improved through use of a casting rig, which prevents arm rotation and twist. Parametrically defined flaps position the mould within an outer frame, thereby ensuring the correct angles between the arms. The addition of a triangular piece spanning the centre of the open side of the mold also added fidelity. The cumulative outcome of these refinements is that a dimensional tolerance of +/−1 mm was achieved across the components in the described case study pavilion.

The research demonstrates that the construction method has approached a state where it can be used for larger, more permanent structures. Important issues such as cladding and further structural testing, however, still have to be addressed before the system is completely functional.

7 Future Work

There are numerous directions for future work. The most interesting and important is to continue to scale up the method's application to larger spans. If the components themselves increase significantly in size folded aPET moulds will cease to be appropriate and so an alternative low material investment casting method that supports the production of unique parts will need to be found. The folding of moulds remains laborious and as such is a serious barrier to the methods adoption. There remains much room for streamlining either with aPET or a replacement material. Additional work can be done to take maximum advantage of each component's uniqueness by adding for example Finite Element Analysis into the workflow to optimize material distribution in relation to live load cases, and not only to the dominant self-weight case, through the various geometric parameters of the component. As this research is targeted at architectural implementation, the design of an associated cladding system and the integration of any necessary connection details into the components remains a necessary development.

References

Block, P., Veenendaal, D.: Design process for prototype concrete shells using a hybrid cable-net and fabric formwork. Eng. Struct. **75**, 39–50 (2014)

Block, P., Lachauer, L., Rippmann, M.: Form finding to fabrication: a digital design process for masonry vaults. In: Proceedings of the IASS, Shanghai (2010)

Block, P., Davis, L., Rippmann, M., Pawlofsky, T.: Efficient and expressive thin-tile caulting using cardboard formwork. In: Proceedings of the IABSE-IASS Symposium, London (2011)

Dallinger, S., Kollegger, J.: Thin post-tensioned concrete shell structures. In: Tailor Made Concrete Structures, pp. 703–706. Taylor & Francis, London (2008)
Day, A.: An introduction to dynamic relaxation. Engineer **219**, 218–221 (1965)
Isler, H.: New shapes for shells. Bull. Int. Assoc. Shell Struct. **8**, c-3 (1960)
Kilian, A., Ochsendorf, J.A.: Particle spring systems for structural form finding. Int. J. Shell Spat. Struct. **46**(2), 77–84 (2005)
Larsen, N.M.: Generative algorithmic techniques for architectural design. Ph.D. thesis, Aarhus School of Architecture, Aarhus, pp. 123–133 (2012)
Larsen, N.M., Pedersen, O.E., Pigram, D.: A method for the realization of complex concrete gridshell structures in pre-cast concrete. In: Synthetic Digital Ecologies: proceedings ACADIA, pp. 209–216. The Printing House, San Francisco (2012a)
Larsen, N.M., Pedersen, O.E., Pigram, D.: Realisation of complex precast concrete structures through the integration of algorithmic design and novel fabrication techniques. In: Proceedings AAG, pp. 161–174. Springer, Wien (2012b)
Pedersen, O.E.: The tectonic potentials of concrete. Ph.D. thesis, Aarhus School of Architecture, Aarhus, pp. 78–90, 131–177 (2013)
West, M.: Thin shell concrete from fabric molds. Available At: http://www.umanitoba.ca/cast_building/assets/downloads/PDFS/Fabric_Formwork/Thin-shell_Concrete_From_Fabric_Forms_SCREEN.pdf (2009). Accessed 16 Apr 2012

Modular Fibrous Morphologies: Computational Design, Simulation and Fabrication of Differentiated Fibre Composite Building Components

Stefana Parascho, Jan Knippers, Moritz Dörstelmann, Marshall Prado, and Achim Menges

Abstract The paper presents a bottom-up design process based on the transfer of biomimetic design principles and digital fabrication strategies for modular fibre-based structures, as demonstrated on a full-scale prototype pavilion. Following the analysis of the structural principles of the beetle elytra, the material differentiation and the morphologic principles of the biological role model are transferred into design and fabrication strategies. Simultaneously, developments of a coreless robotic winding method for glass and carbon fibre reinforced composite elements are incorporated into the design process. The computational set-up developed for the entire workflow is presented, showing the integration of structural analysis with digital simulation, which enables the automatic generation of the robotic winding syntax for individually differentiated components. The investigations, simulation, fabrication and assembly process, which led to the realisation of a highly efficient lightweight architectural prototype, are explained in the current paper.

1 Introduction

Traditionally, architectural design has been prevailed by top-down design methods, which usually subordinate material and fabrication considerations for apre-conceived geometry. While bottom-up strategies have been increasingly explored in design processes, as for example with biomimetic approaches, these are still often followed by a top-down fabrication solution. Contrary to conventional design methods, both the design development and the materialisation process can

S. Parascho (✉) • J. Knippers (✉)
Institute of Building Structures and Structural Design, University of Stuttgart, Stuttgart, Germany
e-mail: s.parascho@itke.uni-stuttgart.de; j.knippers@itke.uni-stuttgart.de

M. Dörstelmann (✉) • M. Prado (✉) • A. Menges (✉)
Institute for Computational Design, University of Stuttgart, Stuttgart, Germany
e-mail: moritz.doerstelmann@icd.uni-stuttgart.de; marshall.prado@icd.uni-stuttgart.de; achim.menges@icd.uni-stuttgart.de

P. Block et al. (eds.), *Advances in Architectural Geometry 2014*,
DOI 10.1007/978-3-319-11418-7_3

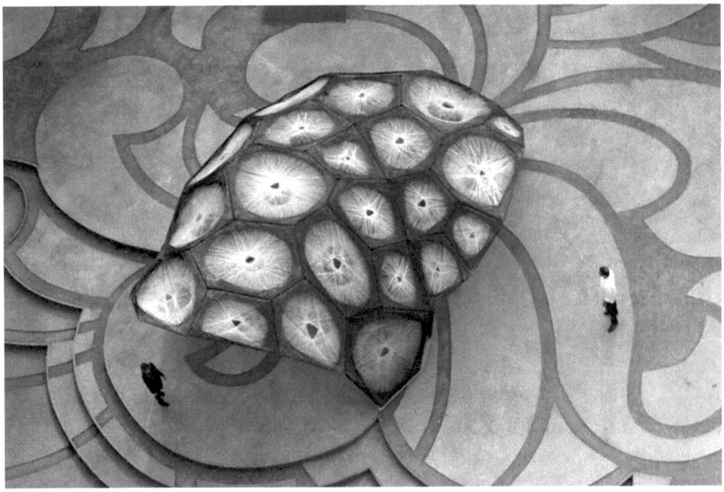

Fig. 1 ICD/ITKE Research Pavilion 2013–2014, top view

be regarded as equal design drivers through the use of biomimetic design principles and the simultaneous development of novel fabrication methods (La Magna et al. 2013; Menges 2013) (Fig. 1).

Biomimetic approaches have proved to have significant potential for design implementations through their systemic complexity and manifold logics (Gruber 2011). Morphological principles of natural organisms are abstracted and transferred into architectural applications for their performative geometries and functional integration. Biological evolutionary processes offer a remarkable example for the integration of multiple requirements into the morphogenetic process. As observed by biological research, natural systems do not evolve as singular optimisation solutions, but rather negotiate between numerous and often conflicting factors, providing a compromise that can fulfil a multitude of requirements, contrary to conventional optimisation procedures (Knippers and Speck 2012).

Similarly, advancements in fabrication methods have led to a constant growth of the geometric design space for architectural applications but rarely to new systems beyond existing typologies of building construction. The development of fabrication strategies based on material behaviour allows an early integration of manufacturing logics into the design process and an exploration of novel structural typologies (Pottmann 2013).

The ICD/ITKE Research Pavilion 2013/2014 presents a design approach based on multiple, concurrent bottom-up investigations, illustrating how parallel research of both biomimetic principles and novel fabrication techniques lead to a highly integrative design process. Biological principles, material properties, structural performance and fabrication constraints are translated into parameters of architectural morphology, allowing the development of a computational design setup as a point of confluence.

2 Context

The presented project builds on previous research conducted by the authors, through the collaboration of the Institute for Computational Design and the Institute of Building Structures and Structural Design which focuses on integrative design methods and their application in architecture, tested and implemented in full scale prototypical research pavilions. The research conducted at ICD/ITKE concentrates on design processes that incorporate biomimetic investigations as a design input from natural systems providing morphological as well as process logics for the fabrication of efficient lightweight structures.

The Research Pavilion 2011 represented a direct application of biomimetic design strategies into a built architectural prototype. The morphology is derived from the structural and connection principles found in the Echinoides' plate arrangement, which led to a remarkable structural performance of the system (Schwinn et al. 2013). The Research Pavilion 2012 focused on the scale of material arrangement, providing differentiated stiffness and strength through the variation of fibre disposition, as found in many biological fibre-based systems like the exoskeletons of arthropods (Reichert et al. 2014).

In order to implement biological principles of fibre-based systems, composite materials such as fibre-reinforced polymers (FRP) are used to locally differentiate material properties through position, orientation, density and type of fibres (i.e. glass or carbon) in order to enhance the structural performance of the system.

FRPs are widely used in industrial applications such as the aerospace industry, due to their high strength to weight ratio and unrestrained mouldability. Standard fabrication techniques rely on the manual or automated application of fibres to pre-fabricated formwork, constraining the otherwise formable material to a logic of serial production or the use of adaptable moulds for niche markets. Neither of these seems suitable for unique large-scale architectural applications; therefore the development of a novel fabrication technique would be required for the production of freeform FRP building components of the lot size one.

In the Research Pavilion 2012, a coreless winding technique was developed through robotic fabrication that allowed the generation of a complex double-curved monocoque shell without the need for a positive mould (Waimer et al. 2013). The formwork was replaced by a minimal steel frame, which enables double curved geometries to emerge through fibre-fibre interaction between the frame elements. In order to produce a single shell through a continuous winding sequence, an on-site fabrication method was developed through the use of a robotic winding set-up. The coreless winding approach reduces the fabrication effort, as it does not require the production of a positive mould; on the other hand, the analysis and simulation effort increases as the complexity of the system is influenced by the constantly changing stress in the fibres.

Further challenges based on the previous research include the expansion of the geometric solution space by eliminating the limitations linked to the robotic set-up (Reichert et al. 2014) as well as improvements in transportability and manufacturing logistics, both achievable through the use of a modular system. A further development of the coreless winding method allows the minimisation of the necessary formwork and the generation of an adaptable and reusable frame for a component-based system. On a structural level, the development of a double-layered system leads to an enhancement of the construction's structural capacity.

3 Investigations

3.1 Bottom Up Biological Investigation

The ICD/ITKE Research Pavilion 2013–2014 continues the successful series of multidisciplinary research projects through collaboration with biologists from Tübingen University and imaging specialists from the Karlsruhe Institute of Technology, resulting in an open-ended biomimetic bottom-up process. A strategic preselection of biological role models, which was based on natural lightweight fibre structures with specific features like material anisotropy, materialisation processes and resulting functional morphologies, preceded the project. The selected role models were investigated by interdisciplinary teams of architects, structural engineers and biologists.

Throughout these investigations the natural fibre composite structure in the beetle Elytra, the protective shell of the abdomen and wings, provided a versatile role model for performative lightweight structures. The comparison of ground beetles and flying beetles revealed how the negotiation of antagonistic performance criteria, e.g. structural rigidity and weight reduction, within an evolutionary morphogenetic process led to the development of a multi-functional lightweight material system. The material efficiency of the system is based on anisotropic organisation of chitin fibre composite material which forms a double layered shell and is coupled by trabeculae, internal bracing structures (Fig. 2).

The geometric articulation of this lightweight construction was studied through microcomputer tomographic scans which were conducted at the ANKA/Institute for photon science and synchrotron radiation and electron microscopy scans by the Department of Evolutionary Biology of Invertebrates in Tübingen. Both methodologies operate on different scales and provide various insights into the structure of the role model. While material organisation and fibre arrangements were analysed based on the 2d SEM scans (Fig. 2b), the micro CT scans were processed to very detailed 3d mesh data and allowed insight into the global as well as the internal structural morphology (Fig. 2a). Like the biological role models, which come in a wide variety of shapes and sizes, an extensive array of performative architectural forms, which follow these structuring principles and fabrication parameters, are possible.

Comparative studies of eight flying beetle species allowed for the abstraction of the underlying functional principles. On the material scale, seamless structural integration of the upper and lower shell is achieved through continuous fibre

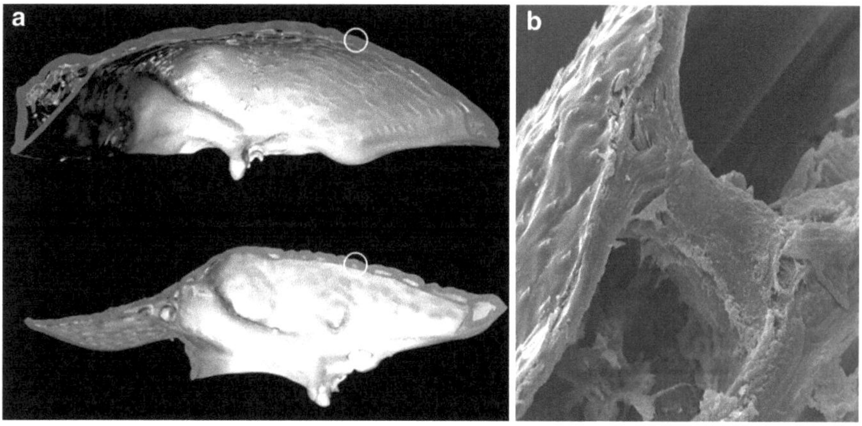

Fig. 2 Beetle elytron, (**a**) cross section Chrysomela (*Top*), Cassidia (*Bottom*), 3d model constructed from Micro CT data by Dr. Thomas Van de Kamp; (**b**) section through Trabecel, Leptinotarsa decimlineata, SEM scan by Prof. Oliver Betz

arrangements within each trabecula. Structural morphologic differentiation can be described on two levels; the local geometric articulation of each trabecula and their arrangement within the global structural system. A clear interdependency of loading situations and the respective morphologic response of the system was identified and abstracted through the description of geometric rules. These relations between structural capacity, fibre arrangements and geometric articulation were transferred into a component based lightweight construction system. The transfer into a technical application requires a level of abstraction, which allows the use of specific functional features without recreating the complexity of the entire natural role model. In the present case this becomes visible through the favouring of the advantages of a modular construction logic over the structural benefit of a seamless continuous shell.

3.2 *Bottom Up Winding Process Investigation*

Parallel to biomimetic investigations in an integrated design process, extensive research on fabrication strategies explore geometric potentials inherent in a material system. This requires a further bottom up investigation of fibre winding techniques and their latent geometric possibilities.

In a bottom-up fabrication process, material is not applied to a pre-defined form, but actively influences the design (Fleischmann et al. 2012), by defining the possible solution space and potential of the global design outcome. As a result, the developed geometry of the system and fabrication strategies are a direct

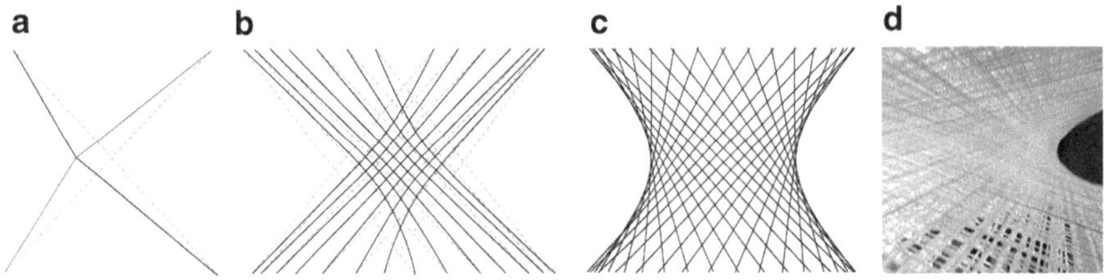

Fig. 3 Coreless winding; (**a**) two fibres connecting four non-planar points; (**b**) fibres deforming under the tension of subsequently wound fibres; (**c**, **d**) anticlastic curvature induced by helicoidal winding

consequence of material capacities, which requires a simultaneous development of both architectural morphology and fabrication.

The geometry of the fibre-based elements is strongly defined by material properties since it is the result of the interaction between subsequently applied fibres. For the generation of the characteristic column-like structure found in the beetle elytra role model a coreless winding method was chosen and further developed, in order to continually minimise the necessary formwork for the production of double-curved FRP geometries. For the transfer of the trabeculae morphology into component geometry, fibres are helicoidally wound around the component frame, which is determined by an adaptive computational model allowing fabrication and material constraints to actively control the size, shape and typology. The resulting fibres form a double curved surface configuration (Fig. 3). The anticlastic geometry is the immediate result of the linear fibres being tensioned once a proper fibre bond is achieved through continuous subsequent winding of multiple fibre layers. With each winding step fibres are pressing on each other, requiring the material to constantly negotiate a state of equilibrium throughout the fabrication process. At the same time, the use of a robotic manufacturing method and the set-up of a parametric robotic code generation system enable the differentiation of each fibre layout.

Through the sequential fibre-laying process both the geometry and the stress state of the fibres are constantly changing in a dynamic process, which leads to the necessity of a complex, constantly updating simulation process.

4 Integration and Verification

4.1 Integrative Design Tool

The co-development of biomimetic insight, material experimentation and fabrication strategy is a reciprocal explorative process where new insight immediately affects the criteria of evaluation and direction of investigation for the adjacent fields. This initial intuition driven process needs to be formalised as the interrelated insight becomes more complex. Computational design strategies offer a suitable platform to integrate and mediate between parallel bottom-up investigations, thus a meaningful synthesis is possible.

The biomimetic investigation led to a structurally driven local geometric variation of the components and their global arrangement, while the investigation of fabrication techniques and material experimentation explored the achievable geometric solution space. These reciprocal investigations need to be mediated to find their confluence in the component geometry as well as the global design outcome. The biological input on the components geometry is to ensure a transfer of the role models structural capacity into the technical implementation of the abstracted principles. Not only local morphologic parameters like structural depth and component diameter are defined through geometric rules in response to loading conditions, but also global structural situations like cantilevering and internal wall structures are affected by biomimetic design principles.

At the same time the specific robot set-up used for fabrication of the components sets geometric constraints, which are necessary to avoid collisions, or out of reach errors during the robotic winding process. The robot tool sets minimal angles between the polygon edges and defines how much non-planarity can be geometrically achieved.

Equally important is the information of the geometry through material behaviour. The winding logic depends on certain geometric conditions to ensure fibre-fibre interaction, which is required to achieve a structurally stable laminate. Therefore the height to width ratio of the component and specific length ratios between neighbouring edges have to be ensured.

This shows how each part of the project equally contributes to the integrated design process and potentially affects mayor spatial characteristic and ultimately the appearance of the global structure.

To negotiate these various inputs onto the local and global geometry, a constraints-based modelling approach was utilised. In a first step, possible cell layouts are negotiated through a planar arrangement of cells with variable dimensions and numbers of edges. Biomimetic as well as structural principles are integrated achieving a more dense and stable structure at edges and bifurcating areas. The initial topological connections of the components are described as a particle-network, which can resolve into one of multiple possible states of equilibria. The initially defined constraints of various process parameters for the control of component geometry can be systematically described which allows an interaction within the constraint-based computational framework. This framework is hierarchically organised, therefore robotic fabrication constraints and material behaviour can set the absolute boundaries of the geometric range, while the biomimetically derived structural principles drive geometric differentiation within these boundaries. Through a first FE Simulation incorporated in the geometry generation tool, dimensions of the connecting members are optimised for structural performance. Global and local geometric characteristics are analysed and regulated within each iteration. This guarantees the greatest possible geometric adaptation while ensuring fabricability of each component. Parameters such as element thickness, out of plane maximum distance, component dimensions and angles

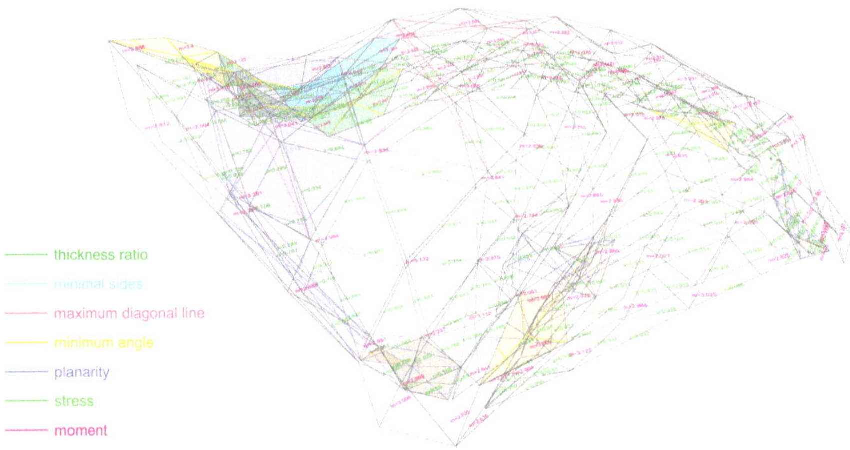

Fig. 4 Intermediate state of the computational form generating tool showing active parameters that are negotiated during the form finding process

as well as structural stability are translated into geometric factors and negotiated until a state of equilibrium is achieved (Fig. 4). Such a computational process of morphogenesis mediates various system influences and results in complex, emergent geometries, which are formed through system information rather than predetermined geometric solutions.

4.2 *Material Behaviour*

Component geometries emerge through the tensioning of the subsequently wound fibres, inducing deformations as a consequence of the fibre-fibre interaction. The overall geometry is not pre-defined, but is rather the result of the fabrication process and material properties. Due to this form-finding process, the material assumes its own geometry, requiring an adaptive winding sequence, which takes into account structural and material constraints. As a result, a simulation of the winding process and the consequences on material behaviour was needed by developing formation rules for the generation of the hyperbolic geometry.

In order to generate the robotic code and proper fibre composite structure for various differentiated components, a winding syntax – a systematic sequence of fibre winding – was developed. The generated FRP material system requires three different layers of fibres to be wound consecutively for the generation of a structural composite for the double curved components (Fig. 5). The first layer of glass fibre serves as form giving and defines the geometric features of the component, while the subsequent carbon fibre layers structurally reinforce both the component and the global geometry.

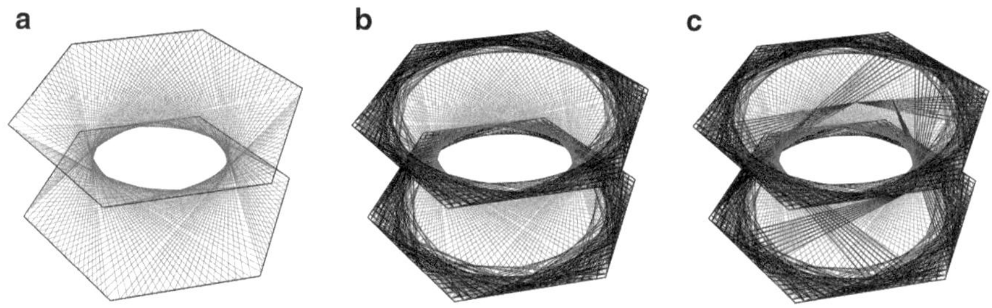

Fig. 5 different fibre layers; (**a**) glass fibre; (**b**) carbon fibre; (**c**) carbon fibre differentiated reinforcement

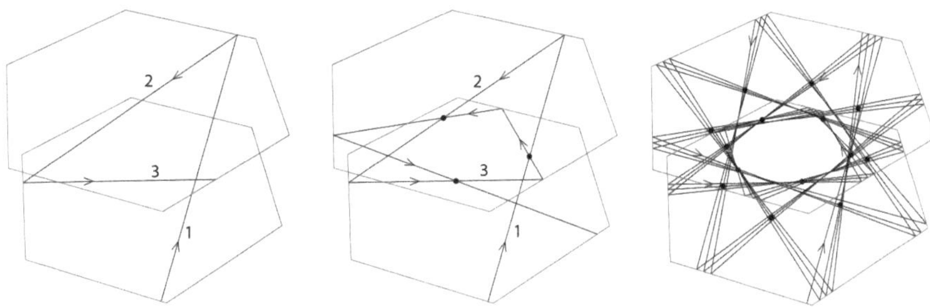

Fig. 6 Glass fibre winding sequence

Simulating the behaviour of the material layout and subsequently developing the winding syntax requires defining a set of rules that outlines the constraints set by the system. These rules were generated from numerous hand winding tests, digital simulations and robotic test windings in order to set the geometric range of possibilities through the pursued robotic winding method. Parameters were developed and tested, such as topology and component dimensions, angles between edges and distance between inner and outer layer, maximum degree of non-planarity and opening degrees of the components, leading to a robust set of boundary conditions that implicitly define the range of the possible global geometric outcomes. The parametric set-up enables the automatic generation of the winding syntax for each component producing a polyline that represents the exact sequence for the generation of the robotic code.

The main challenge for all winding layers is ensuring a proper fibre-to-fibre bond that generates a structurally performant hyperbolic geometry. The form-defining glass layer thus requires each subsequent fibre to tension the previous ones, stabilising the structure with each winding iteration. This is achieved through a fibre laying sequence that loops around the component frames and through the shifting of connection points until all fibres are tensioned (Fig. 6).

The winding syntax for both carbon layers further ensures the fibre tensioning through the definition of the wrapping sequence. The first layer, which locally reinforces the component geometry, is based on a general set of rules adapted to

Fig. 7 Single component and fibre detail

each element, which allows for the differentiation of fibre density and efficient use of material while the second carbon fibre layer responds to global force distribution and component geometry (Fig. 7).

4.3 *Structural Integration*

The geometric outcome is strongly informed by structural analysis data throughout the design process. In this type of integrated design process the structural analysis is used not only as a means of validation of the results, but is included as one of the design drivers on various scales. Structural considerations are partly derived from the biomimetic investigations, since the beetle elytron offers a number of structurally efficient morphological principles. The double-layered shell enhances the stability and structural performance compared to a single layered system. Simultaneously material investigations and Finite Element Analysis supply additional data that is transferred into design logics and embedded in an adaptive geometric generation process.

On a material level, the fabrication method needs to ensure a strong fibre-to-fibre bond in order to achieve the necessary stiffness and resistance. Material performance can be programmed through the differentiation of the angle in which fibres intersect, leading to a variation from an anisotropic material behaviour for small angles to a quasi-isotropic behaviour for angles close to 90°.

The carbon fibre layers are specifically developed to structurally reinforce the individual components as well as the global form. The first carbon fibre layer locally reinforces the component geometry around the edge and allows forces to be transmitted in plane. The geometric configuration of this layer strongly depends on the component frame geometry requiring the integration of fabrication constraints and frame collision avoidance. Through this arrangement, the component's edges are stiffened by generating a local structural height while enabling forces to be transferred throughout the element.

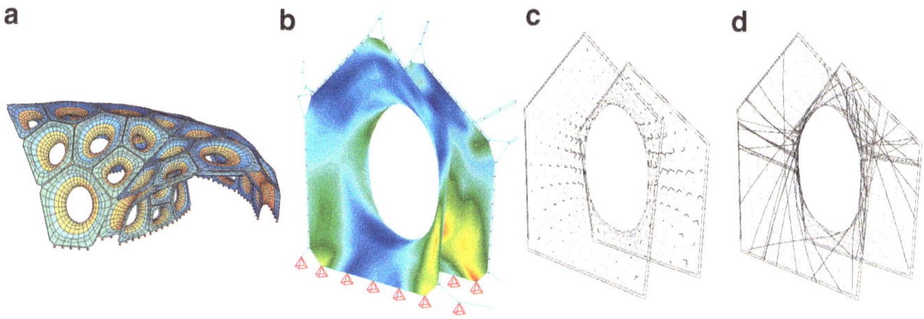

Fig. 8 (**a**) Model for FE Analysis; (**b**) resulting forces in component; (**c**) translation of forces into vectors; (**d**) fibres laid according to force vectors

Contrary to the first layer, the differentiated carbon fibre layer facilitates force transmission between the two structural layers and from component to component, taking into account global parameters such as neighbouring elements and connection points. Furthermore fibres are specifically placed to support the global structure where high forces emerge. For this purpose the global geometry is first analysed as a continuous shell (Fig. 8) with Finite Elements, and the resulting stress state is extracted from the analysis. This is translated into geometrical vectors in order to allow for an interchange between the different working environments. The strength and orientation of each force provide information about where the structure can be efficiently reinforced, enabling the generation of winding paths from the derived force vectors. Simultaneously, the resulting winding paths are informed by the general fabrication logic, ensuring that only feasible winding paths are generated. The density of these additional fibres is varied depending on the strength of the forces, leading to an individual fibre layout for each component that altogether forms a global structurally active construction.

5 Realisation

5.1 Fabrication Set-up

One goal of this research was to develop a fabrication system with a broad range of geometric freedom for the materialisation of fibre composite structures in architectural applications. The fabrication set-up, which continuously develops throughout the research, therefore becomes an integral constituent of the design and materialisation process of complex, distinct fibre composite geometries.

The set-up used for the fabrication of the FRP components relies on two synchronised 6-axis industrial robots, each holding an adaptable steel frame that defines the two boundary polygons of the components. The fibres are wound around

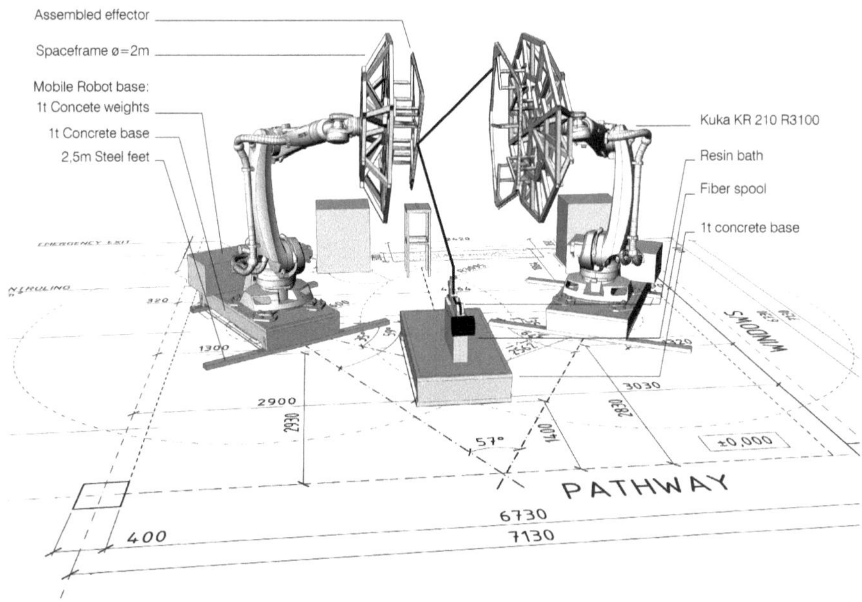

Fig. 9 Fabrication set-up

the control points of the effectors through the parallel movement of the two robots while the fibre source and resin bath are stationary (Fig. 9).

Defining a fabrication system for non-standardised components presents several unique challenges. Namely, providing an adaptive framework with a high level of geometric flexibility, which subsequently must maintain the rigidity necessary for winding tensioned fibres; while minimising the materials and complexity required by the use of a standardised kit-of-parts. Each component is defined digitally by a set of two non-planar closed polygons containing 4–7 vertices each. Each vertex has 8 degrees of freedom (Fig. 10), which defines the frame geometry upon which the fibre composite material is wrapped. In total each component requires the definition of 64–112 variables to entirely constrain the wrapping frame. To minimise the complexity of the system that would be required, industrial robots are used which provide precise control of 6 degrees of freedom per vertex (i.e. x, y, z position and a, b, c rotation). Through a robotically assisted assembly process, vertices of the component polygons are defined in reference to the orientation of the robotic end effector. Adjustable stands are attached to the effector manually, which physically describe the component geometry.

Tolerances are determined by the precision of the robot rather than the manual procedures. Milled guides control the two remaining variables including: the in-axis rotation and alignment of the frame segments (Fig. 10). This assembly system minimises material use and complexity of parts by embedding multiple fabrication constraint into a single construction guide while providing a high degree of reconfigurability by employing the inherent flexibility in a robotic set-up.

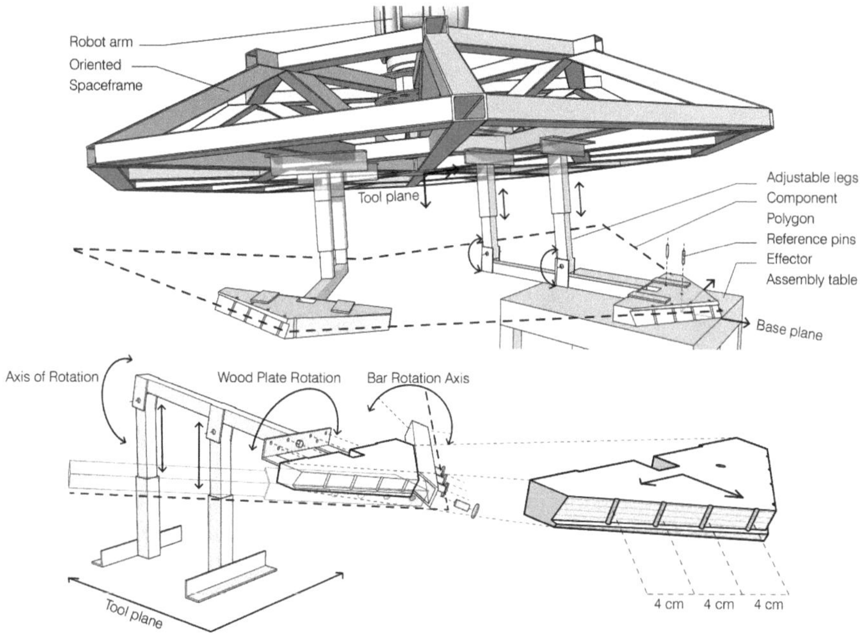

Fig. 10 Diagram of adjustable effector stand and milled guides that define component vertices

The use of two corresponding robots benefits the fabrication process in several ways. Both robots, each with an attached reconfigurable frame, function as one cooperative robotic system. This enables the robotic movements to be synchronised, maintaining the relative frame orientation without the need for interstitial structure, which could possibly obstruct the winding process. The two robots can also equally divide the dead loads placed on the system and resist tensile stresses accrued by sequentially tensioning the fibres. Feedback from integrated sensors in a robotic system allows for adaptive control to compensation for accumulated geometric inaccuracies. As a result, large-scale fabrication of component geometries, which includes smaller tolerances for inter-component connections, higher tensioning of fibres for structural composites and error correction throughout the winding process, become systematically possible.

5.2 Results

The presented integrative design process, based on the two bottom-up investigations, resulted in a full-scale prototype, built at the *****. The pavilion consists of 36 individually shaped components that are assembled into a highly material efficient construction system. Through the use of automated robotic fabrication, a high

Fig. 11 Assembly setup; lightweight and stability of the pavilion

degree of geometric freedom was achieved, resulting in differentiated diameters, thicknesses, non-planarity and number of control vertices. The components ranged in size from 50 to 250 cm in diameter. The lightweight performance of the construction is illustrated through the minimal weight of the prototype, with a total weight of 593 kg of 58 km of glass fibres and 44 km of carbon fibre as well as the related polymer matrix. The individual components weigh between 11.5 and 24.1 kg, and can thus be easily lifted into place by one or two people, allowing for a simple and fast assembly process (Fig. 11).

The pavilion consists of a double layer fibre composite shell, which encloses an area of 50 m^2. The system bifurcates creating an internal structural wall, a principle derived from the biological role model, which divides the space and creates two openings facing the university and the surrounding park.

The pavilion demonstrates the potential of fibre composites for lightweight construction through the stability and efficiency of the structure and illustrates the biologically abstracted principles, from material organisation, i.e. the fibre arrangements, to the structural capacity of the overall construction. The custom development of fabrication methods and the integration of material, fabrication, biomimetics and structural analysis into one multi-informed design process, allowed the development of new types of architectural geometry and related novel spatial qualities, as well as the detachment from existing building types and structural typologies. The overall form, spatial arrangement and material expression contribute to a unique architectural experience while at the same time being extremely efficient with employed material resources (Fig. 12).

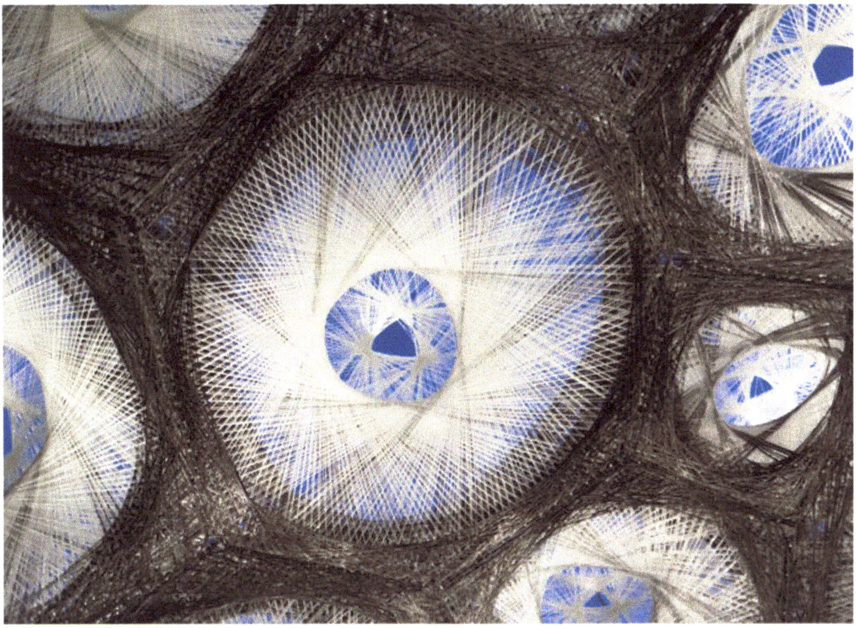

Fig. 12 Various components assembled

Conclusion

The presented research successfully integrates both bottom-up explorations, on a biological level as well as on a material and fabrication level, into a coherent and highly integrative design process. The numerous influencing factors were translated through computation into a common geometric language and could thus be simultaneously considered throughout the design process. At the same time each parameter strongly affected the geometric outcome widening or limiting the solution space.

The use of FRPs and individually adaptable fibre placement enabled a straightforward production of numerous individual geometrically complex but material efficient elements, which was facilitated by the development of a robust computational set-up that allowed an effortless transition from the design intention to the robotic winding code and production.

The biomimetic investigation provided numerous morphological principles that primary defined the components geometries. Furthermore it supplied structurally efficient spatial arrangements that increased the geometrical complexity of the global construct.

(continued)

Through the development of a modular system, the dimension of the resulting structure was no longer limited to the robot reach, which demonstrates how the core-less winding method is applicable in large scale building implementations. In addition the geometric complexity of the global structure was no longer limited by the robotic space, allowing spatial structures such as the internal wall and higher freedom in the global topology of the structure.

Future research is currently concentrating on the further integration between the structural, geometrical and morphological features of the built prototypes. As the core-less winding approach has proved to be extremely efficient for the fabrication of large scale and custom components, further reducing the requirement of formwork elements is currently under investigation. Finally, the biomimetic approach may offer interesting developments, which may open future perspectives in the design and implementation of innovative structural systems (Fig. 13).

Fig. 13 ICD/ITKE Research Pavilion 2013–2014

Acknowledgements This project was possible only through the work of the students participating in the "Fibrous Morphologies" Studio 2012 from the University of Stuttgart, especially Leyla Yunis and Ondrej Kyjanek, as well as the collaboration on structural issues with Vassilios Kirtzakis.

References

Fleischmann, M., Knippers, J., Lienhard, J., Menges, A., Schleicher, S.: Material behaviour: embedding physical properties in computational design processes. Archit. Des. **82**(2), 44–51 (2012)

Gruber, P.: Biomimetics – Materials, Structures and Processes Examples, Ideas and Case Studies. Springer, Berlin/New York (2011)

Knippers, J., Speck, T.: Design and construction principles in nature and architecture. Bioinspir. Biomim. **7**, 015002 (2012)

La Magna, R., Gabler, M., Reichert, S., Schwinn, T., Waimer, F., Menges, A., Knippers, J.: From nature to fabrication: biomimetic design principles for the production of complex spatial structures. Int. J. Space Struct. **28**(1), 27–39 (2013)

Menges, A.: Integral formation and materialization: computational form and material gestalt. In: Manufacturing Material Effects: Rethinking Design and Making in Architecture. Routledge, New York (2013)

Pottmann, H.: Architectural geometry and fabrication-aware design. Nexus Netw. J. **15**(2), 195–208 (2013). doi:10.1007/s00004-013-0149-5

Reichert, S., Schwinn, T., La Magna, R., Waimer, F., Knippers, J., Menges, A.: Fibrous structures: an integrative approach to design computation, simulation and fabrication for lightweight, glass and carbon fibre composite structures in architecture based on biomimetic design principles. Comput. Aided Des. **52**, 27–39 (2014)

Schwinn, T., Krieg, O., Menges, A.: Robotically fabricated wood plate morphologies. In: Brell-Çokcan, S., Braumann, J. (eds.) Rob | Arch 2012 SE - 4, pp. 48–61. Springer, Vienna (2013)

Waimer, F., La Magna, R., Knippers, J.: Integrative numerical techniques for fibre reinforced polymers – forming process and analysis of differentiated anisotropy. J. Int. Assoc. Shell Spat. Struct. **54**(178), 301–309 (2013)

Application of Hybrid Glass-Timber Elements in Architecture

Semi Continuous and Self-generative Glass Layering Structural System

Philipp Eversmann, Paul Ehret, Christian Louter, and Manuel Santarsiero

Abstract The following paper explores the application of hybrid timber-glass elements on semi-continuous architectural structures. The use of glass as a structural material opens multiple fields of investigations. Beyond structural matters and safety issues, architectural questions as functionality and spatiality are briefly addressed, since they are paired with the structural layout. Furthermore, the potential of a glass plate system of overlaying, but yet discontinuous glass elements is addressed in more depth. Geometrical specifications on the structural glass application are elaborated on and generalized into a 'card house' algorithmic discretization model. Through the design and fabrication of a 'case study', the parallel use and digital simulation and empirical physical testing are discussed. A further potential use in the construction industry of the system is debated.

1 Introduction

Spatial liberation of the architectural plan has been an achievement in the use of reinforced concrete technology. The structural efficiency provided by it enabled the design of floor plans with the structure and architectonical elements such as a wall being disconnected.

The use of structural glass, as illustrated for a multitude of new applications by Henriksen (2012), could aim towards a similar design dialectic. It has implications on the functionality but also in the perception of space. The rules implied by the piling of structural glass can determine the spatial arrangement of each floor, generating a different plan for every level.

P. Eversmann (✉) • P. Ehret • C. Louter • M. Santarsiero
EPF Lausanne - ICOM, GC B3 505, Station 18, CH - 1015 Lausanne, Switzerland
e-mail: philipp.eversmann@epfl.ch; paul.ehret@epfl.ch; christian.louter@epfl.ch; manuel.santarsiero@epfl.ch

P. Block et al. (eds.), *Advances in Architectural Geometry 2014*,
DOI 10.1007/978-3-319-11418-7_4

The following sections explain the basic mechanisms of such a system in order to get an understanding of the constructive feasibility. The main structural principles constraining the design also reveal the flexibility of the system and show the possibility to use it as a design driving principle.

2 Structural Glass

In contemporary architecture there is an increasing demand for transparency. To fulfill this quest for transparency, glass is – besides its common use as a façade cladding – increasingly applied as a load bearing structural material. Early applications can be found in the 1980s and 1990s, for example the glass pavilion for the Sonsbeek art exhibition in Arnhem in the Netherlands in 1986 (Nijsse 2003). Ever since, structural glass technology has rapidly and significantly evolved, as is illustrated for instance by Nijsse (2003) and Wurm (2007).

Glass can be applied for various load-carrying components (Henriksen 2012). The most common application of glass as a structural component is its use as a structural beam, where the glass is loaded in in-plane bending. Other applications are columns and (shear) wall systems. The latter application is exploited in the current study. As revealed in Sect. 5, a case study pavilion was constructed based on a hybrid timber-glass structural system. In this system the glass is placed vertically, as a wall system, thereby carrying both the vertical and the horizontal forces acting on the structure. The glass functions both as a vertical column and a horizontal stiffener and stabilizer, in-between horizontal sheets of timber. Furthermore, the overall structure is pre-tensioned by means of vertical steel rods, to further enhance the structural interaction of the timber- glass structure. The cables are placed according to required maximum distances between the structural members and maximum heights of each cable.

3 Semi Continuous Structural System

Simple card houses structures are taken as a reference for the structural development. If we were to consider this system structurally we would enumerate the following principles:

- Postulate: the punctual link between two overlapping panels has sufficient friction to avoid side sliding effect
- One panel needs two points of support to be lifted (Fig. 1)
- Pairing up panels two by two per floors isn't sufficient to hold: either cross bracing or the addition of a 3rd panel enables structural integrity of the system (Fig. 2).

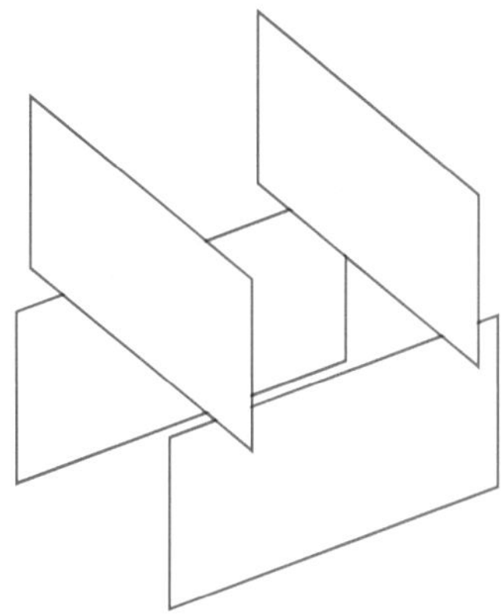

Fig. 1 One panel needs two points of support to be lifted

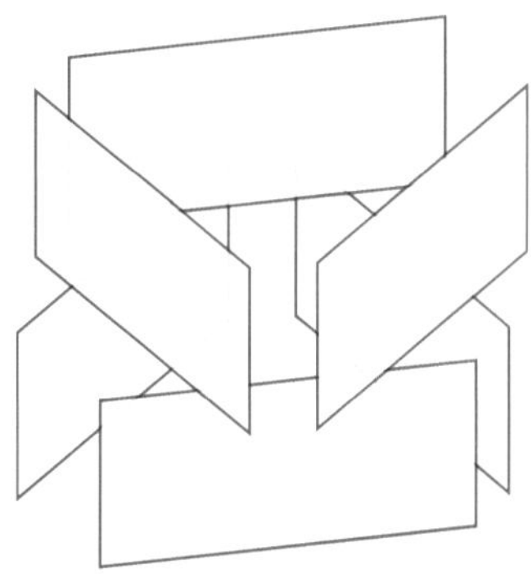

Fig. 2 Wither cross bracing or the addition of a 3rd panel enables structural integrity of the system

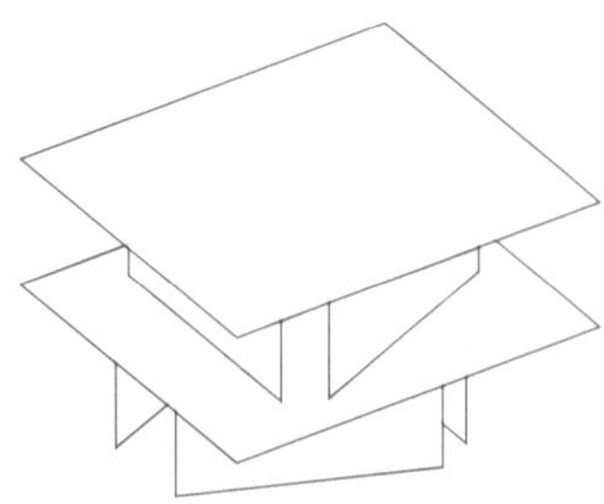

Fig. 3 Floors can stabilize the supporting panels

- The introduction of floors in which there is a link with the panel stabilizes it: it does it for the panel below and the one above (Fig. 3).
- This system involves a chain of panels all connected two by two; on each floor each panel is connected to two panels from the precedent floors, which as a whole can be considered as continuous (Figs. 4 and 5).
- Each floor has at least three panels/walls with at least two different intersections of their axes for rotational stability (Fig. 6).

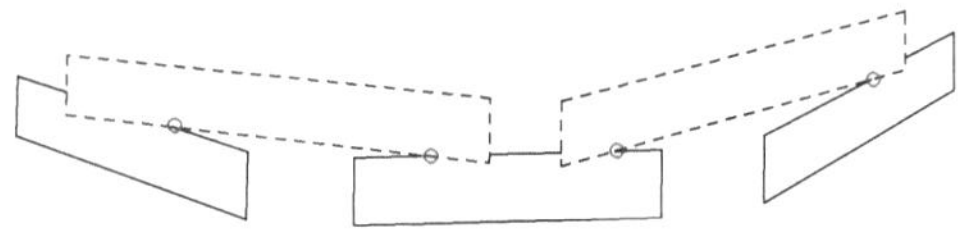

Fig. 4 Two supports/panel

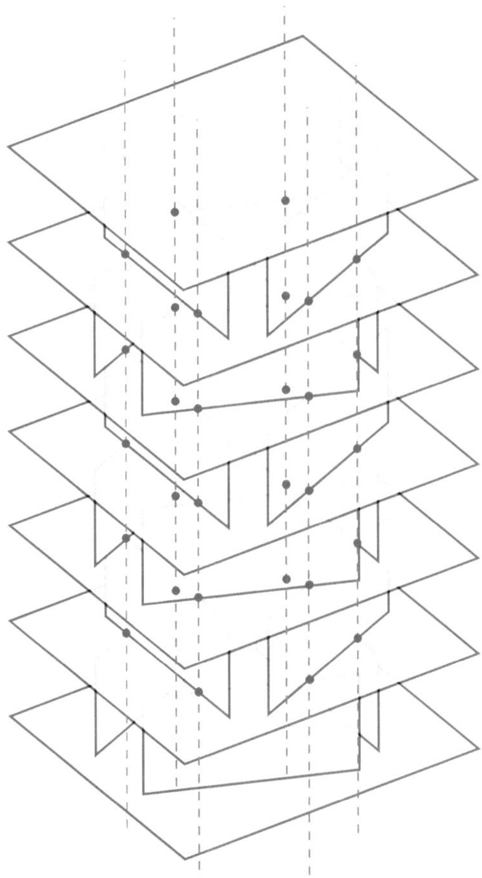

Fig. 5 Support points

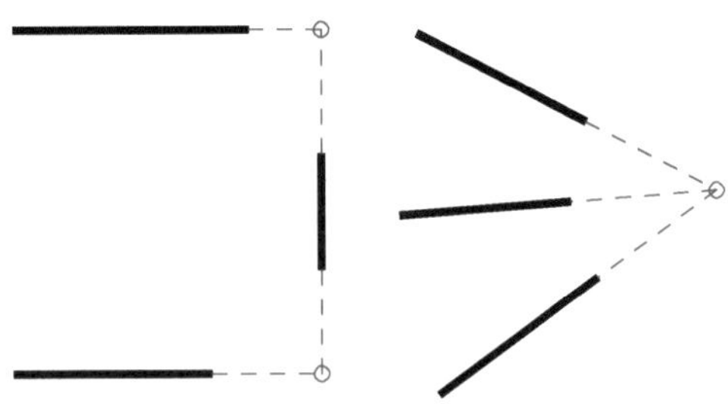

Fig. 6 Floor layout, axis configuration

4 Algorithmic Design: A Self-generative System

The general concern of this study is to describe surfaces following a stacking method. For curved geometries, in particular vaults, 'Trulli', stacked stone vaults in southern Italy (Fig. 7), and nineteenth century glass structures (Lauriks 2008) were taken as a reference.

The goal of the algorithmic modeling is to provide an automated design tool which can accommodate to various surface types (Fig. 8). The choice of double curved surfaces was made to ensure that the system accommodates to a large variety of surfaces, flat and single curved surfaces being considered as particular cases of double curvature ones. The aim is to adjust to the curvature changes while leaving some freedom in the arrangement of the panels for the designer.

The design system takes a reference surface S as an input. The first part consists in slicing S into a multitude of layers with adjustable heights. The resulting curves are offset on the same side of S to secure a complete overlapping of the horizontal panels. These curves can then be fragmented in multiple segments following the local curvature of the input curve. Starting from the bottom curves every new floor verifies the consistency with the points announced in the 2nd section, in particular

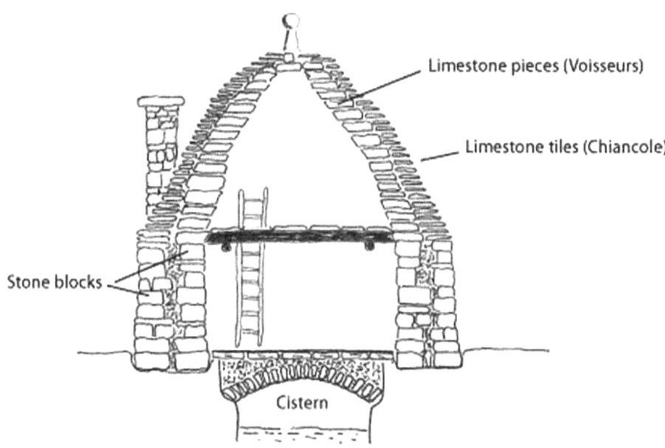

Fig. 7 Section of traditional 'Trullo' (http://www.understandingitaly.com/puglia-content/trulli.html)

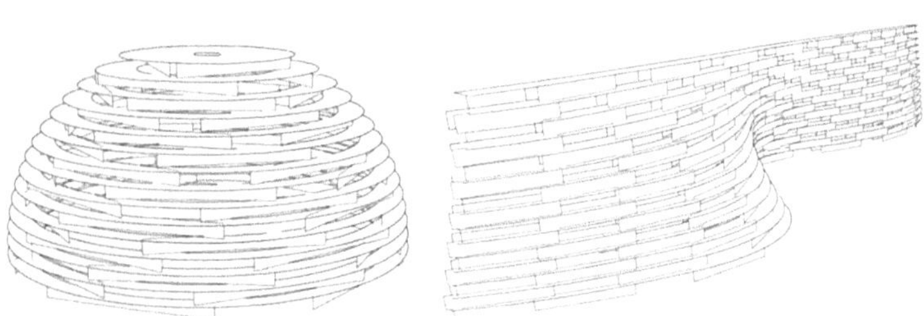

Fig. 8 Stacked dome structure, wall with an inflection point in the generative curve

one glass panel always having at least two support points (Fig. 4) and the axis orientation of the panels (Fig. 6). The solution of this problem can obviously have large number of valid results. For fabrication ease and architectural consistency a minimum and maximum length of each panel can be defined. The designer has the flexibility to change the seams of the curve to adjust the start division points and to correct defects as illustrated in a brick pattern study by Bärtschi and Bonwetsch (2012).

In the realm of a further study for the elaboration of a building it could also be imaginable to integrate different user 'wishes' for the panel layouts to obtain a result by a projected functional use of each 'floor plan'. Each floor can be manually adjusted and be tested by the verification algorithm. Once all panels are positioned they can be optimized in different length families.

5 Case Study: The Pavilion 'Slicing Opacity'

5.1 Architectural Concept

The pavilion Slicing Opacity is the application of the system simplification enunciated in Sect. 3 and 4. It is demonstrating the structural efficiency obtained by the slicing of freeform surfaces into structural planes equally distributed, following a variation of only two different heights.

It illustrates the new potentialities in the use of structural glass for architecture. As the scale of the pavilion was closer to urban furniture the idea of organic shapes was chosen to maximize possibilities of interactions with the visitor (Fig. 9). The design system explained in Sect. 4 has been applied on these freeform surfaces showing the capacities of the system, despite the drastic curvature changes.

Fig. 9 Case study 'Slicing Opacity'

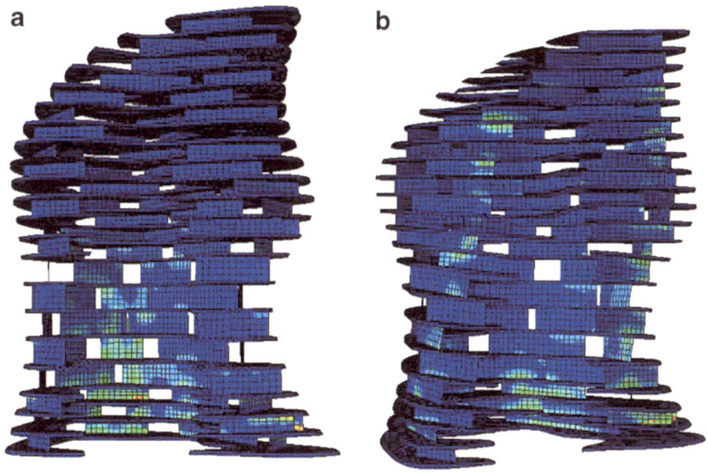

Fig. 10 (**a**, **b**) Fields of maximum principal stress in the glass and timber panels

5.2 Structural Analysis

Preliminary finite element analysis is performed on the structure. Shell rectangular elements are used to model the vertical laminated glass panels. Solid tetrahedral elements are used instead to model the timber horizontal plates.

No sliding is allowed between timber and glass components. In order to evaluate the structure response against lateral load, a horizontal force equal to 0.1 g is applied to the whole structure. Boundary conditions are given specifying displacements and rotations equal to zero to the nodes of the base panel. Figure 10 shows the field of maximum principal stress in the glass plates and timber panels.

Figure 11 shows the field of displacements (displacement scale is $\times 10$).

Results indicate that the maximum stresses are located in the glass panels at the base. This is in agreement to the expectations since the glass panels at the base are subjected to the maximum shear forces.

5.3 Optimization Form and Structure

The algorithmic design described in Sect. 4 is obviously a pure geometric approach to create the structure, using the structural design assumptions of Sect. 3. What if, the structural analysis could directly inform the geometric pattern and look for a structural oriented optimization?

For simulating this approach, a parametric model was set up using the software 'karamba' (http://www.karamba3d.com/). In the model, a set of parameters was

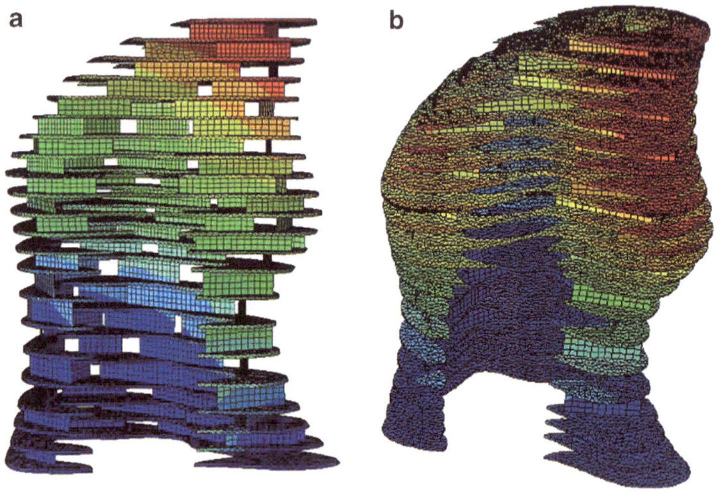

Fig. 11 (**a**, **b**) Visualization of the field of displacements

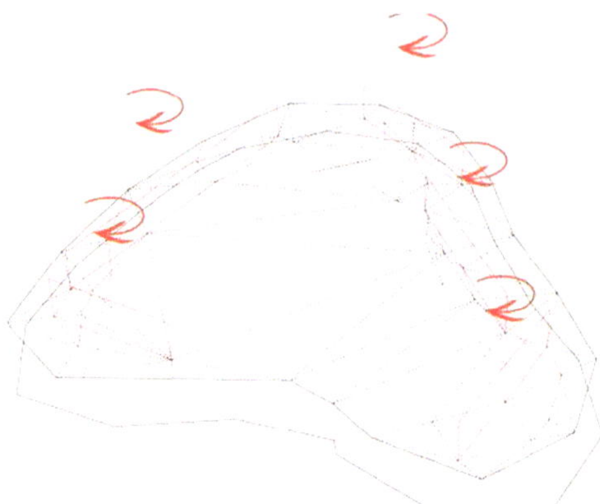

Fig. 12 Glass shells position and resulting meshing

defined to be able to rotate all vertical glass elements separately around their central axis (Fig. 12).

Every position change of the glass shells results in a newly generated mesh geometry for the horizontal elements.

The overall structural comportment was then calculated using different wind load cases with the karamba analysis model. Therefore, all the surfaces were projected into the normal plane of the wind direction to be able to calculate the force per area of every shell element of the glass structure. The resulting forces can then be calculated using the karamba solver (Fig. 13).

Fig. 13 Max displacement of the shell elements in karamba

Using the genetic algorithm 'galapagos' in grasshopper, multiples of variations of the orientation of the panels could be calculated, using the maximal deflection of the whole structure as a fitness criteria.

The movement of all panels or selected ones can be constrained to only certain movements to ensure the architectural integrity of the project.

5.4 *Fabrication*

The idea of the fabrication was to keep the process simple for the wood cutting, which was done at our university's own laboratory, whereas the glass panels were produced by a local glass contractor to ensure quality control and safety requirements are held throughout a manufacturing process which needs very specific tooling and knowledge. Every glass panel is composed of two float glass panels of only 4 mm thickness, laminated with a Polyvinylbutyral film to panels of 11 different lengths and two height families with a total of around 400 panels. In order to gain stiffness and meet the security requirements every single border has been edge treated (Fig. 14).

Physical tests were previously effectuated in order to understand the relationship between the geometry of the glass panels and the structural behavior. The test objects were constructed in real scale and were then structurally evaluated. This

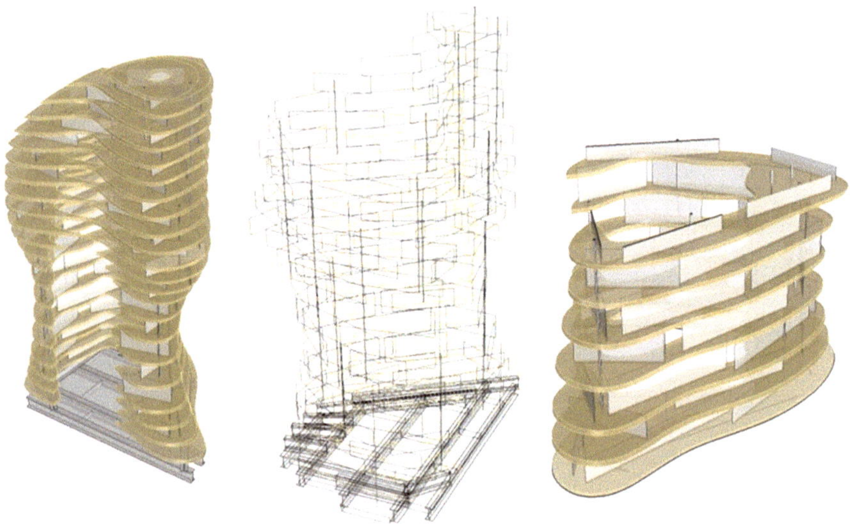

Fig. 14 Tower fabrication model

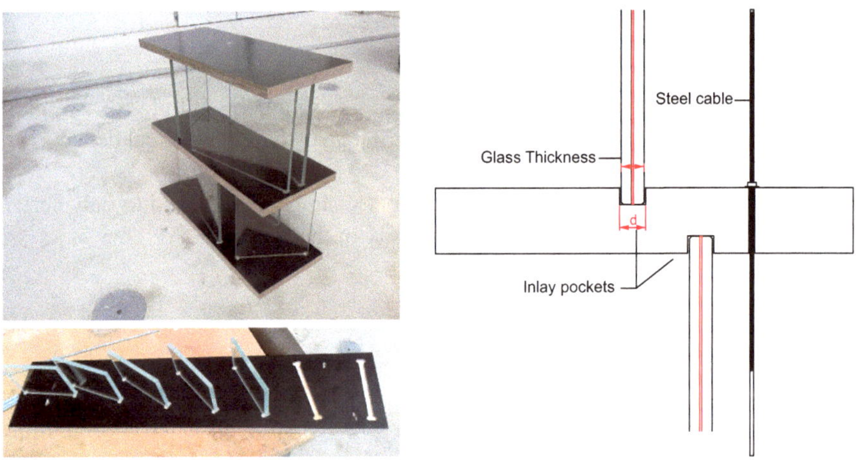

Fig. 15 Physical test objects, glass-timber interaction

method of empirical testing proved a fast way to approach the structural feasibility of the project and to get reliable figures needed for structural evaluation.

A major problem was to accommodate the tolerances of the different glass thicknesses due to the lamination process to the pocketing of the wood panels in a way to provide sufficient holding of the panels on side loads. If the cuts were too large in the tested prototypes, the whole structure became slippery since the wood panels had too much possibility of movement (Fig. 15). If on the other side cuts

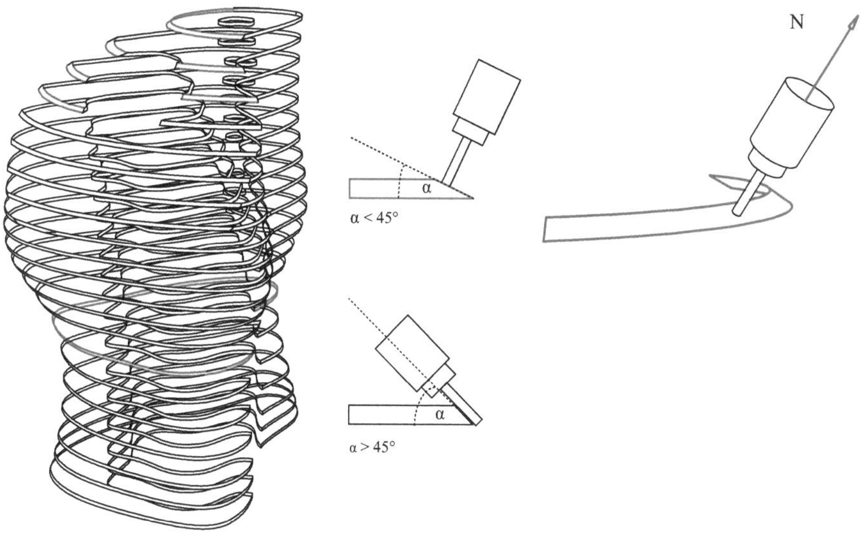

Fig. 16 Classification of surfaces on angles for direction of milling operations

were too thin, it was impossible to insert easily the glass into the wood without extra pressure on its edge, risking unnecessary stress on it.

Since this parameter was neglected in the structural simulation, we chose to pretension the whole system with cables to get more friction between the elements, preventing them from slipping. All wood elements could be cut out of 21 panels of Sperracolor© 1.5 m × 3.5 m plates, a dense plywood with a serigraphy coating for weather protection.

The production consists of only three different processes: drilling for fixation and cable holes and two sided inlay pockets (Fig. 15) to define the inserts for the glass panels for the vertical assembly. Then in order to get the overall shape closer to the original reference surface S the wood edges were chamfered with an angle following the local tangent to S at every single point of the curve C. Due to the machine head size it was cut using the five axis following two different positions (Fig. 16).

6 Architectural Application

The material capacities of reinforced glass as a primary structural material coupled with the proposed stacking system offer new potentials for a maximum of transparency as well as a functional generation of floor plan layouts.

The system can be imagined in a multitude of scales, daylighting can be modulated through the variation of the stacking geometry (Fig. 17).

Though the case study prototype (Fig. 16) could be only effectuated on a relatively reduced scale, structural reinforcement of the glass panels by laminated

Fig. 17 Scale: inhabitable object – full scale floor heights

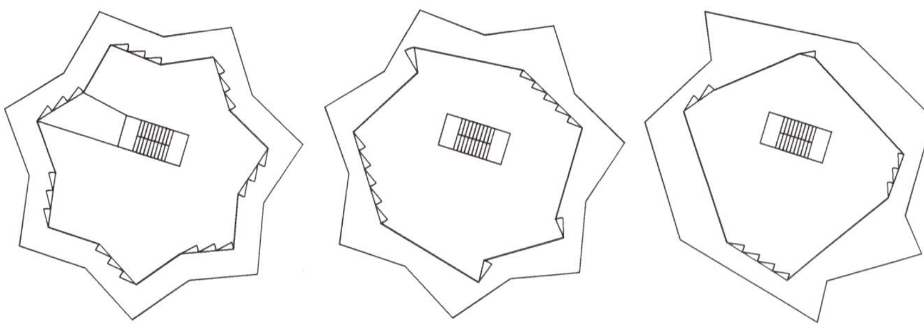

Fig. 18 Case study house

Fig. 19 Case study house

tension elements in the base parts of the panels makes a much larger scale imaginable (Louter 2007). If we were to take "Slicing Opacity" as a mockup of what a new generation building could be, we could pursue towards a functional optimization of floor plan layouts within the limits of certain shape types. Once the user typologies are identified in adequacy with the corresponding floor plan typology and proportions the plans and the structural layout could be generated (Figs. 18 and 19).

Conclusion

This paper presents a system of hybrid timber-glass elements which can be applied on semi-continuous architectural structures. A fabrication case study project shows the algorithmic generation, structural analysis and fabrication detailing in the scale of urban furniture. Following current research in glass reinforcement, the system is applied on bigger scale structures to show future possibilities in an architectural application. The more complex detailing and structural simulation involved in the application on larger structures was not part of this paper, but should be subject to further investigation.

Within its own set of rules this logic opens new fields of investigation for the designer. The glass structural constrains are stretching the modern language (column/beam) to a semi continuous logic as it is explained in the earlier chapters. Further studies could explore the repercussion of this new dialectic of the floor occupation: functional matters as well as visual and perception questions could be addressed, based on the fact that the structural module is transparent.

References

Bärtschi, R., Bonwetsch, T.: A stretcher bond with defects applied to a hyperboloid. In: Advances in Architectural Geometry, pp. 37–42. Springer, Wien/New York (2012)

Henriksen, T.: Future applications of structural use of glass. Challenging Glass 3, In: Conference on Architectural and Structural Applications of Glass, pp. 67–74. Delft University Press, Amsterdam (2012)

Lauriks, L., de Michael, B., Quentin, C., Ine, W.: 19th century iron and glass architecture: common construction details of cylinder and crown glass on iron sash bars. In: Ignatiadou, D., Antonaras, A. (eds.) Annales du 18e congrès de l'Association Internationale pour l'Histoire du Verre, pp. 469–474. ZITI Publishing (2008)

Louter, C.: Adhesively bonded reinforced glass beams. Heron **52**(1/2), 31–58 (2007)

Nijsse, R.: Glass in Structures: Elements, Concepts, Designs. Birkhäuser, Basel (2003). ISBN 3-7643-6439-4

Wurm, J.: Glass Structures: Design and Construction of Self-Supporting Skins. Birkhäuser, Basel (2007). ISBN 978-3-7643-7608-6/978-3-7643-8317-6

Gaudi's Puffy Jacket: A Method for the Implementation of Fabric Slump Casting in the Construction of Thin-Wall Funicular Vault Structures

Iain Maxwell and Dave Pigram

Abstract This paper describes a method for the production of thin-wall funicular (compression-only) structures from unique double-curved concrete components via a novel slump casting technique. The technique deploys fabric formwork within simple two-dimensionally cut frames to enable the efficient production of the unique parts necessary to tessellate form-found funicular geometries. Through the realisation of a high-tech / low-tech ecology of production, the paper seeks the reestablishment of generative pathways between each domain in the design-to-production cycle: architecture, engineering and fabrication. The method and resulting case study pavilions are situated within the historical trajectory of architectural form finding, specifically, the realisation of masonry vault structures.

I. Maxwell (✉)
Architectural Association School of Architecture, London, UK

University of Canberra, Bruce, Australia

Architecture & Technology, University of Technology Sydney [UTS], Ultimo, Australia

Sydney Visiting School, Architectural Association School of Architecture, London, UK

Supermanoeuvre, 42/56-60 Foster Street, Surry Hills, Sydney, Australia
e-mail: maxi@supermanoeuvre.com

D. Pigram (✉)
University of Canberra, Bruce, Australia

Supermanoeuvre, 42/56-60 Foster Street, Surry Hills, Sydney, Australia

Graduate School of Architecture, Preservation and Planning [GSAPP], Columbia University, New York, NY, USA

University of Technology Sydney [UTS], Ultimo, Australia
e-mail: dave@supermanoeuvre.com

P. Block et al. (eds.), *Advances in Architectural Geometry 2014*,
DOI 10.1007/978-3-319-11418-7_5

1 Introduction

Through the realisation of good structural form, poor materials transcend their apparent limitations. This paper describes a novel method for the production of thin-wall funicular (compression-only) vault structures from discrete double-curved precast concrete tiles. In the method and case studies that follow, the deliberate pursuit of a high-tech / low-tech workflow ensues. The method is high-tech with regard to the adoption and development of algorithmic approaches to the production of geometry, one that digitally computes form through the mechanical simulation of material to ensure optimal structural shape and material placement. It is low-tech due to the establishment of a bespoke production method that seeks an efficient approach to the fabrication of unique, thin wall, doubly curved, resilient tile elements. Critically, the coupling of design, engineering and fabrication strategies allows for feedback between each domain and thereby increases the number and nature of design influences providing opportunities for innovation. The method is demonstrated through two case study pavilions.

2 Masonry: Material, Structure and Construction

2.1 Form: Innovation Through Material Constraint

Good structural form (geometry) reduces the existence of stresses within the materials that compose it. Equally, good structural form must also resist bending under live loading. Throughout the history of architecture and building, we find many examples where limited material resources have set the context for significant innovation in order to meet evolving spatial (span) demand. The theory of arch stability via the development of the catenary thrust line (hanging chain) in the seventeenth century serves as a primary historical example, while the work of Eladio Dieste in the twentieth-century exemplifies the imperative for structural innovation borne out of resource limitations. For Dieste, it was the predominance of extruded clay blocks and the rarity and expense of concrete and steel in his native Uruguay that led to the development of his seminal Gaussian vaulting technique (Fig. 1). Dieste's method, primarily used form to resist buckling, where the *"Gaussian vault can be thought of as a series of connected catenary arches with varying rises that share a common springing. At any longitudinal cross section the vault is subjected to axial compression, but the magnitude will vary between sections as the rise of the catenary changes. The variation in axial compression generates shear stresses between the segments, resisted by steel reinforcement placed between the brick units."* (Pedreschi and Larrambebere 2004). Dieste built many such vaults, the largest example of which, the Port Warehouse, spanned 50 m longitudinally.

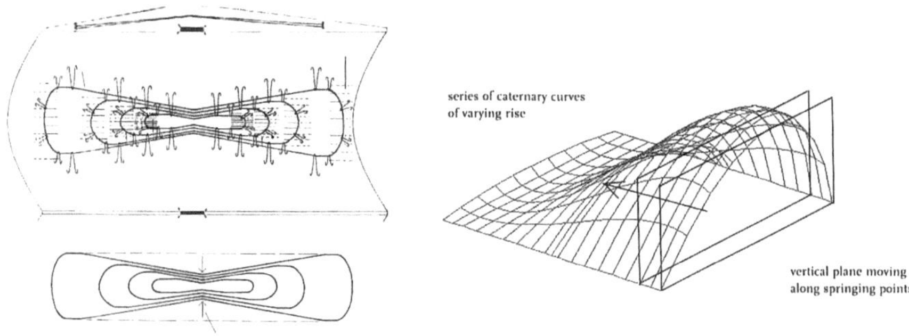

Fig. 1 Eladio Dieste's Gaussian vaulting technique (Pedreschi and Larrambebere 2004)

2.2 *The Stability of Masonry Structures*

Robert Hooke (1635–1703) is generally credited with the theoretical establishment of the thrust-line (catenary) method through the principle of inverting a uniformly loaded suspended rope or chain: *"As hangs the flexible line, so but inverted will stand the rigid arch."* However, as Karl-Eugen Kurrer substantiates, the first mathematical description of the catenary-shape is attributed to the English mathematician David Gregory (1659–1708). Jacques Heyman provides the most extensive review of arch theory and in doing so outlines the historical and theoretical development of the discipline of civil engineering more broadly (Kurrer 2008).

The contemporary rise of computational (algorithmic) modelling, analytical and optimisation techniques has vastly empowered the engagement of design with matters of material and structural stability and sets the platform for an expanded conversation on the nature of architectural form. An extension of architecture's rich history of form-finding that speculatively couples space, geometry, structure and material. A paradigm as familiar with the work of the Master Builders and Gothic masons as it is with the work of Antonio Gaudi and Robert Maillart, Heinz Isler and Frei Otto. Key contemporary figures include: Mike Barnes and Chris Williams (University of Bath), David Wakefield (Tensys), Mutsuro Sasaki (SAPS), John Ochsendorf (MIT), Philippe Block (ETH) and Sigrid Adriaenssen (Princeton University).

2.3 *The Challenge of Thin-Walled Masonry Vaulting*

The fundamental trait of any stable arch or vault is the containment of the thrust-line within the thickness of the structural surface. In confirming Hooke's method David Gregory elaborated: "*none but the catenaria is the true figure of a legitimate arch*...*And when an arch of any other figure is supported, it is because in its thickness some catenaria is included*" (Heyman 1972). Yet, throughout history we find many examples where a masonry arch does not take the ideal form of the catenary. Their stability, as Gregory observed, is only ensured by the fact that within their massiveness the true thrust-line is captured. T.M. Charlton relays, "*since arches carried heavy dead load, due to superstructure, in comparison with which live load was small, elementary statics was sufficient to ensure that the distribution of dead load and arch shape were such that the locus of the resultant of the total shear force at any section and the horizontal thrust, nowhere passed outside the masonry*" (Charlton 1982). Thus, assuming basic mechanical action (element slippage) was accounted for; a number of less-than-perfect shapes may have been tolerated owing to the massiveness of stone construction.

Within a lightweight and thin-walled paradigm such implicit redundancies do not exist. Simply, without the excessive thickness (depth) or latent benefits of over-compression (mass) inherent to historical masonry construction, there is neither margin nor material to spare a thin-shell vault from exposure to non-uniform externalities and ultimately catastrophic failure. As such, shape control and the realisation of better techniques of shape production are paramount!

2.4 Historical Methods of Masonry Construction

There are two primary historical approaches to the construction of masonry arches and vaults: Voussoir and Catalan. Irrespective of the approach, neither method is considered completely stable until the entire vault is complete. The Voussoir method is exemplified by the arches and vaults of the Roman, Byzantium and Gothic periods and denotes the assemblage of individual cut masonry elements. A necessary demand of Voussoir construction is the erection of temporary falsework (centring) to support each block until the overall assembly is complete. The significance of this temporary structuring is further exacerbated by the immense weight of the material used and the limitations its presence places on overall site access.

Conversely, the Catalan method is constructed from small, relatively thin (20–25 mm) and dimensionally similar clay shingles or tiles and produces significantly lighter structures. The use of rapid setting mortar allows Catalan vaults to avoid extensive centring, as such; they are typically built in free-space with minimal guide work. This, together with the use of standardised rather than custom elements, makes them much faster to construct than their Voussoir counterparts. Arising from the brittleness of the tiles and their resulting vulnerability to impact, increased orders of safety – structural redundancy – must be obtained through constructing multiple layers in a cross-laminate manner. Generally three layers are sufficient resulting in an overall thickness of between 75 and 100 mm dependent on curvature and layup. The Catalan method is exemplified by Rafael Guastavino (1842–1908) whose construction company built over a thousand such examples (floor systems and roofs) in North America, the most notable being Boston's Public Library and New York's Grand Central Station (Fig. 2) (Ochsendorf 2010).

Fig. 2 Catalan vaulting employed by the Guastavino Construction Company (Allen 2004)

In the method that follows, a mixed mode approach to masonry assemblies is outlined. One that privileges Voussoir logic in so much as each tile is geometrically unique – and complete falsework is required – yet operates through a lightweight, thin-walled and thereby materially efficient tile strategy. The technique operates on the simple assumption that if the form of the catenary imbues structural capacity and resilience at the scale of the overall structure, then such logics may afford an equivalent resilience at the grain of the tile itself.

3 Form Finding

3.1 Analogue Form Finding

Like traditional design techniques: cutting, carving, folding, weaving; form-finding techniques harness the positive limitations of a given media, here material and physical forces, to resolve formal characteristics in consistent ways. Unlike traditional methods however, form-finding processes embed a considerable level of material and structural intent within active design modelling processes. Robert Hooke's inversion of the suspended rope or chain sets the context for a technique-based approach to computation of funicular geometries, which fundamentally exhibit zero out-of-plane bending moment under self-weight. The quintessential elaboration of such modelling practices is found in the nested hanging-chain models of Antonio Gaudi used to compute increasingly complex load paths.

3.2 Digital Form Finding

Gaudi's models can be reproduced digitally through the use of dynamic relaxation (Day 1965). "The basis of the method is to trace step-by-step for small time increments, Dt, the motion of each node of a structure until, due to artificial damping, the structure comes to rest in static equilibrium." (Barnes 1999). Kilian and Ochsendorf (2005) describe a method for the re-conception of architectural geometry (surfaces, volumes, etc . . . .) as a network of weighted particles (nodes) tethered to one another by variable springs (members). Within such a computational approach, each constituent particle negotiates its immediate neighbourhood of connecting springs (topology) towards a state of equilibrium or equal residual force (i.e. towards a given spring length). Dynamic relaxation (DR) is therefore the iterative application of Hooke's laws of elasticity: that for elastic deformations of an object, the magnitude of its deformation (extension or compression) is directly proportional to the deforming force or load. Or algebraically: the applied force F equals a constant k multiplied by the displacement (change in length) x, thus: $F = kx$. Paul Bourke (1998) provides an example of such a data structure and its implementation.

In support of the research, a DR algorithm was developed in Java and deployed in the open-source 'sketching' environment Processing (Fig. 3). The bespoke

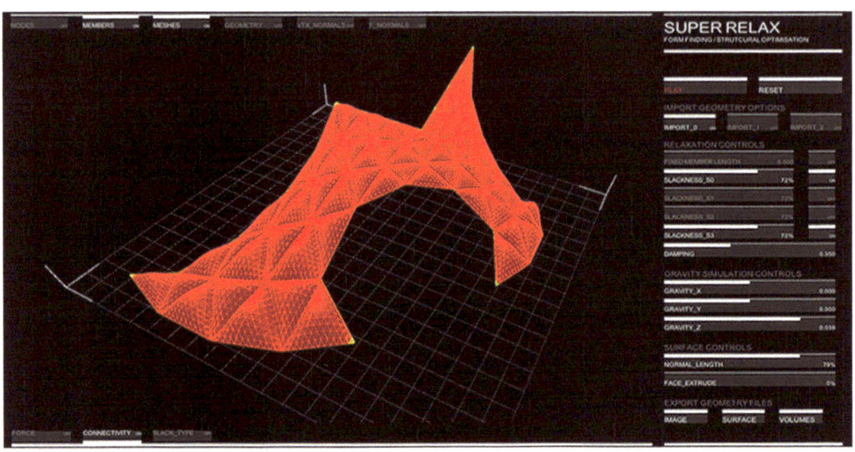

Fig. 3 Implementation of DR algorithm with bespoke graphical interface in processing allowing for multiple member types and representation modes (Maxwell 2013)

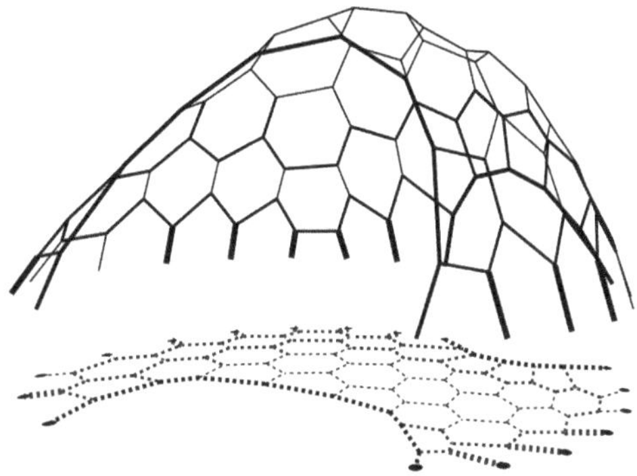

Fig. 4 Tuneable form finding through supply of multiple member types (via rhino layers) that aid overall shape control and boundary definitions (Maxwell 2014)

library – SuperRelax – employs a familiar node-member data structure as described above. The platform interfaces with Rhinoceros3D via a suite of import and export scripts written in Python.

This multi-platform approach enables initial system topology (network of connections) to be quickly established within a native CAD environment (Rhino3D) using geometric primitives such as mesh and line elements. Essentially, the designer can intuitively locate anchor points or specify variegated member types (slack lengths, damping and stiffness rates) through the use of layers in order to define overall shape or to control specific edge conditions (Fig. 4). Conversely, the more readily customisable OpenGL and event-handling context of Processing is harnessed for increased interaction, graphical representation and feedback of simulation data. The software has been validated through the realisation of numerous small-scale pavilions across a suite of material and production techniques (Larsen et al. 2012).

4 Method: Fabric Formwork

4.1 Context

The use of fabric formwork to aid the production of complex – doubly curving – concrete geometries has been used as early as the late nineteenth century for the construction of numerous objects from ferro-cement canoes to patented building systems. Gustav Lilienthal (1849–1933) filed the first known patent for a fabric formed suspended concrete floor system in the USA in 1899 (Block et al. 2011). The more recent work of Mark West at The University of Manitoba's Centre for Architectural Structures and Technology (C.A.S.T.) further demonstrates the application of fabric methods to construct complex column and beam elements. Heinz Isler's formative investigations of shell structures utilized fabric-forming techniques to construct scaled models. As Bill Addis writes, *"[Isler] either soaked the material in plaster and allowed it to harden or soaked it in water and literally freezing the shape by leaving it outside overnight in the Swiss winter.*" (Addis 2005) The consistency across these examples is the material efficiency of the woven textiles to form complex doubly curved moulds for casting. Thus the properties of plasticity unique to concrete can be harnessed for equally significant structural or formal motivations.

4.2 Method Overview

The method described here seeks the establishment of a low-tech mode method for producing non-standard thin-walled (8 mm) Voussoir elements. The latter an important requirement of any masonry vault structure as each element is sure to be unique. Each tile is slump cast within a plywood frame that clamps a fabric surface (flexible formwork). The resulting shape of the tile is the negotiation of the elasticity of the fabric and the mass of the concrete present in the casting frame at the moment of de-centring. Each slumped tile, is a planar facet (triangular mesh face) extracted from a principle parent geometry computed using the digital form-finding software described above. The edges of the facets constitute the primary load paths of the vault while the slumped tile surfaces – also catenary forms – constitute secondary patches. As such, the geometry and structural hierarchy of the resulting vault is akin to a ribbed vault (refer Sect. 4.6).

The intent of the production method is manifold:

(a) To establish a simple, cheap, fast, and robust method for the production of mass-customised lightweight and thin-walled masonry elements (Fig. 5).
(b) To increase the resilience of a thin-walled masonry tile element via double curvature, in order to test the need for a multi-laminate approach to redundancy as exemplified by Catalan construction.
(c) To have the ability to increase the size of each tile without the presence of bending stresses that would otherwise give rise to failure (shear) within a flat plaster tile of an equivalent area and thickness.
(d) To maintain planar tile boundaries to both simplify casting-frame production and on-site scaffolding. The insistence on triangular facets ensures each frame could be constructed from cheap and readily available plywood sheet stock. Equally, commonly available 2/3-axis machinery (CNC routing, laser cutting) or hand tools would be sufficient to produce each frame.
(e) To overcome the loss of formal character associated with the reduction of part population. Simply, reducing the quantum of parts permits a more economical construction approach, however this reductive approach often leads to a loss of geometric detail or global shape. Here, the introduction of a unique second-order formal system allows for a reduction of parts but not fidelity or character.

Fig. 5 Simple approach to uniquely cast tiles: plywood fabric formwork frames (Maxwell 2013)

4.3 Casting Frame Design

Each casting frame is constructed from readily available 8 mm non-structural plywood sheet stock. The complete frame incorporates: a removal centre piece that is the precise boundary shape of the tile; a variable shoulder offset that allows for additional material to be placed in the frame to account for the slumped geometry; a margin which also serves as a clamp for the fabric formwork and a datum to ensure consistent tile thickness (Fig. 6). Scripts were developed in Python (within Rhino3D) to automate the template for each frame. All routine tasks such as facet extraction from the global digital model, geometric offsetting (fabric clamp and casting margin), part naming, neighbouring part notation and labelling were automated and each part nested for optimal digital fabrication (laser-cutting or CNC routing).

4.4 Casting Process

Hydrocal, a gypsum-based plaster with a compressive strength of 5,000 PSI was chosen for its superior strength to traditional plaster and its fast setting time. At this stage, no steel or fibre reinforcement, aggregates, admixtures or plasticisers have been explored.

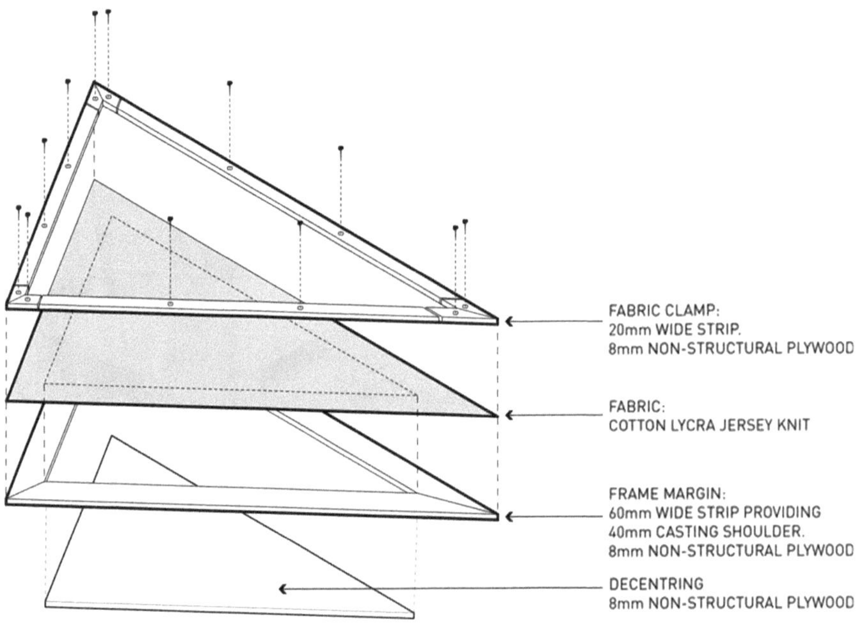

Fig. 6 Casting frame assembly (Maxwell 2013)

The casting process involved three basic operations: placing the Hydrocal, slumping and wire-cutting. Using the plywood margin that clamps the fabric as a datum ensures consistent tile thickness. At approximately 3 min the Hydrocal mixture starts to go 'green' and the casting frame is decentred thus allowing the tile to slump. Through the slumping process, additional material located within the shoulder of the casting frame ensured complete coverage of the tile especially at the boundary. Without this innovation, material cover at the boundary was insufficient leading to cracking or failure. At approximately 12 min the tiles were trimmed using a crude wire cutting approach. Notches in the casting frame allowed for relatively accurate sizing (Fig. 7). After 25 min the tiles were at full green strength and could be moved without altering their shape.

4.5 Relationship to Digital Form Finding Method

The SuperRelax workflow described permits an intuitive and topologically open approach to form finding. Here, the supply of triangular mesh facets and multiple spring types (slack lengths) was sufficient to ensure all resulting vault tiles were planar, within known dimensional limits (sheet size) and therefore capable of slump casting. Importantly, the DR method operates directly on the incoming topology (curves and mesh edges) so that it constitutes the computed – primary – load paths.

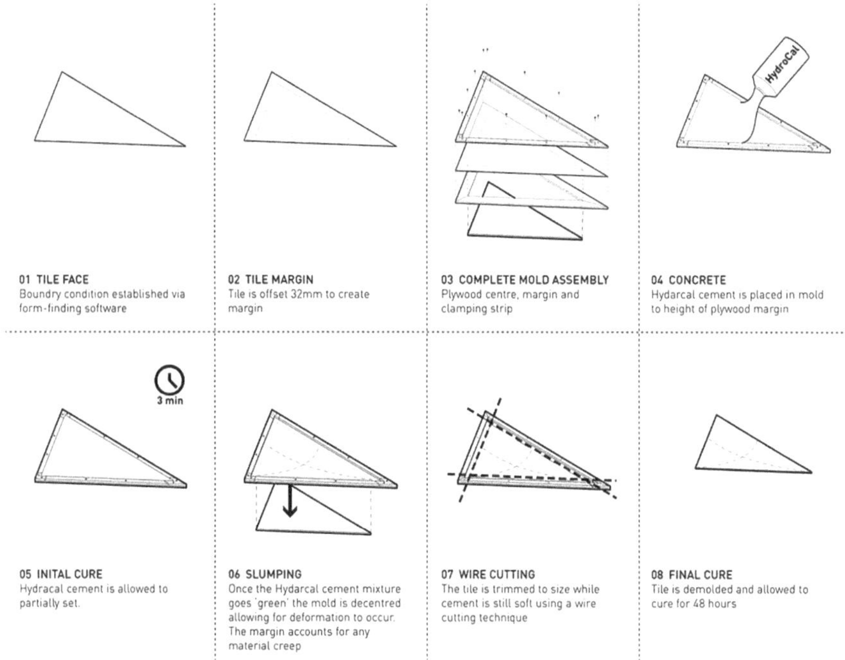

Fig. 7 Diagram outlining casting sequence (Maxwell 2013)

Therefore, no post-processing, namely penalisation, is necessary. This is clearly critical given the thinness of the resulting structural lamina (tiles) as any secondary tiling operations would no longer lie of the idealised thrust plane. Seen in this way, the method offers a novel solution to the dilemma of maintaining formal character albeit with lower part numbers.

4.6 Jointing and Redundancy

The DR method described is limited to the computation of funicular geometries under self-weight. Given the relative lightweight nature of the resulting assemblies, it was assumed they would not benefit from the orders of safety enjoyed by the massiveness of historic structures. Thus, to increase material and structural capacity along the primary load paths, the method of tile jointing was seen as an opportunistic moment of reinforcement. Essentially, each tile serves as a sacrificial formwork with the valley filled with between 20 and 30 mm of additional material (Fig. 8).

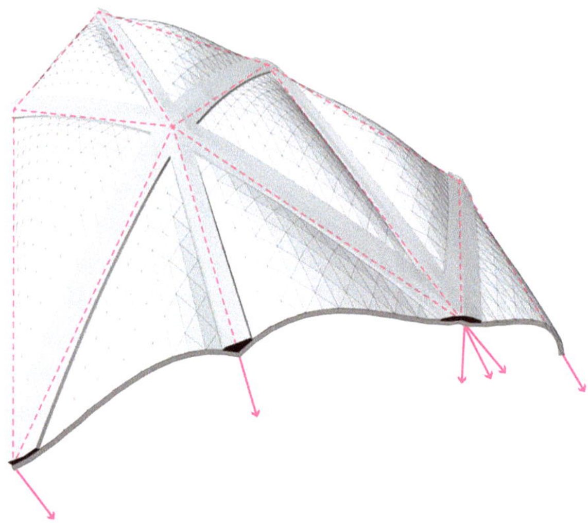

Fig. 8 Use of precast tiles as sacrificial formwork where jointing method provides additional material along primary computed load paths (Maxwell 2014)

5 Case Study Pavilions

5.1 Preamble

Two case study projects are presented, both undertaken within the context of intensive student workshops coordinated by the authors. The pavilions served as timely opportunities to validate the digital and casting techniques previously described.

5.2 Pavilion 01: Institut d'Arquitectura Avançada de Catalunya (IAAC), Barcelona, Spain

At the invitation of Areti Markopoulou, a pavilion was constructed upon an inner-city rooftop during a weeklong intensive workshop undertaken by post-professional students of IAAC's Masters of Advanced Architecture in April 2013 (Fig. 9). A playful yet respectful tribute to the seminal research on form finding conducted by Antonio Gaudi. The pavilion consists of 83 unique fabric formed tiles and achieves a span of 4.85 m with a typical wall thickness of 8 mm. At the time of writing (June, 2014) the pavilion remains in place.

Fig. 9 Completed IAAC vault (Markopoulou 2013)

5.3 *Pavilion 02: University of Technology (UTS), Sydney, Australia*

In June 2013, a second pavilion was realised during a 2-week intensive workshop undertaken by students of UTS' Masters of Architecture (Fig. 10). Here, a nested approach to the vault design was investigated via the introduction of a second lamina that both strikes the ground and rests upon the primary surface (Fig. 11). Again, the pavilion acknowledges Gaudi's nested chain models, only here play out through surfaces. The intuitive and open-ended nature of the *SuperRelax* tool permitted a departure from the dominant character of single lamina vault typical to masonry pavilions via the stitching (welding) of two mesh input geometries.

The absence of digital fabrication machinery meant a slight alteration to the design of the casting frame used previously was necessary. Here, simple strips of plywood were cut to length in lieu of laser cutting in order to produce the fabric clamps and margin. Despite increasing manual labour, an increase in the number of tiles to 193 was still achieved. A maximum span of 6 m was achieved with a typical tile thickness of 8 mm.

5.4 *Evaluation of Slumping Method at the Pavilion Scale*

It was discovered during the casting of the larger tiles, i.e. those predominantly located at the base of the vault where the most forces collect, that the casting frame

Fig. 10 Completed vault, University of Technology, Sydney (Wibowo 2013)

Fig. 11 Drawing demonstrating nested lamina approach to UTS vault (Maxwell 2013)

Fig. 12 Construction: use of cardboard scaffolding to position and support fabric-formed precast panels prior to joining (Maxwell 2013)

deflected significantly and the fabric stretched too much. A simple solution was to double-up both the fabric and the frame resulting in a tile that was 16 mm thick.

5.5 *Pavilion Assembly*

Given the thinness of the tile elements used (8 mm), it is extremely important that the built structure closely matches the computationally found-form so that all load paths remain within the computed thrust line. As is typical in all non-Catalan approaches to vault construction, the structure is not stable in an incomplete state and it was necessary to use extensive construction falsework to ensure the exact positioning of each tile and to support them during assembly. Like the precast tiles, each scaffold element was unique (Fig. 12). The scaffolding was produced from cardboard, laser cut and assembled into triangular tubes, reflecting the geometry of the digitally form-found 'parent' geometry. Again, simple Python scripts were developed to automate this routine task.

5.6 *Destructive Testing*

Destructive testing of the second pavilion prototype was conducted via non-uniformly loading the upper lamina (Fig. 13). The vault demonstrated complete stability under a sustained 1,000 kg + point-loading of sand bags. Catastrophic

Fig. 13 Successful point loading (1,000 kg +) of 8 mm thin-wall funicular vault structure (Maxwell 2013)

failure was only achieved through violently severing the load paths at the vault's leading edge using a metal bar.

6 Future Work

Future work will concentrate on two areas: firstly, material testing of the doubly-curved tiles in order to numerically validate their increased performance over an equivalent planar tile of the same area and thickness. Secondly, to employ finite element analysis in order to digitally examine the stability of the construction method under dynamic and non-uniform loading cases such as wind or impact. The purpose of the latter would afford a direct point of comparison to the cross-lamina approach to redundancy exemplified by traditional Catalan vaulting techniques.

7 Discussion

The pursuit of good structural form permits poor construction materials to transcend their apparent limitations. The traditions of form-finding within architecture trace our discipline's efforts to embed material as an active agent within the design process. Enacted digitally, form-finding methods afford an interactive context for the accelerated evolution of design. When developed in parallel with methods of production, a real opportunity exists to establish generative feedback loops across the allied domains of architecture, engineering, fabrication and construction. Thus increasing the number and nature of design influences enabling extended opportunities for design confidence and ultimately innovation.

The method described permits the low-tech construction of simple unique molds and in turn unique parts that can be used to produce novel, thin and lightweight compression-only structures as demonstrated by the two case-study pavilions.

References

Addis, B.: A history of using scale models to inform the design of structures. In: Huerta, S. (ed.) Essays in the History of the Theory of Structures: In Honour of Jacques Heyman. Instituto Juan de Herrera, Escuela Técnica Superior de Arquitectura de Madrid, Madrid (2005)

Allen, E.: Guastavino, Dieste, and the two revolutions in masonry vaulting. In: Anderson, S. (ed.) Eladio Dieste: Innovation in Structural Art. Princeton Architectural Press, New York (2004)

Barnes, M.: Form finding and analysis of tension structures by dynamic relaxation. Int. J. Space Struct. **14**(2), 89–104 (1999)

Block, P., Veenendaal, D., West, M.: History and overview of fabric formwork: using fabrics for concrete casting. Struct. Concr. **12**(3), 164–177 (2011)

Bourke, P:. Particle system example. Available at: http://paulbourke.net/miscellaneous/particle/ (1998)

Charlton, T.M.: A History of the Theory of Structures in the Nineteenth Century. Cambridge University Press, Cambridge/New York (1982)

Day, A.: An introduction to dynamic relaxation. Engineer **219**, 218–221 (1965)

Heyman, J.: Coulomb's memoir on statics: an essay in the history of civil engineering. Imperial College Press, London (1972)

Hooke's law. Available at: http://en.wikipedia.org/wiki/Hooke's_law

Kilian, A., Ochsendorf, J.A.: Particle spring systems for structural form finding. Int. J Shell Spat. Struct. **46**(2), 77–84 (2005)

Kurrer, K.: The History of the Theory of Structures: From Arch Analysis to Computational Mechanics. Ernst & Sohn, Berlin (2008)

Larsen, N., Pedersen, O., Pigram, D.: A method for the realization of complex concrete gridshell structures in pre-cast concrete. In: ACADIA 12: Synthetic Digital Ecologies Proceedings of the 32nd Annual Conference of the Association for Computer Aided Design in Architecture, pp. 209–216, 18–21 Oct 2012. San Francisco (2012)

Ochsendorf, J.: Guastavino Vaulting: the art of structural tile. Ernst & Sohn/Princeton Architectural Press, Berlin/New York (2010)

Pedreschi, R., Larrambebere, G.: Technology and innovation in the work of Eladio Dieste. In: Anderson, S. (ed.) Eladio Dieste: Innovation in Structural Art. Princeton Architectural Press, New York (2004)

New Opportunities to Optimize Structural Designs in Metal by Using Additive Manufacturing

Salomé Galjaard, Sander Hofman, and Shibo Ren

Abstract An initial research has been carried out, exploring the opportunities of using the design freedom created by Additive Manufacturing in metal on structural elements. The application of Additive Manufacturing (AM) bears the potential of increasing efficiency and shortening lead time by reducing processing steps, material use and labor intensity.

The aim of this research is the exploration of AM technology potentials as a feasible and robust design and manufacturing solution for structural building elements. This research is based on the redesign of an existing node, applying the new production opportunities of AM. This resulted in insights into the different design steps involved and knowledge on process, costs and structural characteristics. The start was a topology optimization on an existing design for a structural node, followed by a further rationalized design for production. A comparison in aesthetics, structural behavior and costs has been carried out between the new design and the conventional design. The first results look very promising and the work will be continued to enable the use of these new opportunities in structural designs in the near future.

1 Introduction

From 2008 till 2013 Arup was involved in the design of the tensegrity structures for the Grote Marktstraat in The Hague, the Netherlands, which were to be used as street lighting. Figure 1 shows a rendering of one of the lighting elements.

Tensegrity – tensional integrity – is a structural principle based on the use of isolated components in compression inside a net of continuous tension, in such a way that the compressed members do not directly connect and the tension cables define the system spatially.

Due to the irregular shape of the structure, the structural nodes, connecting the cables to the struts within the tensegrity, all have slightly different shapes. There are

S. Galjaard (✉) • S. Hofman (✉) • S. Ren (✉)
Arup, Amsterdam, The Netherlands
e-mail: salome.galjaard@arup.com; sander.hofman@arup.com; shibo.ren@arup.com

P. Block et al. (eds.), *Advances in Architectural Geometry 2014*,
DOI 10.1007/978-3-319-11418-7_6

Fig. 1 Rendering of one of the original tensegrity structures in The Hague. This design will not be built as such (Architect: ELV Architecten © Studio i2)

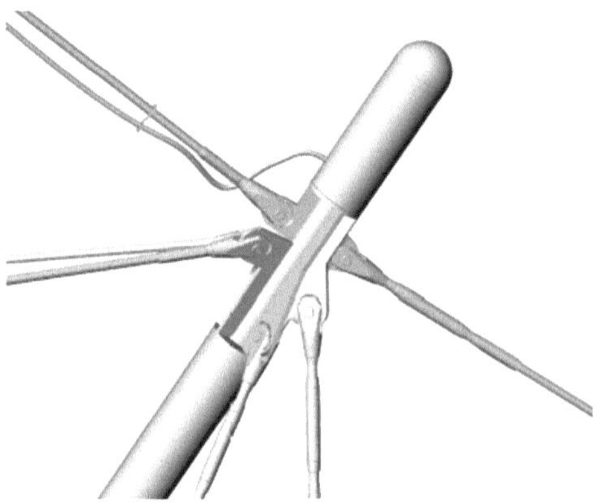

Fig. 2 Model of the structural node and light fixture, linking six cables to a strut

1,200 variations in angle and position of the attached cables. Figure 2 shows one of the nodes from the original design.

Conventional production of the nodes includes high labor intensity as it is literally a 'one of' production a 1,200 times over with very little rationalization. Each node will be made from six to seven unique machined steel plates, welded on a central tube in varying directions (ASTM International, Designation, Direct Manufacturing Research Center 2013; Galjaard 2013; Snelson 2012; Strauss 2013).

Funded by Invest in Arup we were able to explore novel methods of production in order to research opportunities to produce the nodes in a faster and smarter way. This research project has been executed separately from the original project and only after our work on it was finished.

1.1 Additive Manufacturing

The term Additive Manufacturing refers to a whole set of different production techniques, able to process different groups of materials like plastics, metals and ceramics. The American Society for Testing and Materials (ASTM) defines it as a process of joining materials to make objects from 3D model data, usually layer upon layer, as opposed to subtractive manufacturing methodologies, such as traditional machining.

During this research project we focused specifically on Selective Laser Sintering (SLS) for metals, also called Direct Metal Laser Sintering or Selective Laser Melting. During the production process, powdered metal material is selectively melted layer by layer via lasers. Layers typically have a thickness of 20–100 μm. The metal is fully melted into a solid homogeneous mass. There is a wide variety of alloys available and new ones are continuously being developed.

Products produced by SLS can have thin walls, deep cavities or hidden channels, but sometimes need support structures due to the weight of the solidified material and heat dissipation. The technique is mostly used in industries that benefit from weight reduction and where products are complex and produced in low quantities, like the aviation, aerospace and automotive industry.

2 Method

The set-up of our research consisted of the following three steps:

1. Topology optimization
2. Production of the optimized design
3. Detailing of the original design for comparison

2.1 Topology Optimization

With the rapid development of computer aided design technique, topology optimization as a design approach has been widely implemented to optimize material layout when designing load-bearing structures. In the classical problem of a Michell truss (1904), the optimal design concerns a minimum-weight for the planar truss was generated for a given load and boundaries. In the earlier 1960s, a modern optimization theory was further developed based on mathematical programming and sensitivity analysis (Schmit 1960). This has triggered the intensive development of various mathematical methods for topology optimization and numerous commercial products to solve the design problem with remarkable complexity (Bendsφe and Sigmund 2004). By assuming a design domain subject to a set of boundary condition, the optimal form can be predicted subject to the prescribed performance targets.

The definition of the design domain is based on the traditional node design, following a set of simplified starting points regarding loads, material and other boundary conditions for one of the nodes. The starting points for analysis were as follows:

1. 3D geometry of the original node.
2. To create an isolated model, limited to the part to which the cables are connected. The connection to the strut and light fixture and pin & fork of the cable are not included.
3. The node itself will be connected to a strut on one end and a lighting fixture will be attached on the other end.
4. The node has six cables attached. Loads per cable have a maximum of 109 kN in tension. Design to load path.
5. The cables have a fork and pin that need enough space to be connected. This requires the positions of the connection openings to be some distance apart.
6. The diameter and finishing of the node to be similar to the connecting strut for aesthetical reasons, no design space constraints.
7. Reference material is a suitable steel/stainless steel or aluminum alloy.
8. Minimalize material, weight reduction.

The software package Within Enhance has been used by WithinLab to perform the topology optimization. It features an optimization engine and an integrated FEA solver. Based on the given starting points and boundary conditions WithinLab designed a geometrically optimized node using the state-of-the-art freedom of AM. Figure 3 shows the original node with the direction of the loads from the connecting cables.

The topology optimization method is used to generate an optimal form. The initial condition is modified to make the node self-supporting given the manufacturing process. Based on that, a design space is defined as the maximum volume within which the optimization can occur by removing material where it is less necessary.

A single-target optimization process is then performed with a predetermined target volume percentage and the design objective of minimizing the total material weight. The Von Mises stress is analyzed during the optimization process as one of the main design parameters.

2.2 Production of the Optimized Design

The AM node was produced by CRDM/3D Systems to gain insight in production requirements and costs. For this, possible improvements of the design for production are considered in more detail. Issues to consider were:

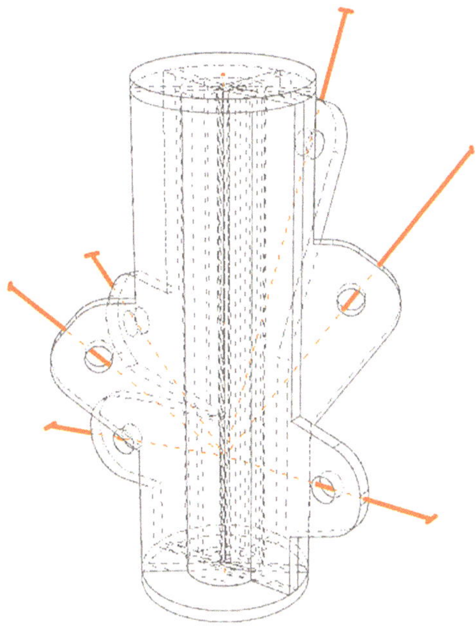

Fig. 3 Direction of loads from connecting cables

1. Size of the node in relation to the laser printer compartment.
2. Number of parts to fit in printing chamber.
3. Layer thickness in relation to printer capacity (Wattage) and production time.
4. Printing direction.
5. Amount of support structure or form changes to prevent support structure.
6. Wall thickness and gaps.
7. Heat dissipation.
8. Availability of printing material (powder).
9. Post-processing, heat treatment, polishing, surface treatment, &c.

Figure 4 illustrates some of the above mentioned issues.

Given the printing direction and geometrical definition as indicated in Figure 5, a hangover analysis has been performed to analyze the self-supporting feature of the node. A general design rule is that overhangs with an angle less than 45° to the horizontal need support structure. This can however vary depending on the production method and machine brand used.

Support structures use up material, increase production time and require several post-processing steps. Eliminating support structure therefore has a positive influence on the cost and quality of the end product.

The product that resulted from the topology optimization has been redesigned to include the AM features so as to speed up the production process and reduce the post-processing, such as support removal or machining.

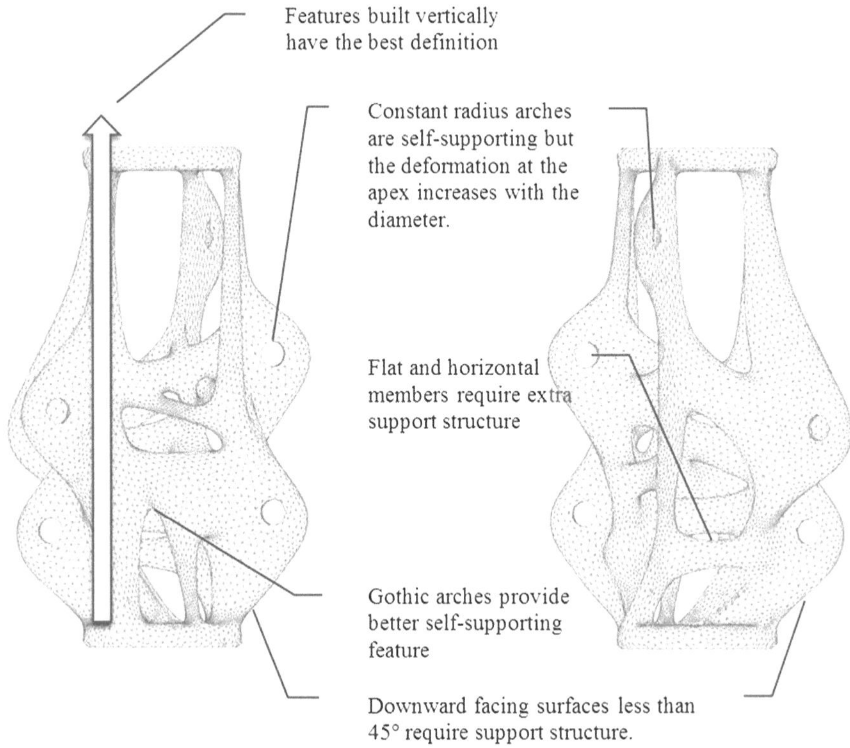

Fig. 4 Key aspects related to design for AM production, *left*: front view of the AM node, *right*: back view of the AM node

The original node was designed for stainless steel but was in the project for budget reasons later changed to galvanized S355 steel. In consultation with CRDM/3D Systems, Maraging steel 1.2709 of ultra-high strength (yield stress of 1,000 N/mm^2) in fine powder is selected for the AM node.

2.3 Detailing of the Original Design for Comparison

The original design has been further engineered and shop drawings were made in order to produce it in a traditional way and to verify and compare the structural characteristics, costs and aesthetics.

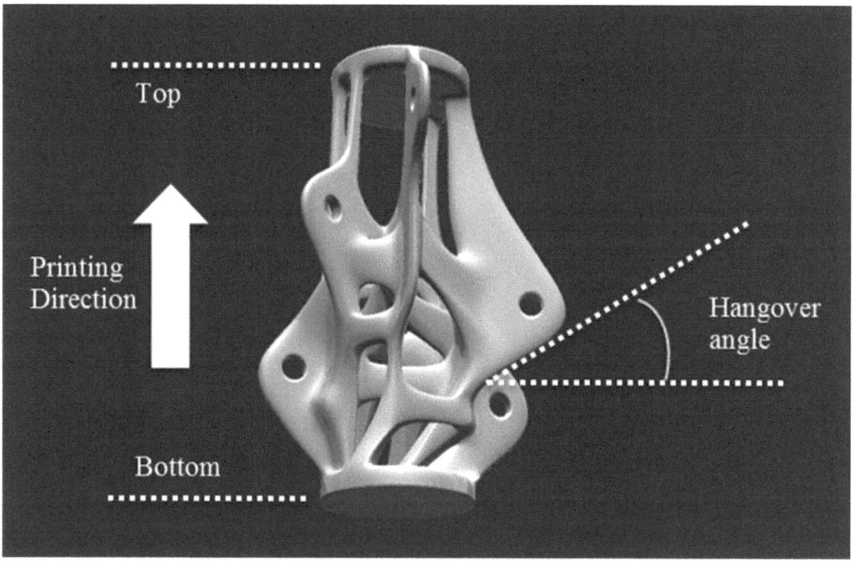

Fig. 5 Printing direction and hangover angle

3 Results

3.1 Topology Optimization

With the given boundary conditions, the optimized design features an efficient organic form which follows the flow of the internal forces within the structure. It is noted that the constraints that have been used for the optimization are set at the top of the node, which is opposite to a realistic situation where the steel node is fixed with the tube at its bottom. Figure 6 shows several steps in the optimization process as done with the WithinLab software.

3.2 Production of the Optimized Design

In order to produce the AM node, the design of the node is further analyzed and optimized by taking into account the production features of the AM process.

Areas with hangover angles smaller than the selected angle of 45° are highlighted in red in Fig. 7 to evaluate the node's self-supporting features.

With this knowledge the node has been further optimized for production. Some of the changes are only meant as an example of the possibilities as they would normally require to be implemented earlier in the design process. Figure 8 shows the design after topology optimization (left) and after design for production (right).

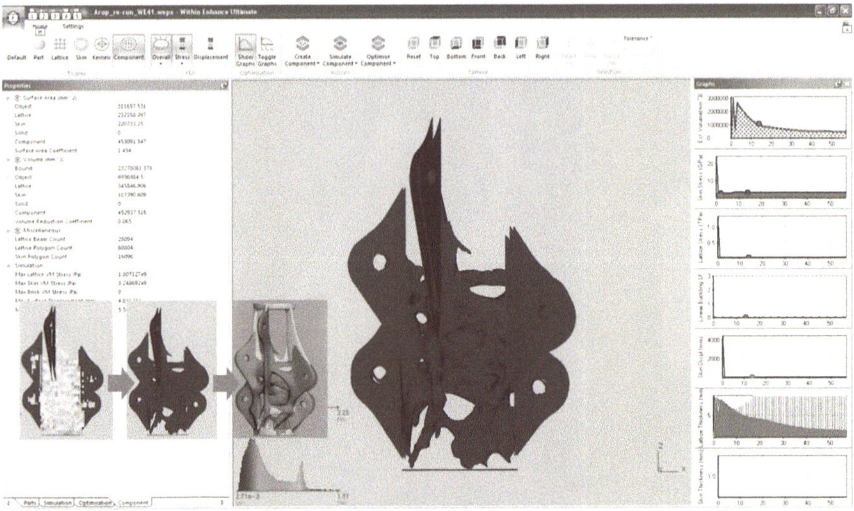

Fig. 6 Optimization process with within Enhance, by WithinLab

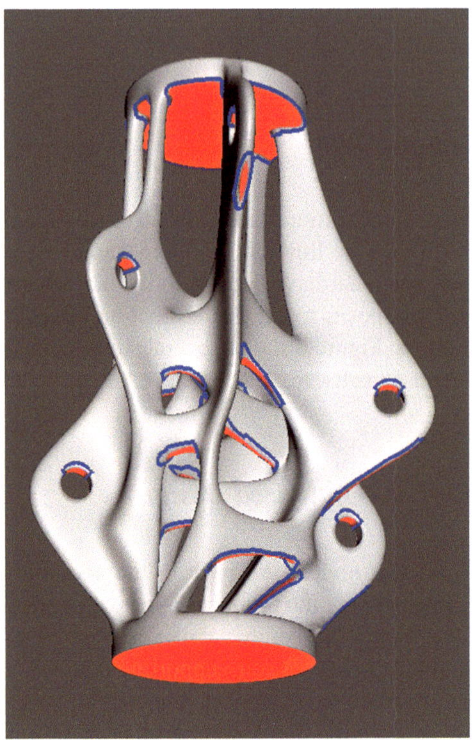

Fig. 7 Hangover analysis with in *red* areas that exceed a 45° angle

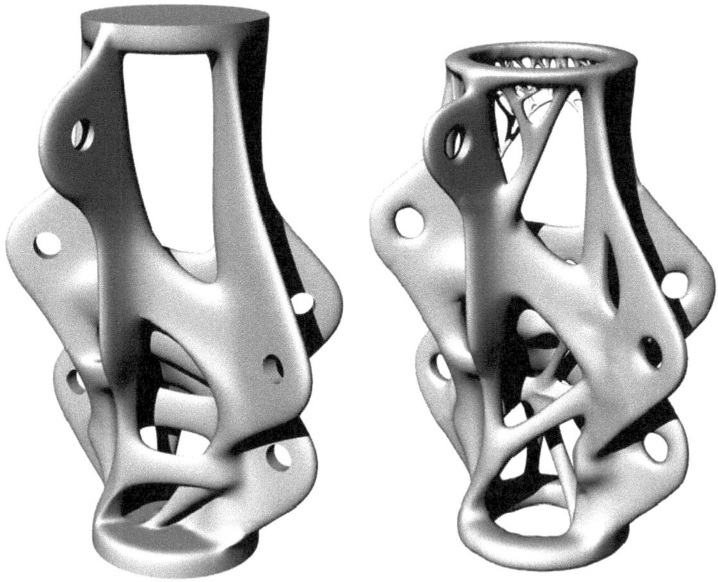

Fig. 8 The optimized node after topology optimization (*left*) and the further optimized-for-production model (*right*)

Adaptations to the design made after the topology optimization are:

- Move holes for connecting cables closer to the center.
- Reduce the height of the element from 420 to 370 mm.
- Change the support condition at the bottom.
- Replace the top plate for lighting support with a ring to reduce the weight and support structure.
- Prevent support structure by including results from overhang analysis.
- Reshape the constant radius arches to gothic arches and add chamfers to eliminate the support structure.
- Add branch elements near top plate to have the geometry itself become the support structure.
- Further remove material where stress level is low based on the results of the structural analysis as shown in the next paragraph.
- Replace flat and horizontal members to reduce the support structure needed.

3.3 Detailing of the Original Design for Comparison

Both the traditional node and the node designed for AM have been produced on a 40 % scale due to financial and production considerations. The comparison on

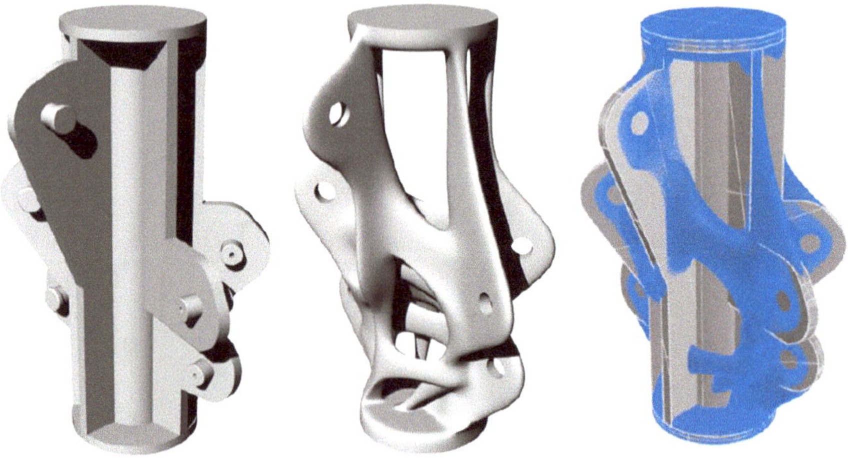

Fig. 9 Comparison of the traditional design and the design after the topology optimization. *Left* to *right*: traditional design, design after the topology optimization, comparison with two models overlapping at the same position

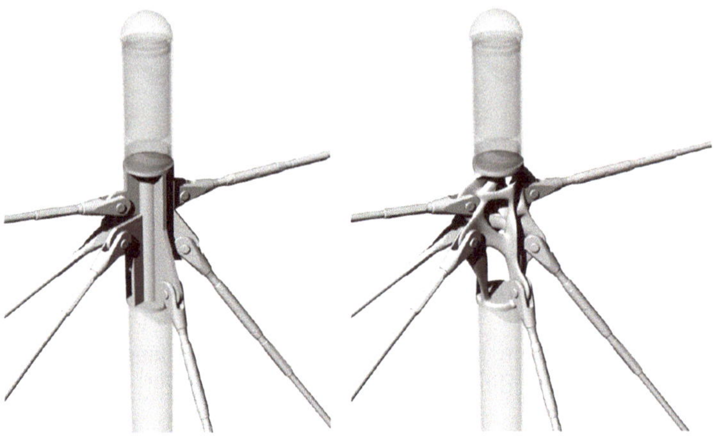

Fig. 10 Comparison of the traditional and the topology optimized design in the structure

aesthetics, structural behavior and costs is presented in this paragraph and is based on the original size of the node.

Aesthetics Figures 9, 10, 11, and 12 show different versions of the original and optimized node for reasons of comparison.

Structural Comparison Both structures are able to take the design loads. Figure 13 shows the calculated stresses in the two designs. The Von Mises stresses are lower than the yield stress with the exception of local peak values which would be acceptable given the plastic behavior of the steel and the ultimate tensile stress

Fig. 11 A model of the traditional node, scaled to 40 %, in galvanized steel. It is produced in the traditional way by cutting and welding (*left*). A model of the node, scaled to 40 %, produced by Additive Manufacturing in Maraging steel (*right*)

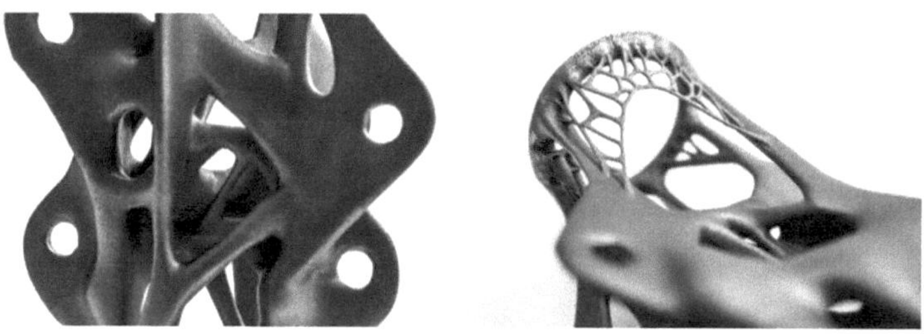

Fig. 12 Two close-up images illustrating the complex structure in the middle of the AM node (*left*) and showing the integrated support structure of the upper ring (*right*)

(510 MPa) for S355. Note that the design load forms an unrealistic combination that is in equilibrium only when a shear force at the support at the bottom plate is accepted. More realistic combinations would likely lead to lower stresses.

Cost Comparison The costs of several parameters are estimated for the original node and the AM node. Figure 14 shows an overview of the costs for the original node, the AM node when produced in 2–5 years time and the predicted costs for a further optimized node, based on the lessons learnt during this design process.

The costs are based on the original size of the nodes, not the scaled version.

The printing costs of the current design for AM are high at this point in time. This is mainly due to the relatively large size of the elements. Printing with the currently available machines might take up to 15 days with a 200 W laser, printing in 14 μm thick layers. Next generation lasers will have a much higher wattage (1,000 W), printing 10 times faster using fewer layers. Also the size of the building chambers

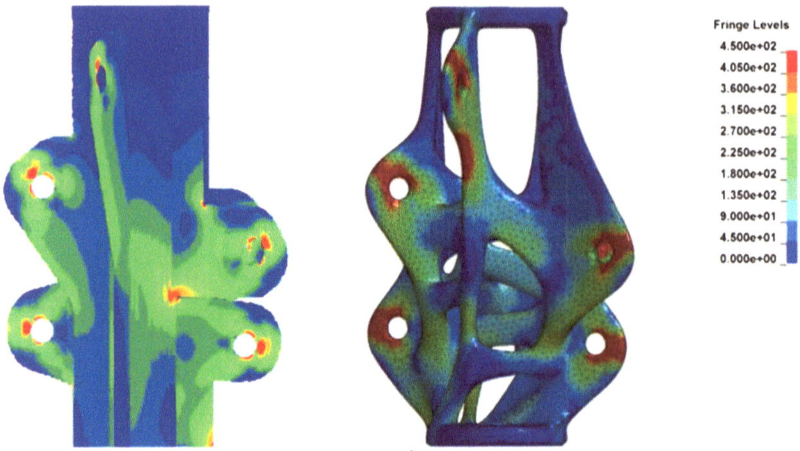

Fig. 13 Plot of Von Mises stresses [MPa] in the original design (*left*) and the design after topology optimization (*right*). Highest stresses can be observed in *red*, lowest in *blue*

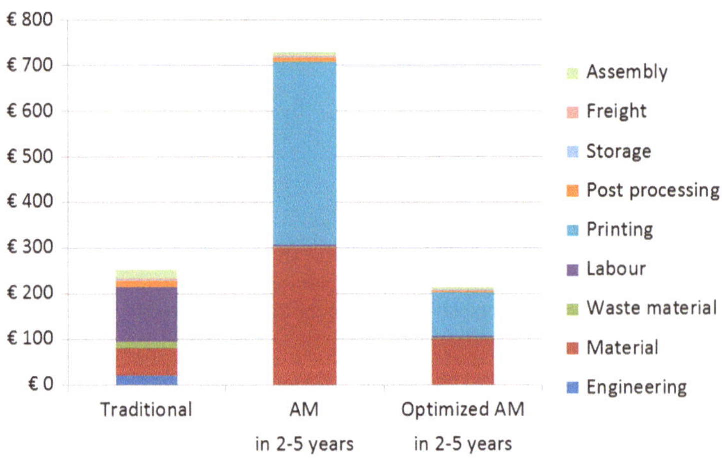

Fig. 14 Cost comparison for traditional fabrication versus AM in the near future, also for a further optimized node

of the AM machines is increasing. In larger printers, more pieces will fit the printing chamber further reducing the cost per piece. These bigger and faster machines can be available after testing of the results as soon as the coming year.

In the figure above a cost comparison is also added for a further optimized node, on which the research team is working at the moment. The node can be reduced in size by redesigning it including the connection points to the cables and light fixture.

Also a possible 60 % further reduction of material by making use of the higher strength of the available Maraging steel powder is assumed.

4 Discussion

The topology optimization resulted in an organic form with less material while the original functions as cable connectors are still ensured. During the process of the research, we realized that the initial boundary conditions for the topology optimization of the node restricted a more optimal outcome:

- Because of the almost three times higher strength of Maraging steel a substantial reduction of the material can still be obtained, further reducing the total weight of the element. This will be in addition to the 15 % weight reduction in the current optimization.
- The structural comparison between the original node and the AM node shows us that although both structures can take the design loads, the AM structure does not yet utilize its ultra-high strength. Some simple tensile tests on printed test samples from the same material confirm the potential applicability of this additional strength.[1] Additional testing on the printed steel parts will be necessary to safely incorporate this production method in the building industry.
- The AM design can be optimized to reduce the peak values thus making more efficient use of the material. Eliminating these stress concentrations would also be highly beneficial for structures with fatigue issues.
- Lowering the weight would have a significant effect on the forces in and size of the cables in the tensegrity structure as a whole, attributing to further reductions.
- The loads and the connection of the cables can be optimized to make a more compact node
- Incorporating the connection of the lighting fixture could lead to additional advantages.
- We also expect that a more integrated design approach, in which the topology optimization and optimization for production are executed in parallel, will result in a more optimal solution.

The studied node would today not be an economic alternative for the traditionally machined and welded node. The large size would require other parameters as maintenance or aesthetical motives to govern, even when printed in a few years. However for a further optimized version reducing size and material use, the printing costs may soon be less than the high labor cost of the traditional node.

Even though there is still a lot of work and testing to be done, we believe the initial results are very promising. Our work gets close to the essence of *form follows function* and besides the different aesthetics, which should be judged subjectively,

[1] No strength-enhancing post-processing operations were applied to the test samples after printing.

the design freedom, material savings, limited storage and transport costs and possible installation benefits should be enough of a stimulus to continue the research to see whether we can add Additive Manufacturing to our list of possible production techniques in the Building Industry.

If we manage to get further proof of the safe applicability of this production technique, we believe that it could change the way we design. Not only on a structures level, but for buildings and architecture as a whole. We believe this technique should be used next to the existing ones that have proven their value over many decades of construction. But from the moment that we believe Additive Manufacturing can safely be used, our imagination will be the only limiting factor in designing for the future.

5 Future Steps

Our follow-up research will be two-fold: to process the insights mentioned in chapter “Application of Hybrid Glass-Timber Elements in Architecture” involves a complete re-evaluation of the starting points and redesign of the node, which will be the next step in our work. We will also try to integrate the functionality of other products in the tensegrity structure in the node – increasing the complexity of the node, but limiting the amount of products in the structure and making installation easier.

The second part of our research will focus on material testing. We are setting up a complete list of tests, necessary to get the right input and feedback on reliability and potential limitations of the powdered materials in combination with the production technique.

Acknowledgements We are grateful for the support of Arno Held from EOS e-Manufacturing Solutions in the early stages of the research project. Dr. Siavash H. Mahdavi and his software company Within have provided invaluable input for the topology optimization. Daniel Kirk and Simon Hammond have been very supportive during the AM production phase. We learned a lot from their knowledge and expertise and hope to continue to work with them in future AM related projects.

References

ASTM International, Designation: F2792-12a, Standard Terminology for Additive Manufacturing Technologies, p. 2

Bendsφe, M.P., Sigmund, O.: Topology Optimization; Theory, Methods and Applications, 2nd edn. Springer, Berlin (2004)

Direct Manufacturing Research Center.: Thinking Ahead the Future of Additive Manufacturing – Innovation Roadmapping of Required Advancements, Paderborn (2013)

Galjaard, S.: Tensegrity structures with integrated street lighting. IABSE Future of Design, IABSE UK Newsletter, 35th edn. (2013)

Michell, A.G.M.: The limits of economy of material in framed structures. Phil. Mag. **6**, 589–597 (1904)

Schmit, L. A.: Structural design by systematic synthesis In: Proceedings of 2nd ASCE Conference Electronic Computation. ASCE, New York (1960)

Snelson, K.: The art of tensegrity. Int. J. Space Struct. **27**(2 & 3), 71–80 (2012)

Strauss, H.: AM Envelope – The Potential of Additive Manufacturing for Façade Construction. Delft University of Technology, Faculty of Architecture, Architectural Engineering + Technology department, Delft (2013)

Interactive Modeling of Architectural Freeform Structures: Combining Geometry with Fabrication and Statics

Caigui Jiang, Chengcheng Tang, Marko Tomičić, Johannes Wallner, and Helmut Pottmann

Abstract This paper builds on recent progress in computing with geometric constraints, which is particularly relevant to architectural geometry. Not only do various kinds of meshes with additional properties (like planar faces, or with equilibrium forces in their edges) become available for interactive geometric modeling, but so do other arrangements of geometric primitives, like honeycomb structures. The latter constitute an important class of geometric objects, with relations to "Lobel" meshes, and to freeform polyhedral patterns. Such patterns are particularly interesting and pose research problems which go beyond what is known for meshes, e.g. with regard to their computing, their flexibility, and the assessment of their fairness.

1 Introduction

Architectural projects of high geometric complexity greatly benefit from the incorporation of essential aspects of function, fabrication and statics into the shape modeling process. This integrated view is one of the major goals of Architectural Geometry. One has to avoid detailed physical simulation as this would hardly be compatible with interactive shape manipulation. Instead, we aim at developing shape modeling tools which employ simplified mathematical models in order to respect manufacturing and structural constraints. Still, one needs to develop efficient numerical solvers for the arising systems of constraints. In the present paper, we briefly address a few core ideas on the mathematical formulation and mainly discuss a class of numerical algorithms which are well suited for interactive design.

C. Jiang (✉) • C. Tang • H. Pottmann
KAUST

M. Tomičić
Waagner Biro Stahlbau

J. Wallner
TU Graz

H. Pottmann
TU Wien

P. Block et al. (eds.), *Advances in Architectural Geometry 2014*,
DOI 10.1007/978-3-319-11418-7_7

From a mathematical perspective, the problems to be solved are not typical constrained optimization problems. Constraint-aware design is about the solution of under-determined systems of usually nonlinear, and frequently redundant, equations and inequalities and about ways of guiding the user towards preferred solutions. This guidance is provided by additional targets related to the aesthetics of a design and other goals which depend on the specific application scenario and task.

1.1 Prior Work

The methodology presented in our paper is applicable to a wide range of problems in Architectural Geometry, but we confine our survey of prior work to the integration of structural constraints and to interactive constraint-aware design.

Statics-aware design The incorporation of structural constraints into shape design is probably best accomplished by aiming at force equilibrium. This approach is taken within the *thrust network method* in connection with the design of self-supporting structures (Block and Ochsendorf 2007; Block 2009; Block and Lachauer 2011; Vouga et al. 2012; Panozzo et al. 2013; de Goes et al. 2013; Liu et al. 2013), as well as closely related form-finding methods for compression support structures (Lachauer and Block 2012), fabric formwork for concrete shells (Van Mele and Block 2011) and tension structures (Barnes 2009).

The design of structures which are built from simple elements (such as planar quad panels) and which are structurally sound at the same time, is much less investigated. Schiftner and Balzer (2010) do not simultaneously consider statics and the rationalized geometry. Vouga et al. (2012) show how to rationalize any self-supporting shape by a self- supporting planar quad mesh, but do not directly design it. Form-finding with polyhedral meshes in static equilibrium is the topic of Tang et al. (2014), of which our paper is a follow-up.

Interactive design tools for freeform architecture Most early work in architectural geometry was already aimed at fabrication-aware design and realized this goals by means of constrained optimization algorithms. The most successful ones are already incorporated in commercially available software (e.g. "*EvoluteTools*").

A lot of recent progress comes from the research group of M. Pauly at EPFL. Their *shape-up* algorithm (Bouaziz et al. 2012) uses projections onto individual constraints, but has the disadvantage that each constraint requires a separate treatment. Local shape modifications of constrained models using ideas from sparsity have been presented by Deng et al. (2013). Finally, Deng et al. (2014) extended *shape-up* within an augmented Lagrangian formulation to a framework which aims at real-time manipulation and a careful distinction between hard constraints and soft targets. Poranne et al. (2013) introduced plane coordinates as auxiliary variables for modeling polyhedral surfaces. In this way, the planarity constraint is quadratic. This approach motivated the systematic avoidance of constraints of degree ≥ 3 by Tang et al. (2014) which forms the basis of the present paper. Extensive comparative tests showed that it clearly outperforms previous work in terms of the combination of speed and high accuracy of constraint satisfaction. Moreover, it is well suited for the integration of statics based on the thrust network method.

Remark Tools based on evolutionary optimization such as *Galapagos* attracted a lot of attention in the architectural community. Their simplicity comes at the cost of low efficiency and accuracy and thus they are not suitable for the interactive modeling tasks we are interested in here.

1.2 Contributions and Overview

In Sect. 2, we briefly outline the surprisingly effective method of Tang et al. (2014) which is suitable for interactive design of constrained meshes and more general structures. Section 3 discusses form-finding with polyhedral meshes. We show how to design meshes with planar faces subject to further constraints such as alignment with a given boundary, force equilibrium, and others. This method is very well suited for modeling structures with repetitive elements. This is demonstrated in Sect. 4 by means of so-called honeycomb structures and *Lobel frames* which are closely related to them. In Sect. 5 we turn to the wider topic of polyhedral patterns and discuss ways for assessing and controlling the aesthetics of such patterns. Along with a discussion of our findings and insights, we address future research directions in section "Conclusion".

2 Computational Setup

The computational setting we are about to describe is very general, and we illustrate it by means of a few examples which also serve to introduce the geometric objects which are the focus of this paper. Their interrelations are the topic of later sections. Each time we describe a set of variables which define a certain geometric structure, provided certain constraints are fulfilled. We aim at constraints which are linear or quadratic equations, and which involve as few variables as possible.

"Lobel" meshes (cf. Fig. 1), are defined by having equilateral triangles as faces. We use the vertices $\mathbf{v}_1, \mathbf{v}_2, \ldots$ as variables, together with the edge length "l". The equilateral property reads $(\mathbf{v}_i - \mathbf{v}_j)^T(\mathbf{v}_i - \mathbf{v}_j) = l^2$, for all edges $\mathbf{v}_i\mathbf{v}_j$. We conclude that the set of constraints imposed on a triangle mesh in order to enforce the Lobel property is one quadratic equation per edge.

Polyhedral meshes A more complex example are meshes whose faces can be arbitrary n-gons, but are required to be planar. In addition to vertices, we here use the normal vectors $\mathbf{n}_1, \mathbf{n}_2, \ldots$ as variables. The condition that face No. k is planar is expressed by $(\mathbf{v}_i - \mathbf{v}_j)^T\mathbf{n}_k = 0$, for all edges $\mathbf{v}_i\mathbf{v}_j$ of that face. It is also convenient to require the normalization $\mathbf{n}_k^T\mathbf{n}_k = 1$. Again, we get a set of quadratic constraints.

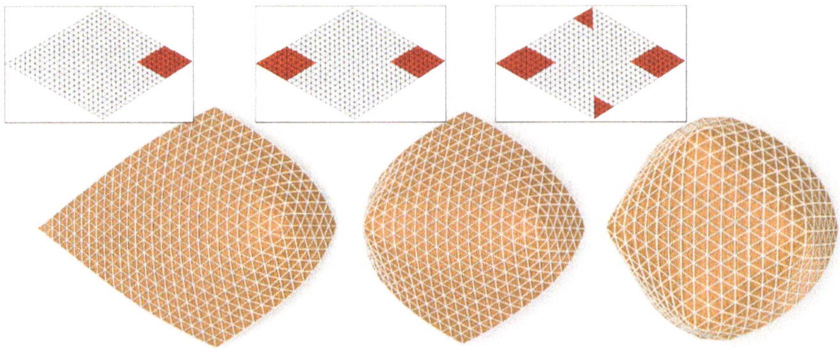

Fig. 1 *Lobel Meshes*. This series of images visualizes surfaces composed of equilateral triangles created by cutting out pieces of a diamond (*red areas* in *inset figures*) and gluing the newly arising boundaries together. Finding such surfaces is one example of a mesh optimization problem with constraints (here: equal edge lengths in a triangle mesh). Structures made from equilateral triangles are called "Lobel" structures in honor of the French architect Alain Lobel who intensively studied them (See Lobel (1993))

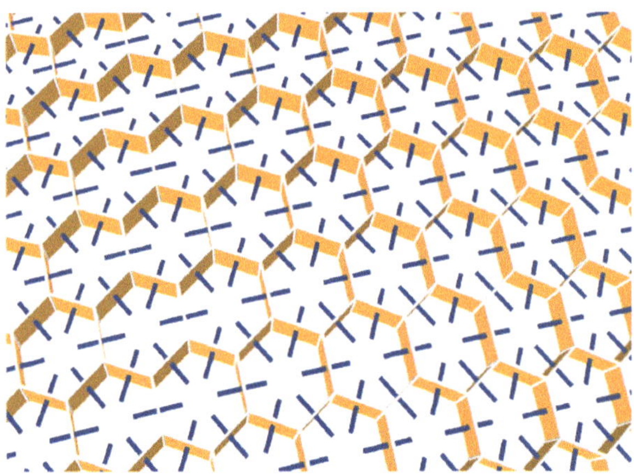

Fig. 2 Designing *honeycomb structures* is an instance of modeling with constraints: A collection of quadrilaterals form the walls of a honeycomb structure, if their combinatorial arrangement is along the edges of a hexagonal-dominant mesh, and their intersection angle is 120°. Constraints are planarity of quads and the correct intersection angles. Note that one can find a Lobel mesh which intersects the honeycomb orthogonally, and vice versa

Honeycomb structures as defined by Fig. 2 are studied by Jiang et al. (2014). They are arrangements of quadrilaterals combinatorially different from the arrangement of faces of a mesh; however the constraints describing their planarity are the same. The required intersection angles of 120° are most elegantly expressed by requiring that whenever faces No. i, j, k meet in a common axis, their normal vectors must form an equilateral triangle, which results in the equation $\mathbf{n}_i + \mathbf{n}_j + \mathbf{n}_k = \mathbf{o}$.

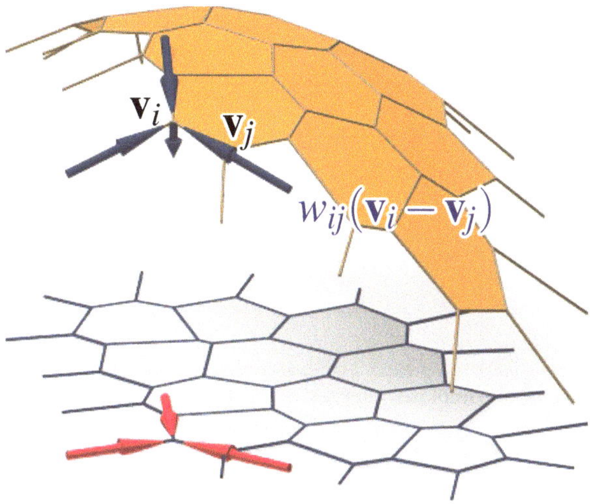

Fig. 3 *Force equilibrium.* The simplest case is that forces $w_{ij}(\mathbf{v}_i - \mathbf{v}_j)$ in edges $\mathbf{v}_i\mathbf{v}_j$ counterbalance the weight of edges, which is modelled by the weight "ρ" per unit length. With $(0, 0, -1)$ as direction of gravity, force balance at vertex $\mathbf{v}_i$ reads

$$(0, 0, -1) \cdot \sum_{j\sim i} \rho l_{ij} = \sum_{j\sim i} w_{ij}(\mathbf{v}_i - \mathbf{v}_j)$$

where summation is over all vertices $\mathbf{v}_j$ connected to $\mathbf{v}_i$ by an edge, and the edge lengths are defined by $l_{ij}^2 = (\mathbf{v}_i - \mathbf{v}_j)^T(\mathbf{v}_i - \mathbf{v}_j)$

Self-supporting meshes We wish to incorporate *forces* in our computational setting, since meshes with compressive equilibrium forces in their edges (Fig. 4) play an important role e.g. in the stability analysis of masonry, see e.g. Block and Ochsendorf (2007). The force which vertex No. j exerts on vertex No. i has the form $w_{ij}(\mathbf{v}_i - \mathbf{v}_j)$. Since $w_{ij} = w_{ji}$, information on forces is stored via one force coefficient w_{ij} per edge. This edge experiences compression if $w_{ij} \geq 0$. The inequality $w_{ij} \geq 0$ is made an equality by introducing a dummy variable ω_{ij} and requiring $w_{ij} = \omega_{ij}^2$ (while w_{ij} represents a force per edgelength, ω_{ij} is there only to assist in a mathematical trick and does not have a physical interpretation).

As to constraints, we must formulate what the forces should be in equilibrium with. The simplest case, which e.g. applies to Fig. 4b, is discussed by Fig. 3. Again, all equations are quadratic. For details and other ways to define the weight, e.g. proportional to area as shown by Fig. 4a, see Tang et al. (2014).

Projection onto the constraint space Tang et al. (2014) describe how to treat constraints of the kind described above (the restriction that constraints should be at most quadratic is not severe, since any equality or inequality expressible in polynomial form can be made quadratic, by introducing products of two existing variables as new variables, until the polynomial degree is down to 2). They describe how to quickly solve the system of constraints, and thus make interactive modeling

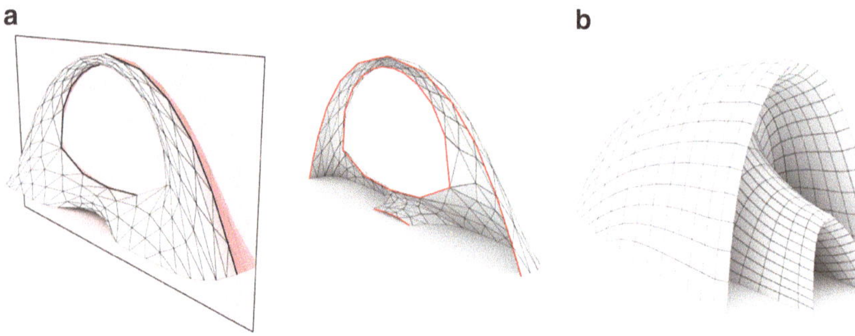

Fig. 4 *Self-supporting meshes* represent a bar-and-joint framework with compressive forces in the edges counterbalancing the deadload. They serve as fictitious "thrust networks", proving stability of freeform masonry via Heyman's Safe Theorem (**a**), or as the basis of structures made of beams and rigid joints (**b**) (*Remark:* These images do not show forces)

of constrained geometry possible. A typical application of their method would be to implement a graphical interface where a user can interactively create and modify a given geometry, while the system performs *projection onto the constraint space*, i.e., searching for variables which fulfill the constraints and which are close to the values they had before, and close to those the user wanted.

For their algorithm, Tang et al. (2014) make use of a fairness energy "$E(\mathbf{x})$" which is a quadratic expression in the collection $\mathbf{x}$ of variables. The definition of E varies depending on the application. The method is a Newton iteration; in the i-th iteration the nonlinear system of constraint equations is converted to a linear system of the general form $A_i\mathbf{x} - \mathbf{s}_i = \mathbf{o}$, whose solution $\mathbf{x}_i$ is computed. The iteration stops whenever the desired accuracy is reached or $\mathbf{x}_i$, $\mathbf{x}_{i+1}$ are equal for all practical purposes. Linearization of quadratic equations is a standard procedure; nonlinear constraints like vertices being confined to a curve or surface are linearized by replacing those curved objects by their tangent resp. tangent plane. Since the system of constraints is both underdetermined and redundant, one cannot solve $A_i\mathbf{x} - \mathbf{s}_i = \mathbf{o}$ directly. Instead $\mathbf{x}_i$ is found via regularization, as minimizer of $\|A_i\mathbf{x} - \mathbf{s}_i\|^2 + \epsilon E(\mathbf{x}_i) + \epsilon'\|\mathbf{x}_i - \mathbf{x}_{i-1}\|^2$, with $\epsilon, \epsilon' \ll 1$. For details we refer to Tang et al. (2014).

3 Interactive Form-Finding with Polyhedral Meshes

Loss of design freedom by geometric constraints Meshes with planar faces are important objects of architectural geometry, since they represent the shapes of steel-glass-constructions. Their modeling is, of course, easiest for triangle meshes, because planarity is automatic there. Quad meshes are more of a challenge. It is known that "fair" quad meshes (whose mesh polylines emulate the isoparameter lines of smooth surfaces) offer little design freedom once the reference surface to be covered is fixed. If edges are required to intersect near-orthogonally, they must already follow the principal directions of the reference surface (Liu et al. 2006), leaving only the density of edges as a design element. This is true also for quad meshes whose edges carry forces counterbalancing vertical loads: The edges must follow *relative* principal curvature lines (with the Airy stress potential assuming the role of unit sphere, cf. Vouga et al. (2012)). With either side-condition it is usually impossible to perform a fair quad meshing of the reference geometry such that the mesh's boundary is nicely aligned with the original surface's boundary.

Interactive modeling of polyhedral meshes If one can solve the constraint equations quickly enough, however, then interactive geometric modeling becomes available. It is no longer necessary to perform surface analysis and subsequent remeshing as proposed e.g. by Liu et al. (2006) and Vouga et al. (2012). Figure 4b shows a self-supporting quad mesh created in this way. For details see Fig. 5.

The computational setup for this projection is as described in Sect. 2. As fairness energy we use the sum of terms of the form $\|\mathbf{v}_i - 2\mathbf{v}_j + \mathbf{v}_k\|^2$, for each possible choice of consecutive vertices $\mathbf{v}_i, \mathbf{v}_j \mathbf{v}_k$ on a mesh polyline.

We do not pursue this topic further, since it is anyway treated by Tang et al. (2014). We mention only a few more constraints easily incorporated into this setup, such as symmetry w.r.t. to a plane, or w.r.t. to rotation about an axis, or inequalities involving panel sizes. Another constraint is a prescribed total area of the mesh, or a prescribed enclosed volume, see Fig. 6. Constraints like the latter however involve many more variables than the ones expressing local properties like planarity. They will slow down computations, since the matrices A_i will no longer be sparse.

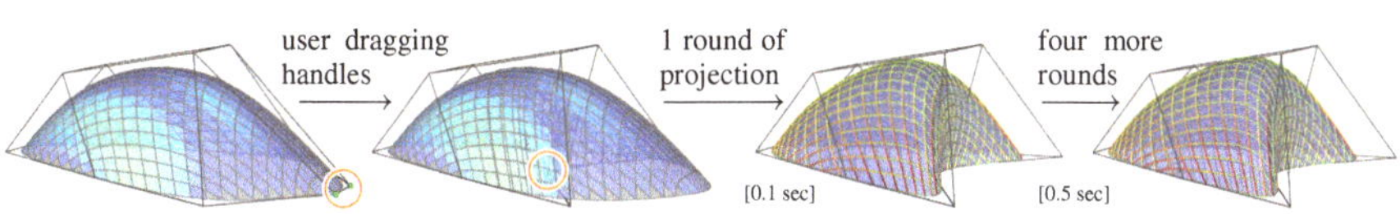

Fig. 5 *Interactive modeling of forces in the edges of a quad mesh under planarity constraints.* We use the method of Tang et al. (2014) to combine subdivision and projection onto the constraint space (defined by planarity of faces, the self-supporting property, and the condition that boundary vertices must not leave the boundary curve of the subdivision surface generated by the control points as set by the user). (**a**) The user works on the control points of a mesh created by subdivision. (**b**) Each time the user moves the control points and releases them, the system preforms projection onto the constraint space. (**c**) and (**d**) Numerical information is conveyed by color coding the faces (*blue* means planar) and edges (color corresponds to magnitude of forces)

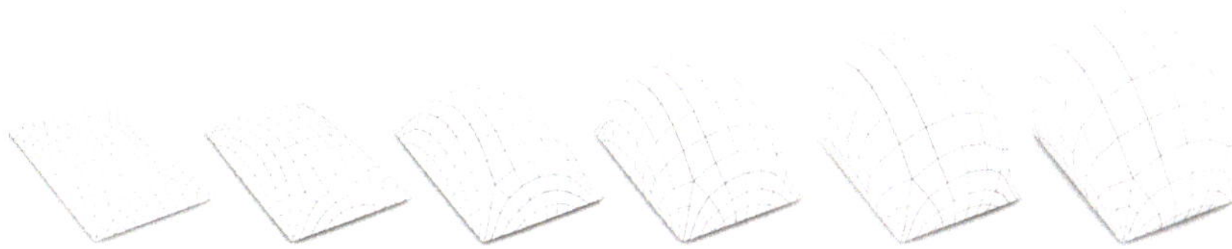

Fig. 6 Interactively modifying a polyhedral mesh by increasing the enclosed volume

4 Element Repetition: Honeycomb Structures and Lobel Frames

For fabrication of freeform skins in architecture, it is very relevant if a substantial number of parts is the same. For example Singh and Schaefer (2010) studied meshes where only a small number of shapes of triangles occur. Here we are very much interested in two kinds of geometry: meshes where all faces are the same, and support structures where all nodes are the same. These two questions lead us to the concept of Lobel mesh on the one hand, and honeycomb structure on the other hand.

Lobel meshes Frameworks built from equilateral triangles are sometimes called "Lobel frames", after the French architect Alain Lobel who intensively studied them. We are here interested in triangle meshes all of whose faces are equilateral triangles and which we would like to call *Lobel* meshes. An equivalent definition of such a Lobel mesh is that all edges have the same length. Yet another definition is that all angles in the mesh are 60°. Examples are shown by Figs. 1, and 7.

The duality Honeycomb – Lobel mesh Objects dual to Lobel meshes are honeycomb structures, defined as an arrangement of open hexagonal cells bounded by quadrilateral walls, such that the cells and walls follow the faces and edges of a hexagonal mesh, respectively. In addition we require that walls intersect at 120° – see Fig. 2. This leads to the important property that any structure made of beams which follow the walls of the honeycomb has *congruent nodes*. For modeling and properties of honeycombs see Jiang et al. (2014).

Given a simply connected honeycomb, one can step by step construct a Lobel mesh whose edges are orthogonal to the walls of the honeycomb, see Fig. 8. Conversely, given a Lobel mesh, one may construct the corresponding honeycomb. The honeycomb, however, carries more geometric information than the Lobel mesh: the latter is uniquely determined, up to scale, by the planes carrying the walls of the honeycomb. The information on the location of vertices of the honeycomb is lost. This duality between honeycomb structures and Lobel meshes is easy to see, but it does not seem to have been mentioned in earlier publications. We also mention that Lobel meshes are a special case of the *circle packing meshes* studied by Schiftner et al. (2009), since placing spheres at the vertices (whose diameter is the mesh's edge length) yields a packing.

Discrete developable surfaces Lobel meshes are *developable*, meaning that every 1-ring neighbourhood of a valence 6 vertex can be mapped isometrically into the plane without stretching or tearing. This follows immediately from the fact that all angles in the mesh are 60°. Depending on the global topology and geometry of the mesh, one can grow this developable neighbourhood by adding more and more triangles. If a vertex has valence different from 6, we must cut the Lobel mesh near that vertex to develop it into the plane, cf. Figs. 1 and 10b.

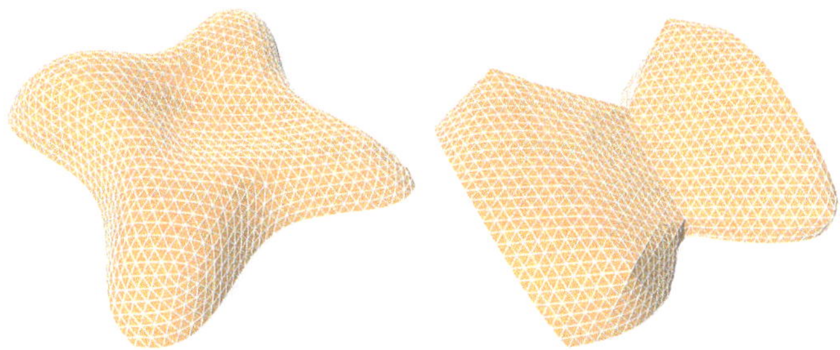

Fig. 7 Projecting a triangle mesh onto the constraint space defined by the "Lobel" property makes all faces equilateral and the mesh developable, apart from cuts necessary to accommodate valence 5 vertices

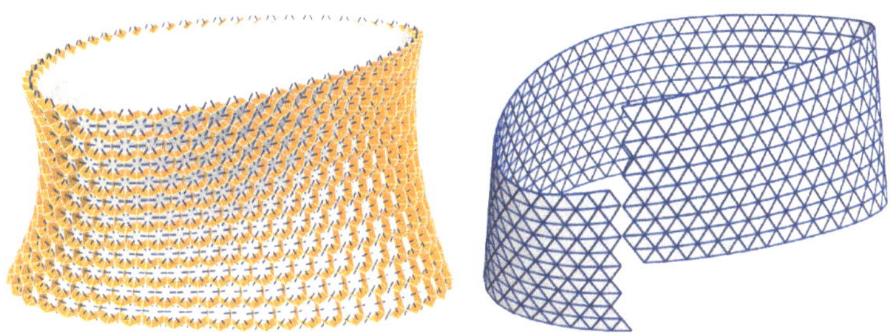

Fig. 8 *Lobel meshes and Honeycombs*. A combinatorial dual to the walls of a honeycomb (*left*) is a triangle mesh where edges intersect at 60° (*right*), implying the Lobel property. Step-by-step construction of the Lobel mesh orthgonal to the honeycomb is possible only for simply connected honeycombs; otherwise the construction does not globally close

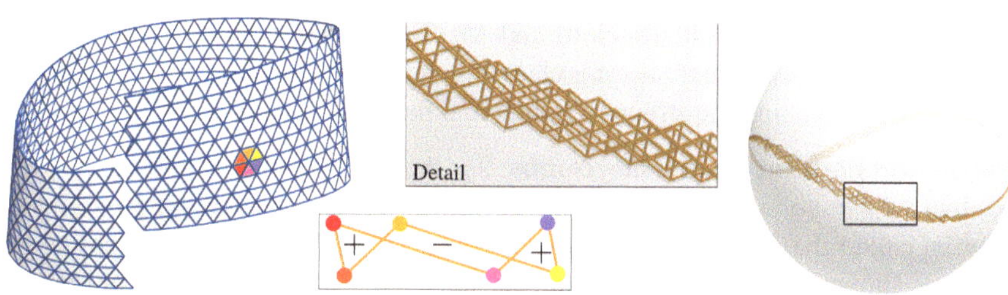

Fig. 9 *Local and global properties of Lobel meshes*. The image at *right* illustrates the curve-like *spherical image* of the Lobel mesh at left. It follows from the discussion in Jiang et al. (2014) that the normal vectors of a 1-ring neighbourhood of any vertex typically form a zero area hexagon on the unit sphere

A Lobel mesh which approximates a *smooth* surface should therefore have properties similar a developable surface. One such property is that developables have only a 1-dimensional variety of normal vectors in contrast to the 2-parameter set which non-developable surfaces have. Lobel meshes exhibit a discretized version of this behaviour, cf. Fig. 9: When constructing a Lobel mesh from a honeycomb, the intersection lines of walls become the normals of faces in the Lobel mesh. We visualize those normals by the *spherical image* of the Lobel mesh. As Jiang et al. (2014) remark in their discussion of honeycombs, the spherical image of the Lobel mesh consists of zero-area spherical hexagons (which follows from the Gauss-Bonnet theorem), and a certain fairness/regularity implies that these hexagons fill a curve-like stripe on the unit sphere. For details we refer to that paper.

5 Novel Forms of Fairness: Polyhedral Patterns

Motivation We already discussed the problematic task of representing a given reference shape by a mesh with planar quadrilateral faces. If the mesh polylines are to mimick the isoparameter lines of a smooth surface parameterization e.g. in the manner of Fig. 6, we have very little design freedom, because such a quad mesh is never far away from the network of principal curvature lines, cf. Liu et al. (2006) and Zadravec et al. (2010). If the reference surface is not convex, the resulting meshes might easily be unacceptable (besides the fact that lack of design freedom is frequently unacceptable in itself). There are several ways out of this situation:

1. One might give up the condition that all faces are quadrilaterals and thus gain design freedom;
2. One might give up the condition that the mesh strictly follows the reference geometry (making use of interactive modeling tools);
3. One might give up the condition of smoothness/regularity.

Solution No. 1 has been sought for the "flying carpet" roof in the Cour Visconti in the Louvre, Paris, by R. Ricciotti and M. Bellini. Even if not visible through the outer skin of triangular shading elements, many of the glass panels below are quadrilaterals, making the structure lighter and have fewer parts.

Polyhedral patterns from honeycombs For the third solution proposed above, it is obviously important that regularity is given up in an aesthetic and regular (if it may be called that) way. What we mean by this is demonstrated by means of the example shown by Fig. 10: Jiang et al. (2014) have proposed to derive polyhedral patterns from freeform honeycomb structures. The honeycomb contains a "top layer" hexagonal mesh, and in this mesh we split hexagons in half by introducing additional edges. The resulting quad mesh is being planarized by means of the computational framework of Tang et al. (2014). We cannot however employ this algorithm in the way in its original form, since it is guided by a smoothness energy which is not appropriate here.

In order to derive an alternative smoothness energy, we have a closer look a the patterns under consideration. Obviously patterns Fig. 10c, d are quite regular, but the edges do not form "straight" mesh polylines. This is even less so for pattern Fig. 10e. It is still possible, however, to select sequences of vertices on a succession of

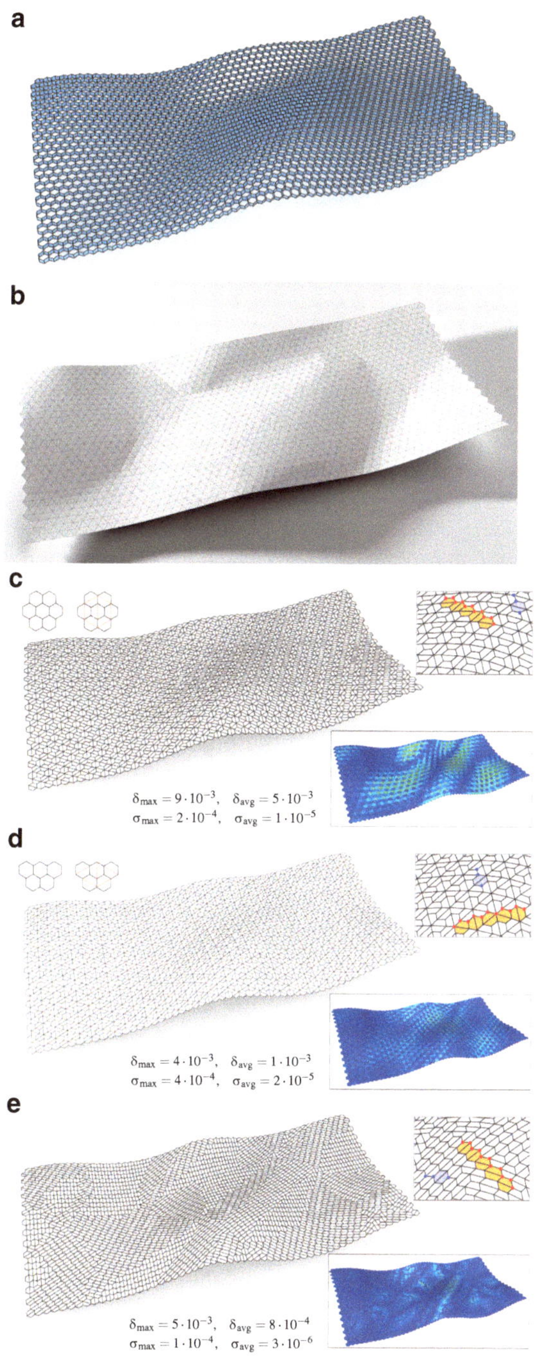

Fig. 10 *Honeycombs and polyhedral patterns.* (**a**) A honeycomb following the "flying carpet" roof in the Cour Visconti in the Louvre, Paris. Its walls are near-orthogonal to the reference geometry in so far as they have been initialized this way, before projection onto the constraint manifold has been applied. (**b**) The Lobel mesh dual to this honeycomb. Developability of this surface is visualized by a rendering with tangential light, recalling crumpled paper. (**c**–**e**) (*opposite page*) Consider the top layer hex mesh of the honeycomb in (**a**), introduce new edges (colored strokes in left hand insets) and seek a nearby polyhedral mesh, guided by an alternate smoothness energy. While the patterns in (**c**) and (**d**) follow a single rule, the one in (**e**) is irregular. It is created by a greedy rule: among the three possibilities to split a hexagon in half we take the one which makes the resulting quads most planar. The *inset figures* are right show a detail and the quality of planarity by color coding faces according to the value of δ, where δ is defined as distance of diagonals, divided by average edge length. We also give the value σ which is the distance of vertices to the reference geometry, divided by average edge length of the mesh

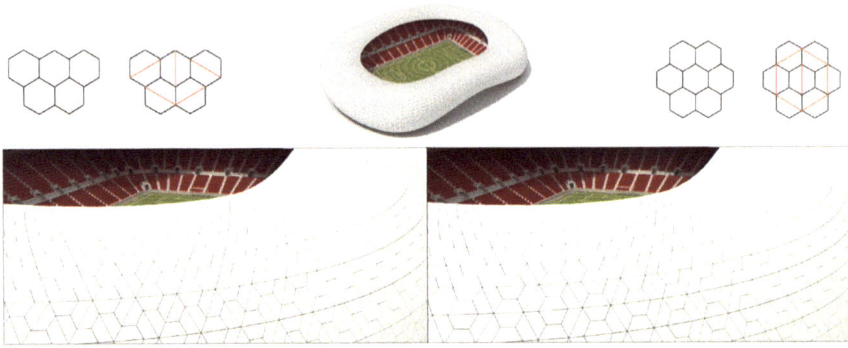

Fig. 11 Different kinds of polyhedral patterns resolve this freeform skin into planar quads. Like for the example of Fig. 10, the pattern has been found via a two-step procedure: (i) compute a honeycomb structure following the given reference surface, and (ii) split hexagons in half and planarize the resulting quad mesh. Both steps use the method of Tang et al. (2014); the 2nd step being guided by the "zigzag" fairness energy mentioned in Sect. 5

hexagons which form a periodic zigzag sequence, such as those highlighted in red in Fig. 10 (right hand insets) and Fig. 11.

A zigzag sequence is considered fair, if $(\mathbf{v}_i - \mathbf{v}_j) - (\mathbf{v}_k - \mathbf{v}_l)$ is small, for any choice of consecutive vertices $\mathbf{v}_i, \mathbf{v}_j, \mathbf{v}_k, \mathbf{v}_l$. We therefore take as a fairness energy the sum of squares of such expressions. Similarly we consider triples $\mathbf{v}_i$, $\mathbf{v}_j$, $\mathbf{v}_k$ of vertices analogous to those highlighted in blue – there are six of them for every hexagon. We would like the kink in the short polyline $\mathbf{v}_i\mathbf{v}_j\mathbf{v}_k$ to be small, so we add as a constraint $(\mathbf{v}_i - \mathbf{v}_j) \times (\mathbf{v}_j - \mathbf{v}_k) = 0$, but multiplied with a small factor to make it a "soft" constraint. With these modifications, the method of Tang et al. (2014) works fine, as demonstrated by Fig. 10.

Conclusion

We have shown how the algorithmic concept of Tang et al. (2014) can be applied to several classes of geometric objects which are relevant to architectural geometry for various reasons (flatness of panels, or repetition in elements). These objects include:

- Polyhedral meshes in general, and polyhedral meshes with equilibrium forces in their edges in particular.
- More general than meshes are honeycomb structures with their interesting relations to Lobel meshes and to developable surfaces.
- An interesting topic are polyhedral patterns. We have in particular demonstrated patterns derived from honeycomb structures.

(continued)

While part of this paper is merely an exposition of the capabilities of Tang et al. (2014) we have also presented new geometry, namely the relations between honeycomb structures, Lobel frames, and developable surfaces, thereby extending recent work on honeycombs, cf. Jiang et al. (2014).

Limitations The algorithmic concept for solving constraint equations we used in this paper has turned out to be efficient and fast in such cases where the constraints can be formulated by linear or quadratic equations each of which involve only few variables. Already Tang et al. (2014) observed that higher-order polynomial constraints cause a dramatic drop in performance. So do constraints which involve many variables (such as total volume of a mesh).

A different kind of limitation is posed by geometric problems which contain a discrete optimization component, such as a change in combinatorics. An example of this is the paneling algorithm of Eigensatz et al. (2010) which contains an assignment problem. Relevant to our work, it would be a challenge to incorporate automatic changes in combinatorics into our modeling system for polyhedral meshes.

Future Work It should not be difficult to extend the methods discussed in this paper to situations where the variables and constraints do not act on "visible" geometry variables directly, but indirectly, e.g. on control points of NURBS surfaces. In this way e.g. developability of NURBS can be incorporated into our computational framework. An interactive system which combines aspects of shape, function, and fabrication is a long term goal of our work. For all types of applications (developables, crumpled paper, ...), comparison with existing methods is needed.

Acknowledgements This research was supported by KAUST base funding, NAWI Graz funding, by the DFG-Collaborative Research Center, TRR 109 *Discretization in Geometry and Dynamics*, through grants I 705 N-26 and I 706-N26 of the Austrian Science Fund (FWF), by FWF project P 23735-N13 and by the European Community's 7th Framework Programme under grant agreement 286426 (GEMS).

References

Barnes, M. R.: Form finding and analysis of tension structures by dynamic relaxation. Int. J. Space Struct. **14**(2), 89–104 (2009)

Block, P.: Thrust network analysis: exploring three-dimensional equilibrium. PhD thesis, MIT (2009)

Block, P., Lachauer, L.: Closest-fit, compression-only solutions for free form shells. In: Proceedings of the IABSE – IASS London Symposium, London, 8pp, 2011

Block, P., Ochsendorf, J.: Thrust network analysis: a new methodology for three-dimensional equilibrium. J. Int. Assoc. Shell Spat. Struct. **48**(3), 167–173 (2007)

Bouaziz, S., Schwartzburg, Y., Weise, T., Pauly, M.: Shaping discrete geometry with projections. Comput. Graph. Forum **31**, 1657–1667 (2012)

de Goes, F., Alliez, P., Owhadi, H., Desbrun, M.: On the equilibrium of simplicial masonry structures. ACM Trans. Graph. **32**(#93), 1–10 (2013)

Deng, B., Bouaziz, S., Deuss, M., Zhang, J., Schwartzburg, Y., Pauly, M.: Exploring local modifications for constrained meshes. Comput. Graph. Forum **32**, 11–20 (2013)

Deng, B., Bouaziz, S., Deuss, M., Kaspar, A., Schwartzburg, Y., Pauly, M.: Interactive design exploration for constrained meshes. Comput.-Aided Des. (2014). doi:10.1016/j.cad.2014.01.004

Eigensatz, M., Kilian, M., Schiftner, A., Mitra, N., Pottmann, H., Pauly, M.: Paneling architectural freeform surfaces. ACM Trans. Graph. **29**(#45), 1–10 (2010)

Jiang, C., Wang, J., Wallner, J., Pottmann, H.: Freeform honeycomb structures. Comput. Graph. Forum **33**(5) (2014). Proc. SGP

Lachauer, L., Block, P.: Compression support structures for slabs. In: Hesselgren, L. et al. (eds.) Advances in Architectural Geometry 2012, pp. 135–146. Springer, Vienna (2012)

Liu, Y., Pottmann, H., Wallner, J., Yang, Y.-L., Wang, W. Geometric modeling with conical meshes and developable surfaces. ACM Trans. Graph. **25**(3), 681–689 (2006)

Liu, Y., Pan, H., Snyder, J., Wang, W., Guo, B.: Computing self-supporting surfaces by regular triangulation. ACM Trans. Graph. **32**(#92), 1–10 (2013)

Lobel, A.: Formes et structures engendrées par des éléments identiques, vol. 1–6. A. Lobel, Bourg-la-Reine (1993). ISBN 2-9514254-0-6

Panozzo, D., Block, P., Sorkine-Hornung, O.: Designing unreinforced masonry models. ACM Trans. Graph. **32**(#91), 1–12 (2013)

Poranne, R., Ovreiu, E., Gotsman, C.: Interactive planarization and optimization of 3D meshes. Comput. Graph. Forum **32**(1), 152–163 (2013)

Schiftner, A., Balzer, J.: Statics-sensitive layout of planar quadrilateral meshes. In: Ceccato, C. et al. (eds.) Advances in Architectural Geometry 2010, pp. 221–236. Springer, Vienna (2010).

Schiftner, A., Höbinger, M., Wallner, J., Pottmann, H.: Packing circles and spheres on surfaces. ACM Trans. Graph. **28**(#139), 1–8 (2009)

Singh, M., Schaefer, S.: Triangle surfaces with discrete equivalence classes. ACM Trans. Graph. **29**(#46), 1–7 (2010)

Tang, C., Sun, X., Gomes, A., Wallner, J., Pottmann, H.: Form-finding with polyhedral meshes made simple. ACM Trans. Graph. **33**(#70) (2014). Proc. SIGGRAPH

Van Mele, T., Block, P.: A novel form finding method for fabric formwork for concrete shells. J. Int. Assoc. Shell Spatial Struct. **52**, 217–224 (2011)

Vouga, E., Höbinger, M., Wallner, J., Pottmann, H.: Design of self-supporting surfaces. ACM Trans. Graph. **31**(#87), 1–11 (2012)

Zadravec, M., Schiftner, A., Wallner, J.: Designing quad-dominant meshes with planar faces. Comput. Graph. Forum **29**, 1671–1679 (2010)

Biomimetic Lightweight Timber Plate Shells: Computational Integration of Robotic Fabrication, Architectural Geometry and Structural Design

Oliver David Krieg, Tobias Schwinn, Achim Menges, Jian-Min Li, Jan Knippers, Annette Schmitt, and Volker Schwieger

Abstract The research presented in this paper pursues the development and construction of a robotically fabricated, lightweight timber plate system through a biologically informed, integrative computational design method. In the first part of the paper, the authors give an overview of their approach starting with the description of the biological role model and its technical abstraction, moving on to discuss the computational modelling approach that integrates relevant aspects of biomimetics, robotic fabrication and structural design. As part of the validation of the research, a full-scale, fully enclosed, insulated and waterproof building prototype has been developed and realized: The first building featuring a robotically fabricated primary structure made of beech plywood. Subsequently, the methods and results of a geodetic evaluation of the fabrication process are presented. Finally, as the close collaboration between architects, structural and geodetic engineers, and timber fabricators is integral to the process, the architectural and structural potentials of such integrative design processes are discussed.

O.D. Krieg (✉) • T. Schwinn • A. Menges
Institute for Computational Design, University of Stuttgart, Keplerstrasse 11, D 70174 Stuttgart, Germany
e-mail: oliver.krieg@icd.uni-stuttgart.de; tobias.schwinn@icd.uni-stuttgart.de; achim.menges@icd.uni-stuttgart.de

J.-M. Li • J. Knippers
Institute of Building Structures and Structural Design, University of Stuttgart, Keplerstrasse 11, D 70174 Stuttgart, Germany
e-mail: j.li@itke.uni-stuttgart.de; j.knippers@itke.uni-stuttgart.de

A. Schmitt • V. Schwieger
Institute of Engineering Geodesy, University of Stuttgart, Geschwister-Scholl-Strasse 24D, D 70174 Stuttgart, Germany
e-mail: annette.schmitt@ingeo.uni-stuttgart.de; volker.schwieger@ingeo.uni-stuttgart.de

P. Block et al. (eds.), *Advances in Architectural Geometry 2014*,
DOI 10.1007/978-3-319-11418-7_8

1 Introduction

New developments in computational design and digital fabrication currently lead to a rethinking of architectural design and delivery processes. In contrast to traditional, hierarchical, and form-driven design approaches, integrative strategies can activate discipline-specific knowledge in early design stages. In the case of the research presented in this paper, this includes the activation of reciprocal dependencies between architectural geometry, biomimetic engineering, structural design, robotic fabrication and geodetic evaluation (engineering geodesy being the scientific discipline concerned with surveying). Specifically, the authors introduce an innovative, biologically informed, computational design approach for the development and construction of a robotically fabricated lightweight timber plate structure (Fig. 1).

The timber construction industry is currently experiencing significant innovations in fabrication technology and structural applications. As a renewable resource and with its negative carbon footprint and low embodied energy (Kolb 2008; Alcorn 1996), wood undoubtedly plays a major role in the current development towards sustainable, carbon neutral construction. Additionally, from a life cycle analysis standpoint, the utilization of locally available building materials becomes crucial. Besides using local wood products, the project also investigates how innovative fabrication techniques, as part of an integrative computational design strategy, can lead to a rethinking of sustainability: Increasing structural performance and resource efficiency through a higher degree of shape variation in the building parts, while at the same time offering new spatial qualities and architectural potentials. The hypothesis is that by following biomimetic principles in form-finding and construction, a performative, structurally efficient, and geometrically intricate plate structure system can be developed, which, following the high-level biomimetic principle (Gruber and Jeronimidis 2012) of functional integration, functions as a materially efficient load-bearing layer and building envelope at the same time.

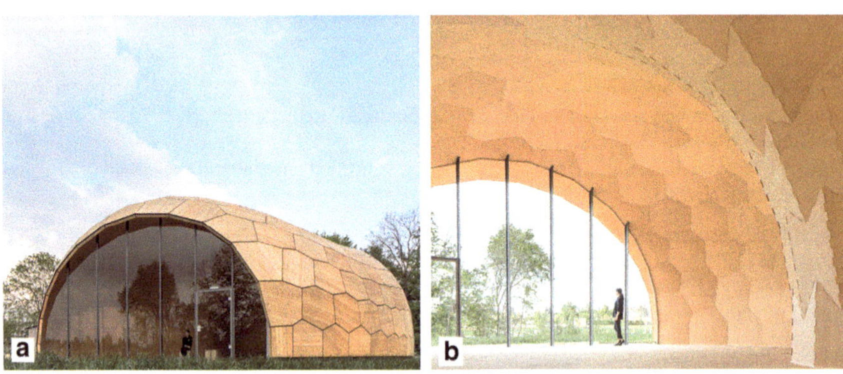

Fig. 1 Exterior (**a**) and interior (**b**) views of the finished prototype building

The goal of the research project therefore is to investigate the architectural and structural potentials of an integrative design approach for timber plate structures, enabled by a close collaboration between architects, structural and geodetic engineers, and timber fabricators. Consequently, this method is evaluated through the design, fabrication, construction, and verification of a full-scale building prototype. Built for the Landesgartenschau Schwäbisch Gmünd, Germany, in 2014 and named the *Landesgartenschau Exhibition Hall*, the case study is a fully enclosed, insulated and waterproof building.

2 Computational Design Strategies for Plate Structures

2.1 Constructional Morphology of Plate Structures in Nature

A distinguishing feature of structural systems in nature is their geometric complexity. In contrast to engineering, living organisms are unconstrained by predefined structural typologies or aspects of calculability. Furthermore, geometric complexity forms the basis for material efficiency, which in turn is the prerequisite for survival. In other words, while in nature material and energy are more expensive than shape or geometry, in technology the opposite was true as of today (Vincent 2009). Current developments in digital fabrication, however, ease many of the economic restrictions for varying the building components' shape. In particular, the kinematic freedom of robotic fabrication opens up the opportunity to introduce geometric complexity in construction systems for more efficient structures in architecture, a development that can be summarized as: more form, less material.

In this context, plate structures are of particular interest as they are a performative construction system made of individual, planar elements. Geometrically, they can be organized such that the individual plates constitute the primary load bearing structure, instead of the joints as in the case of grid shells (Bagger 2010). However, previous research also showed that an integrative, performance driven computational design method is critical for the development of an efficient and adaptive plate structure system (La Magna et al. 2012; Schwinn et al. 2012). Plate structures can be found in many different biological systems, of which the sand dollar species of the class of sea urchins (*Echinoidea*) is most notable for its constructional morphology (Fig. 2). The skeletal shell of a sand dollar is a modular system of polygonal plates composed of calcium carbonate and covered by a thin dermis and epidermis (Barnes 1982). On a microscopic level, the plates are joined by interlocking calcite protrusions (Fig. 2) that can be seen as the biological equivalent to man-made finger joints. Through their topological rule of always joining three plates in one point, plate structures such as the sand dollar's skeleton transfer bending forces mainly through shear forces along the plate's edges.

By transferring biomimetic principles into a construction system, the material efficiency of geometrically differentiated building elements that is inherent in

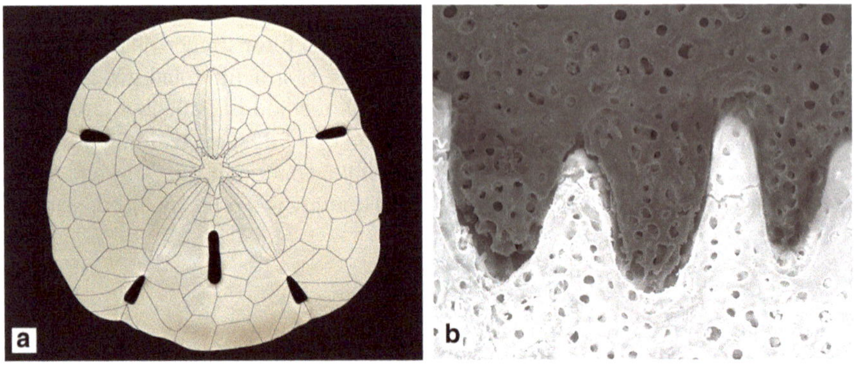

Fig. 2 (**a**) Top view of a sand dollar with highlighted topology. (**b**) Microscopic view of a sand dollar's plate edge

biology can be used for developing a new kind of construction system, which not only incorporates the connection principle within the building part, but also integrates fabrication constraints and material parameters. The research project therefore aims at the development of a modular, lightweight plate structure capable of adapting to free-form surfaces by making use of the extended design space, or Machinic Morphospace (Menges 2012), for plate structures provided by robotic fabrication.

2.2 *Integrative Computational Design for Plate Structures in Architecture*

In architecture, one of the prerequisites for developing an integrative design method for plate structures is solving the planar polygonal approximation of complex double-curved surfaces: not only in positive and negative Gaussian curvature regions, but also in regions close to the so-called parabolic line – the transition zone – where $K=0$. While this is a topic that is actively being researched in the fields of computer graphics (Cohen-Steiner et al. 2004; Zimmer et al. 2012) and architectural geometry (Manahl et al. 2012; Troche 2008; Wang et al. 2008), a novel approach is being investigated as part of this research that synthesizes the principles of tangent plane intersection (TPI), the biomimetic principles of natural plate structures, and the rules and constraints of robotic fabrication in an agent-based modelling (ABM) approach. The promise of this approach is that (1) in addition to the numerical solution of the approximation problem, various sometimes conflicting objectives can be integrated into a design-fabrication feedback loop, where the various goals relating to fabrication, geometry, structure, and architecture can be described and negotiated in a behavioural model. (2) In contrast to the linear hierarchical model of the digital chain, which formalises a deterministic design

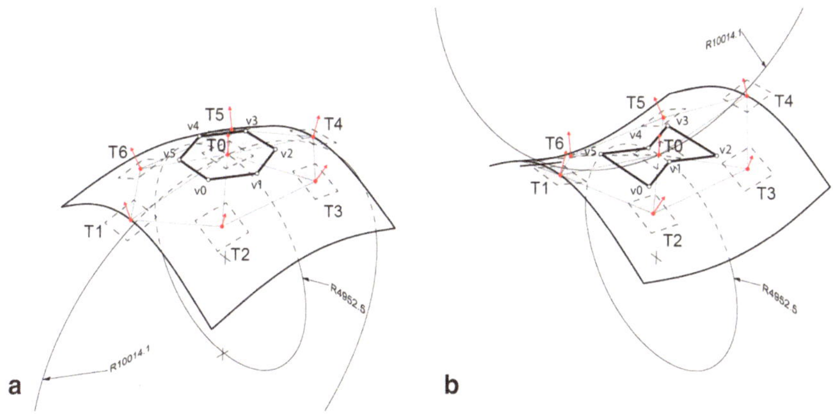

Fig. 3 Tangent plane intersection for doubly curved surfaces. Oscillating *circles* indicate curvature (1/R), principle curvature directions and orientation. A. Synclastic, positive Gaussian curvature $K > 0$. B. Anticlastic, negative Gaussian curvature $K < 0$

process, in this approach the rules and constraints of digital fabrication are activated as design drivers resulting in a non-linear design process.

Tangent plane intersection (TPI) has a number of characteristics that make it applicable in the context of agent-based modelling for plate structures: first, all the intersection points of a tangent plane T_0 with its neighbouring planes lie in the plane of T_0 and can consequently represent the vertices of a planar polygon (Fig. 3); second, TPI produces valid results for double-curved geometry, i.e. synclastic and anticlastic surface regions, which is beneficial for the global stability of a structure, however with one caveat: TPI produces degenerate results in parabolic regions, where Gaussian curvature K approaches 0, requiring additional strategies: in this case the intersection points of three neighbouring planes usually lie far away from the input points, so that they have to be projected back and, consequently, will temporarily form non-planar polygons. Agents with non-planar polygons therefore follow a prioritized behaviour that leads them to minimize the non-planarity by numerically differentiating the measure for non-planarity, i.e. the out-of-plane distance, at that point, thus moving in a gradient descent towards areas of planarity; third, in order to determine which tangent plane intersects with which neighbouring plane the concept of a "neighbourhood" has to be established, which is implemented through the triangulation of the input point set. While the concept of the neighbourhood is integral to ABM where it is defined by distance/proximity, in the case of TPI, it is defined by topology, e.g. a Delaunay triangulation of the u,v-parameter space. Both definitions are consequently incorporated into the ABM approach for plate structures.

Agent-based systems are part of a rule-based computational modelling method whose premise is that through the calculation of the local interactions between individual entities (*autonomous agents*), which are based on pre-defined rule sets (*behaviours*), a global configuration emerges that meets higher-level goals. As such,

ABM is an example of behaviour-based artificial intelligence (Brooks 1986) and is used in a variety of fields ranging from finance to robotics to biology. The underlying models and the related behaviours of the individual agents are based on abstractions of the relevant application context. In the architectural context, ABM is often used for the simulation of pedestrian movements as in building evacuation or urban environments (Aschwanden et al. 2008), but also in design exploration (Snooks 2013). In these cases, the agent models require the capacity for locomotion, which, introduced by Reynolds (1999), allows each agent to adjust position and orientation based on its local interactions with neighbouring agents and the environment. Recently, ABM has been introduced in the context of design integration, specifically focussing on fabrication requirements (Baharlou and Menges 2013).

In the context of the presented research, the behaviours of the agents integrate the requirements of fabrication, biomimetic principles, and aesthetic criteria. Parameters of the model are divided into global parameters, such as number of plates and average plate size, and local parameters. Local parameters, which are the parameters that govern agent behaviour, are directly related to fabrication constraints such as available stock size, working envelope of the fabrication setup (Fig. 4). They are also based on the joint fabrication strategy, which determines the min-max ranges of the connection angles, and the min-max ranges for the edge lengths. Aesthetic criteria (symmetry within the plate, edge proportions) and biomimetic principles (heterogeneity of plate sizes, number of edges, and 3-plates rule) are controlled by the adaptive separation value between agents, the topology of the underlying mesh representation, and by responding to the local curvature at a the agent locations. All behaviours are defined such that the agent system is globally driven towards convergence.

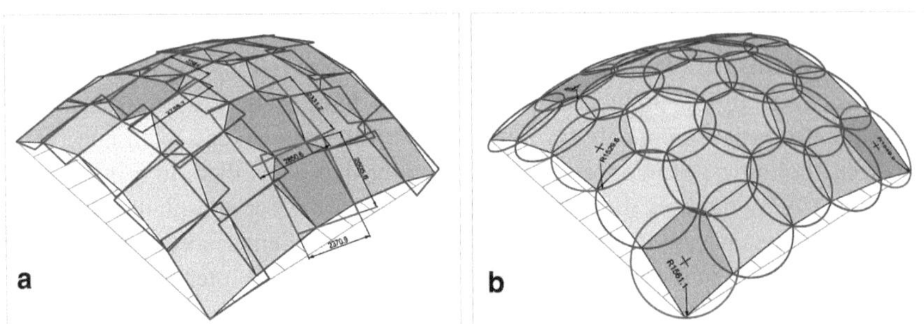

Fig. 4 Global fabrication constraints are represented in the behavioural model and evaluated at runtime. (**a**) The minimal bounding rectangle evaluates the producibility in relation to the available stock material sheet size. (**b**) The circumcircle evaluates the plates in relation to the available fabrication workspace diameter

3 Integrative Structural Design

3.1 Structural Features

Segmented shell structures are common in nature but rare in building construction. At the academic level, a few segmented shells have been built out of panels from glass (Blandini 2005; Bagger 2010; Almegaard et al. 2007) or wood (La Magna et al. 2013), but with only limited use at larger scales. The main reason for this is the lack of applicable joint constructions satisfying conditions of sufficient structural load transfer and the restricted adaptability to various geometric and structural situations. Only a few studies on this subject can be found (Wester 2002; Veer et al. 2003), which either use adhesives for joining or apply combined mechanical/adhesive techniques.

Joints between prefabricated segments nearly always weaken shell structures as they disturb the continuity of their stiffness. Joints can also reduce the capacity to transfer bending moments or membrane forces. The structural capacity of joints not only impacts the overall load-bearing behaviour of the shell, but also determines the efficient arrangement of individual segments on a free-form surface.

The structure of the *Landesgartenschau Exhibition Hall* is designed as a segmented plate shell carrying loads mainly through membrane forces. It generates the global stability by its global double-curved shape and the binding strength in the connections between the segments. The *Exhibition Hall* is supposed to stand on site for at least 5 years and is therefore designed to meet the respective wind and snow load requirements.

Shell structures take out-of-plane loads by in-plane forces such that the out-of-plane bending and shears are minimized. Compared with lattice shells in which internal forces are mainly transferred through axial forces, plate shells transfer internal forces across edges not only through axial forces but also through in-plane shear forces (Fig. 5); the force flows could occur in either ways and their ratio is largely determined by the design of the joint and the joint patterns on the shell geometry. Observing the influence of the joint stiffness to the structural performance gives important clues about how to proceed with the structural design.

Like lattice shells with 6-valent triangular segments, plate shells with trivalent polyhedral segments are also kinematically stable (Wester 2002; La Magna et al. 2012) and connections with bending stiffness are thus not required. However, trivalent polyhedrons will become kinematically unstable if one of the three included angles of a vertex is close to 180°. Additional bending stiffness in connections can effectively improve the structural performance of plate shells, especially in areas where included angles approach 180°. Therefore, bending stiffness is still considered in the connection design of the Exhibition Hall.

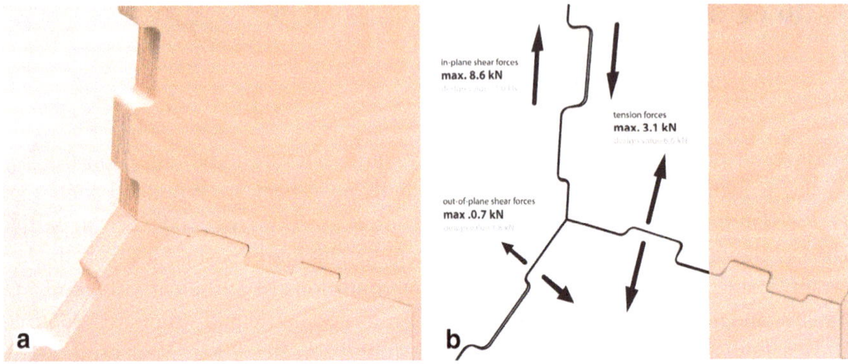

Fig. 5 Visualization of the forces inside the plate structure. The dominant in-plane shear forces are transferred through the plate's finger joints

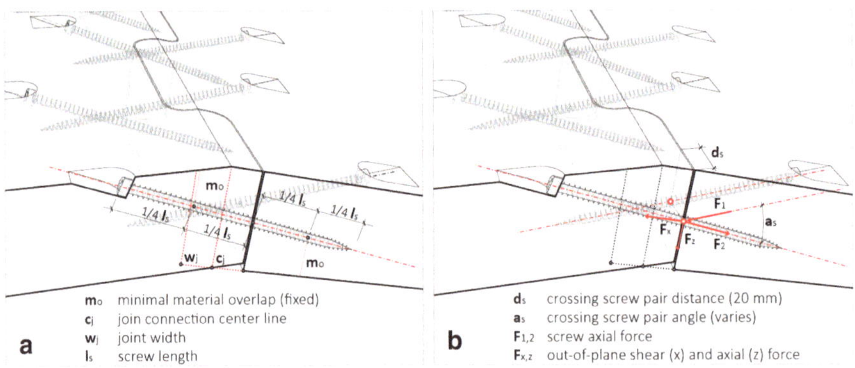

Fig. 6 Visualization of the developed finger joint connection. Each finger joint includes a pair of parallel crossing screws, which intersect from the side view at the connection point between two plates. The crossing screws function as a small truss structure embedded in the plywood. Out-of plane shear F_x and axial force F_z are therefore taken by screw axial force F_1 and F_2

3.2 *Structural Integration of Finger Joint Connections*

Structurally designing a feasible connection, which provides sufficient strength and stiffness between the panels within a thickness of only 50 mm, is an essential aspect of the joint development and was one of the key challenges of the project. The objective that the connection could ultimately be applied to the building industry also means following the building code in the development process. The research therefore leads to a novel connection design for plate structures: by mimicking the skeletal configuration of sea urchins, finger joints are used in connections to resist the decisive in-plane shear forces, while the smaller axial forces and out-of-plane shear forces are taken by pairs of crossing screws (Fig. 6). The concept of crossing screws was first introduced by Blass and Bejtka (2004).

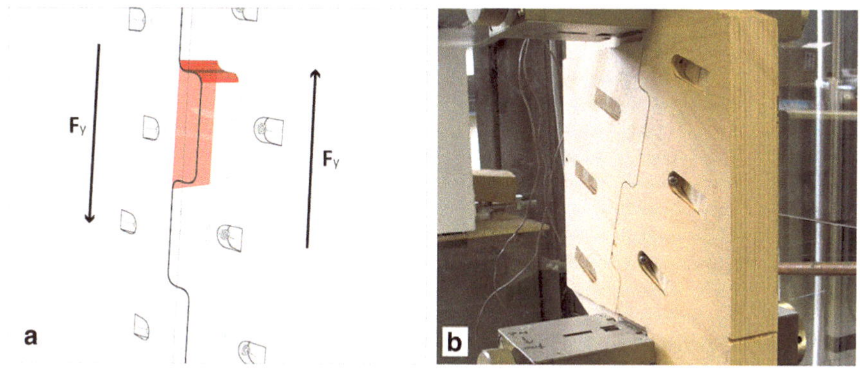

Fig. 7 (**a**) The in-plane shear force F_y is taken by the contact surface and transferred to the shear plane. (**b**) The compression capacity of the contact surface and the shear capacity of the shear plane are evaluated through physical tests

The finger joints function similarly to step joint in traditional truss works. The thrust force will be taken by the contact surface and be transferred to the shear plane. Therefore, two failure criteria need to be examined: the shear capacity against the thrust and the compression capacity of the contact surface (Fig. 7).

The axial forces and the out-of-plane shear forces are taken by the crossing screws, which lie in parallel planes, normal to the plate edge, with distance d_s to each other; the screws are intersecting with each other with a specific included angle in the side view (Fig. 6). Since the in-plane shear forces are taken by the finger joints, the analysis of the internal forces of screw joints can be reduced to a two dimensional force diagram problem; a simplified model is adapted here, in which the crossing screws are treated as a truss structure embedded in the plywood. The axial force and the out-of-plane shear force that are derived in the global structural model are taken as external forces of this truss structure.

3.3 *Global Structural Analysis*

From a structural point of view, an adequate model for structural analysis needs to be developed and fully integrated in the design process. Therefore, an automated sequence has to be developed to generate structural models from trivalent polyhedrons. In this way, the design process is directly linked to the structural analysis and the configuration of connections can easily be varied for finding an optimal solution (Fig. 8).

The structural performance of plate structures is largely determined by the connections between the plates. A proper simulation of the forces inside the connections is therefore an important aspect of the structural design. In the developed structural model the stiffness in connections is simulated by spring elements: Every

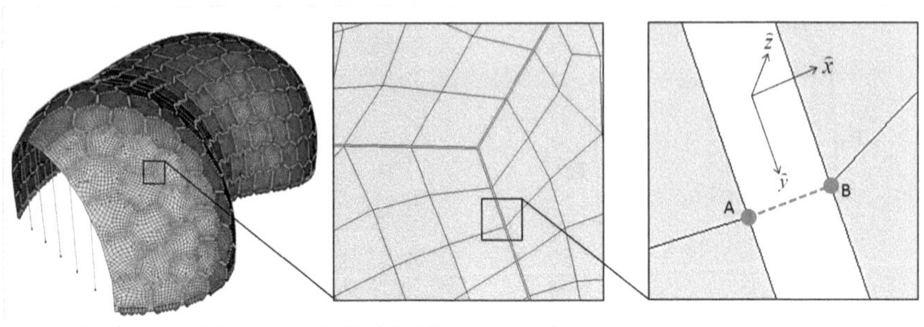

Fig. 8 *Left*: the global FE model of the pavilion. *Right*: spring elements are acting on two nodes, *A* and *B*, situated at the two opposite sides of neighbouring panels. Axes *x*, *y*, and *z* are the directions of the axial force, the in-plane shear, and the out-of plane shear respectively. Axis *y* is also the axis of bending resistance

connection consisting of a finger joint and a pair of crossing screws is simulated by four spring elements which act on two specific nodes situated at the two opposite plates in the finite element model (Fig. 8, right). Three springs are used for simulating the axial resistance; the in-plane shear resistance and the out-of-plane shear resistance respectively, while the fourth spring is used for simulating the bending resistance. Each spring is assigned with a specific direction of action as in Fig. 8.

The spring coefficients are determined as follows: The axial resistance of a single screw is physically tested with various tilting angles in the plywood, and with these test values the spring coefficients for the axial resistance and the out-of-plane shear resistance can be determined by the simplified truss model. The spring coefficient for the in-plane resistance is determined by a simplified step joint model while the spring coefficient for the bending resistance is directly determined by the four-point-bending test of a real size joint.

4 Robotic Fabrication

4.1 Robotically Fabricated Finger Joints for Plate Structures

Through the extended kinematic range of robotic fabrication, complex connections become feasible and open up the possibility to integrate biomimetic principles at a deeper hierarchical level. Robotically fabricated finger joints had been developed in previous research for specific construction techniques (Schwinn et al. 2012; Krieg and Menges 2013). As in the previous research, the joint connection is developed with regards to the specific robotic fabrication setup and material parameters; in this research, construction and assembly constraints play an additional major role. As discussed above, an important driver in the plate structure system's development

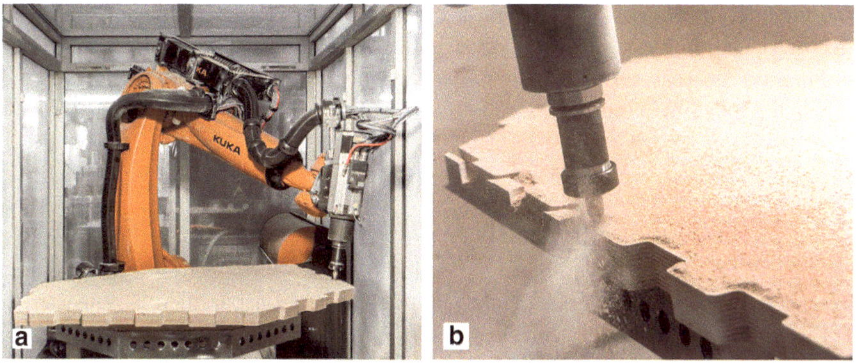

Fig. 9 (**a**) Robotic fabrication setup for milling a plywood plate. (**b**) Close-up of the milling process

Fig. 10 (**a**) Assembly of a structural plywood plate. (**b**) The plates are fixed and held in place through the crossing screws

is the potential applicability in the building industry and therefore, in keeping with the resource efficiency of living organisms, the efficient and economic use of material. While this is achieved globally through an optimal distribution of the individual plate segments, and locally through the structural performance of the plate connections, the joint is also designed to keep production times short (Fig. 9), and to facilitate straightforward assembly (Fig. 10).

The finger joint's parameters are subsequently integrated into the digital design and fabrication workflow. Based on the topology analysis of the plate structure surface model, the various tool paths for the finger joint fabrication can be automatically generated for each individual plate edge based on custom CAM strategies on three different levels. Besides milling the plate's boundary and the finger joints, the robotic fabrication process also ensures the precise localization of the crossing screws. While the joint width is a preset parameter to allow the space for the crossing screws and match requirements of the building code, other parameters such as the crossing screw angles, lengths and numbers, as well as the finger joint assembly vector are a function of the varying edge-specific, local geometric parameters such as angles between plates or plate edge length. This parametric process ensures a consistent connection performance while staying inside the machine's morphospace.

4.2 Strategic Use of Robotic Fabrication

The specific setup that defined many of the fabrication parameters consisted of a KUKA industrial robot arm with a 12KW spindle as an effector mounted on the robot flange, an additional revolute external axis used as a turntable, and an automatic tool changer. These components were preconfigured in a mobile robotic cell that was installed in the timber fabricator's workshop for the duration of the plate fabrication. The plates' formatting process consisted of two main fabrication steps: first, the plate stock material was cut with oversize from beech plywood boards on a Hundegger SPM machine, so as to minimize cut-off; second, the plates were mounted on the turntable in the milling cell, where milling of the finger joints down to the nominal joint dimensions could be accomplished. Both, the machining code for robotic fabrication as well as the code for pre-formatting the plates on the CNC-machine could be automatically generated from the digital plate information model.

Different cycle strategies were implemented in order to automate not only the precise trimming of each plate's boundary, but also construction details where necessary. For example, pockets and holes for the crossing screws are calculated for their optimal material overlap and screw angle in order to make the subsequent assembly on site easier. Pre-drilling is essential in the robotic fabrication process since the low material thickness does not allow for tolerances in the screw positions.

In this production process, robotic fabrication represents the crucial technology enabling the fabrication of the lightweight timber plate structure by solving the fabrication of the complex joints. However, other digital fabrication technologies such as CNC milling or waterjet-cutting, complemented robotic fabrication in order to form a coherent fabrication sequence. Not only the structural layer but also all other remaining layers of the building envelope required for insulation, waterproofing and cladding could be digitally fabricated.

5 Geodetic Evaluation

5.1 Evaluation of Fabrication Accuracy

Quality assurance, e.g. checking accuracy values, is an important task when developing new fabrication and construction processes. Thus a sample of 24 out of overall 243 plates was analysed. For the measurement the laser tracker API

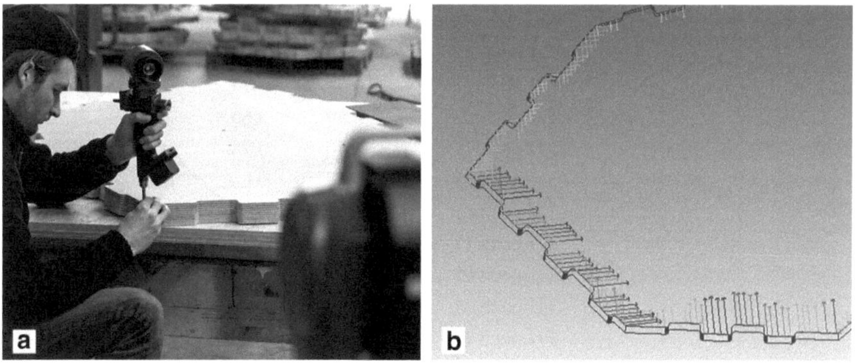

Fig. 11 Measurement of the plate segments (**a**) and presentation of deviations in x and y direction (**b**)

Table 1 Accuracy (RMS) in each direction and 2D

	x [mm]	y [mm]	2D [mm]	z [mm]
Accuracy (RMS)	0.48	0.51	0.70	0.74

Radian™ is used together with an API IntelliProbe 360™, a hand-held probing instrument (Fig. 11). The laser tracker is a polar measuring system, measuring angles and distances and delivering point coordinates with an accuracy / root-mean-square error (RMS) of 50 μm in a distance of about 10 m. The high accuracy requirements are due to the fabrication that should be realized with 1 mm standard deviation, leading to a 10-times improved measurement RMS.

The 24 plates were measured with the help of an especially produced adapter for the API IntelliProbe 360™ to measure the plate's edges. Ten points on each finger joint element are measured. For a statement about the fabrication accuracy, the newly designed adapter for the IntelliProbe 360™ is tested with a reference metal work piece. The fabrication tolerance of that work piece is around 0.1 mm. One of the edges of this work piece is measured ten times. The medium RMS of these 3D-measurements is 0.24 mm.

The values in Table 1 include the combined measurement and fabrication RMS as well as object properties and external influences on this object. The average 2D-RMS is given in Table 1 to 0.70 mm.

Wood is a natural material and its geometry can vary over time, based on its moisture content that results from changes in temperature or relative humidity. This is the reason why tolerances are generally defined for fixed temperature. Therefore, the measurements are repeated for four elements before transportation and before assembly defining three so-called *measurement epochs*: right after fabrication (epoch 1), two weeks later (epoch 2), and before assembly (epoch 3). The RMS of the comparison between the measured and the CAD model for the three epochs are shown in Table 1.

Table 2 shows that the in-plane accuracy (RMS in X- and Y-direction) of a plate is between 0.28 and 1.11 mm. The values in the z-direction register the

Table 2 RMS of comparison between CAD model and measurements for four elements in three epochs

Number of element	dx[mm]	dy[mm]	dz[mm]	3d[mm]
1 (Epoch 1)	0.37	0.64	0.51	0.90
1 (Epoch 2)	0.54	0.73	0.63	1.11
1 (Epoch 3)	0.59	0.66	1.21	1.50
2 (Epoch 1)	0.80	0.78	0.81	1.39
2 (Epoch 2)	0.67	0.85	0.99	1.46
2 (Epoch 3)	1.11	0.55	0.70	1.43
3 (Epoch 1)	0.54	0.52	0.57	0.94
3 (Epoch 2)	0.42	0.28	0.73	0.89
3 (Epoch 3)	0.49	0.29	0.75	0.94
4 (Epoch 1)	0.43	0.73	0.96	1.28
4 (Epoch 2)	0.55	0.71	1.16	1.46
4 (Epoch 3)	0.68	0.80	1.43	1.77

plywood's tendency to buckle and dish over time and are not an effect of fabrication. However, due to the three-dimensional fabrication process the in-plane accuracy will be affected by these differences. The influence of non-specified external influences (differences in RMS between epochs) varies between 0.05 and 0.60 mm. A statistical test can be used to check if these variations are significant. A Gaussian distribution based test with a confidential interval of 95 % is assumed. The test compares the averages of the deviations. It can be shown that no plate changes can be detected significantly.

5.2 *Evaluation of Construction*

After the Exhibition Hall's assembly, the building will be scanned by a laser scanner in at least two epochs to compare the result of the construction process with the CAD model and to detect any global and local deformations. The laser scanner is used for the deformation analysis, since it delivers a high resolution and a 1–2 mm accurate point cloud. For the scans the laser scanner Leica HDS7000 is used. It is planned to scan the pavilion from at least three different laser scanner positions. The different positions will be registered by reference targets, which make it possible to combine the different scans into one 3D model. This process also includes cleaning the point cloud from outliers.

To compare the different epochs of laser scanning, a point network is build up around the pavilion for registration and geo-referencing. The comparison of the epochs will be made by comparing the surfaces and edges of the elements.

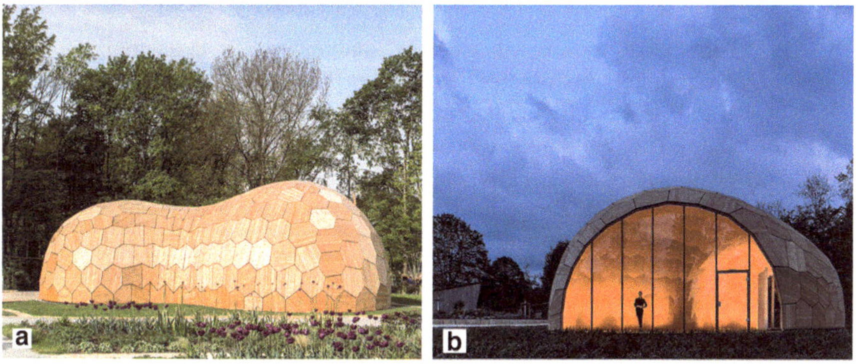

Fig. 12 Photographs of the finished prototype building

5.3 Preliminary Conclusions on Accuracy in Robotic Fabrication

The evaluation of the fabrication process delivers a 2D-RMS of up to 0.70 mm. The average plate accuracy over the 24 elements is 0.48 mm for d*x*, and 0.51 mm for d*y*, which, in the context of timber manufacturing, is exceptionally low. However, measuring the average fabrication RMS proved to be not entirely straightforward since it is influenced by fabrication as wells as object properties and external influences. Slight dimensional changes in the material as well as rough surfaces due to the milling process might also have led to deteriorated deviations. An improvement may be reached by integrating the measurement process more deeply into the fabrication process and to design and implement a sensor-actuator control loop to correct for small fabrication deviations during the fabrication process.

Conclusion and Outlook

In order to be able to evaluate the dependencies and effects of material, fabrication, assembly, and building lifespan / occupation on timber plate structures and the role of robotic fabrication therein, a building prototype has been developed and realized as part of the presented research. The prototype building, which is open to the public since May 2014, is located at the *Landesgartenschau Schwäbisch Gmünd*, Germany. With a usable floor area of 125 m^2 and a shell surface area of 245 m^2, the insulated and waterproof building spans 10 m with a structural layer thickness of just 50 mm beech plywood. The automated machine code generation for all different 243 plates with 7,356 joints was possible through an integrative computational design process (Fig. 12). Through the integration of an assembly order and specific assembly vectors for each plate, the construction process lasted only 3 weeks.

(continued)

In addition to questions of accuracy, automation, or behavioural form finding, the research investigates a new concept of geometry in architecture: shape variation not as a formal feature, but a necessity for achieving structural performance. The building's global shape and the plates' differentiated arrangement enable the development of a material-efficient and lightweight construction system. While enclosing a volume of 605 m^3 the structural layer is only made of 12 m^3 of beech plywood. The plate structure development is also connected to the introduction of a new timber product. The Landesgartenschau Exhibition Hall is one of the first buildings to use beech plywood as the main structural building material. Relating to future foresting strategies, beech will become a more common timber product and also provides a higher structural performance than contemporary softwood timber constructions. The material's strength was essential in the development of both the joint geometry and the crossing screw connection.

Further improvement of the fabrication process with respect to accuracy and tolerances may be reached by integrating the geodetic measurement process completely into a sensor-actuator controlled fabrication process. By strategically using robotic fabrication and other numerically controlled machines throughout the fabrication process, the building prototype showed how robotic fabrication can become a valuable component in a fabricator's tool kit that not only integrates into existing fabrication technologies but, due to its flexibility, also opens up new applications. Given that the traditional discipline boundaries were effectively removed – architects controlling fabricators machinery – the project also demonstrates the viability of possible new alliances in the building industry and, potentially, new and more performative delivery models.

The prospect of this research is that through the integration of the various fields of biology, architecture, structural and geodetic engineering, in one coherent computational design approach, more intelligent contributions to the built environment can be made with respect to resource efficiency, structural material utilization, and architectural quality.

Acknowledgements The work presented in this paper is part of a collaborative research project between the University of Stuttgart and Müllerblaustein Holzbau GmbH. The authors would like to thank their project partners Landesgartenschau Schwäbisch Gmünd 2014 GmbH, Forst BW and KUKA Roboter GmbH. The research project was partly funded by the European Union EFRE fund and the state of Baden-Württemberg through the Cluster Forst und Holz Initiative.

References

Alcorn, A.: Embodied Energy Coefficients of Building Materials. Centre for Building Performance Research, Wellington (1996)

Almegaard H, Bagger A, Gravesen J, Juttler B, Sir, Z.: Surfaces with piecewise linear support functions over spherical triangulation. In: The Mathematics of Surfaces XII, LNCS 4647, pp. 42–63. Springer (2007).
Aschwanden, G., Halatsch, J., Schmitt, G.: Crowd simulation for urban planning. In: Architecture in Computro – 26th eCAADe Conference Proceedings, pp. 493–500. Antwerpen. (2008)
Bagger, A.: Plate shell structures of glass studies leading to guidelines for structural design. Civil Engineering, Technical University of Denmark (2010)
Baharlou, E., Menges, A.: Generative agent-based design computation: Integrating material formation and construction constraints. In: Proceedings of the 31th eCAADe Conference, vol. 2, pp. 165–174. Technical University of Delft, Netherlands (2013)
Barnes, R.D.: Invertebrate Zoology. Holt-Saunders International, Philadelphia (1982)
Blandini, L.: Structural use of adhesives in glass shells. Ph.D. thesis, University of Stuttgart (2005)
Blass, H.J., Bejtka, I.: Selbstbohrende Holzschrauben und ihre Anwendungsmöglichkeiten. In: Holzbau Kalender 2004, Bruderverlag, Karlsruhe (2004)
Cohen-Steiner, D., Alliez, P., Desbrun, M.: Variational shape approximation. ACM Trans. Graph. **23**(3), 905 (2004)
Gruber, P., Jeronimidis, G.: Has biomimetics arrived in architecture? Bioinspir. Biomim. **7**(1), 010201 (2012)
Kolb, J.: Systems in Timber Engineering: Load bearing Structures and Component Layers. Birkhäuser, Basel (2008)
Krieg, O.D., Menges, A.: Potentials of robotic fabrication in wood construction: elastically bent timber sheets with robotically fabricated finger joints. In: Proceedings of the 33nd Annual Conference of the Association for Computer Aided Design in Architecture, pp. 253–260. Waterloo (2013)
La Magna, R., Waimer, F., Knippers, J.: Nature-inspired generation scheme for shell structures. In: Proceedings of the International Symposium of the IASS-APCS Symposium 2012, Seoul (2012)
La Magna, R., Gabler, M., Reichert, S., Schwinn, T., Waimer, F., Menges, A., Knippers, J.: From nature to fabrication: biomimetic design principles for the production of complex spatial structures. Int. J. Space Struct. **28**(1), 27–39 (2013)
Manahl, M., Stavric, M., Wiltsche, A.: Ornamental discretisation of free-form surfaces. Int. J. Archit. Comput. **10**(4), 595–612 (2012)
Menges, A.: Morphospaces of robotic fabrication, from theoretical morphology to design computation and digital fabrication in architecture. In: Robotic Fabrication in Architecture, Art and Design, pp. 28–47. Springer, Wien (2012)
Schwinn, T., Krieg, O.D., Menges, A.: Robotically fabricated wood plate morphologies – robotic prefabrication of a biomimetic, geometrically differentiated lightweight finger joint timber plate structure. In: Robotic Fabrication in Architecture, Art and Design, pp. 48–61. Springer, Wien (2012)
Snooks, R.: Self-organised bodies. In: Architecture in Formation, pp. 264–267. Routledge, London (2013)
Troche, C.: Planar hexagonal meshes by tangent plane intersection. In: Advances in Architectural Geometry 2008, (pp. 57–60). Vienna (2008)
Veer, F.A., Wurm, J., Hobbelman, G.J.: The design, construction and validation of a structural glass dome. In: Proceedings of the Glass Processing Days. Poster 12, Finland (2003)
Vincent, J.: Biomimetic patterns in architectural design. Archit. Des. **79**(6), 74–81 (2009)
Wang, W., Liu, Y., Yan, D., Chan, B., Ling, R., Sun, F.: Hexagonal meshes with planar faces (TR-2008-13). In: HKU CS Tech Report TR-2008-13. Hong Kong (2008)
Wester, T.: Nature teaching structures. Int. J. Space Struct. **17**(2), 135–147 (2002)
Zimmer, H., Campen, M., Herkrath, R., Kobbelt, L.: Variational tangent plane intersection for planar polygonal meshing. In: Advances in Architectural Geometry 2012, pp. 319–332. Springer, Wien (2012)

Form Finding of Twisted Interlaced Structures: A Hybrid Approach

Sina Nabaei, Olivier Baverel, and Yves Weinand

Abstract Our study presents a set of form-finding procedures to explore curved structures made from interlaced panels. Interlacing introduces a particular coupling between assembly components which has to be formulated along with a pertinent flexible body model. We examine here a hybrid approach: panels are simulated a first time using an elastic rod model formulated within a constrained elastic energy minimization where user can virtually buckle, twist and interlace strip assemblies. A thin shell model dynamically integrated comes complementary to the rod approach in order to resolve intersections in case of panels colliding while interlaced. Some conceptual structures are presented to demonstrated the procedure.

1 Introduction

Interlacing can be regarded as a unifying process to create an assembly from individual components: e.g. knitting a tissue from yarns, weaving a basket from wood strips. Interlacing, regardless of scale, inherits some elegant structural notions to be brought beyond the wickerwork scale. It lets assemble flexible slender components into a resistant object, with far more load bearing capacity. It lets span over a distance times more than the cross section dimension of its components. Complex coupling introduced by interlacing also distributes forces over the assembly components to ensure a more homogeneous load take down. While interlacing with tensile elements is more matter of pattern complexity, some key differences impose when assembling flexible components with bending resistance and anisotropic cross section. They are flexible enough to undergo large elastic deformations but would not allow any arbitrary pattern unlike textiles: panels would fail in tight closure or high curvature patterns. Bending and twisting forces influence

S. Nabaei (✉) • Y. Weinand
Chaire of Timber Construction (IBOIS), École Polytechnique Fédérale de Lausanne (EPFL), CH-1015, Lausanne, Switzerland
e-mail: sina.nabaei@epfl.ch; yves.weinand@epfl.ch

O. Baverel
Navier Research Unit, École des Ponts ParisTech (ENPC), Champes-Sur-Marne, France
e-mail: baverel@enpc.fr

P. Block et al. (eds.), *Advances in Architectural Geometry 2014*,
DOI 10.1007/978-3-319-11418-7_9

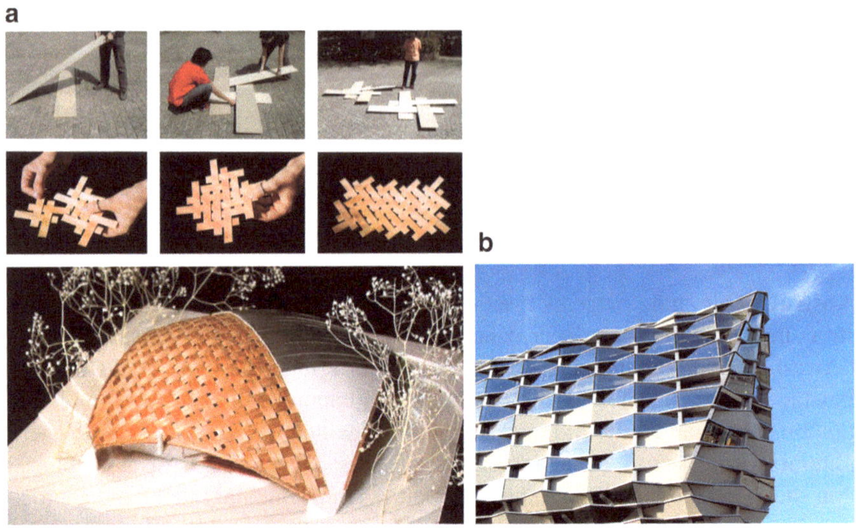

Fig. 1 Interlacing: structural versus visual influence. (**a**) Wickerwork house, Nagano, Japan, 2002, Shigeru Ban architects. (**b**) Facade of the Pabellón de Aragón, Expo 2008, Zaragoza, Spain

how the interlaced relaxed geometry would look like. The fundamental interest of reproducing a pattern with components resistant in bending and torsion (unlike tensile-only ones) is that the assembly will relax into an actively curved geometry which can span over an architectural ambiance. Curved spanning (versus being a tissue) is the key to a structural regard toward interlaced assemblies rather than a facade motif or an ornament (see Fig. 1). This forward implies that the geometry for an actively curved interlace of components with bending and torsional resistance has to be form-found. This form-finding problem with all implied complexities is explored in this manuscript. We have to adopt a nonlinear structural model to simulate panels with bending and twist degrees of freedom, have to come up with a formulation to impose an interlacing pattern as coupling constraints and perform collision detection/handling if required. In terms of technical details, the Euler Elastica (Love 2013) (see Fig. 2) is the building block of the specific family of structures highlighted in our study. We use an extended energy functional to take into account also for twist contributions and append three further structural design potentials on top of a simple Elastica: we explore twisted Elastica made from flexible panels and couple these individual Elasticas with proximity and/or alignment constraints coming from our desired interlace design pattern. We also deal with collision resolving using a pseudo-dynamic shell solver in case interlace and twist boundary condition cause panel intersection.

The manuscript is organized in following order: the related work is briefly reviewed spanning over form-active structures, physics-based rod and shell models. Our hybrid form-finding approach is then detailed, rod-wise and shell-wise and is

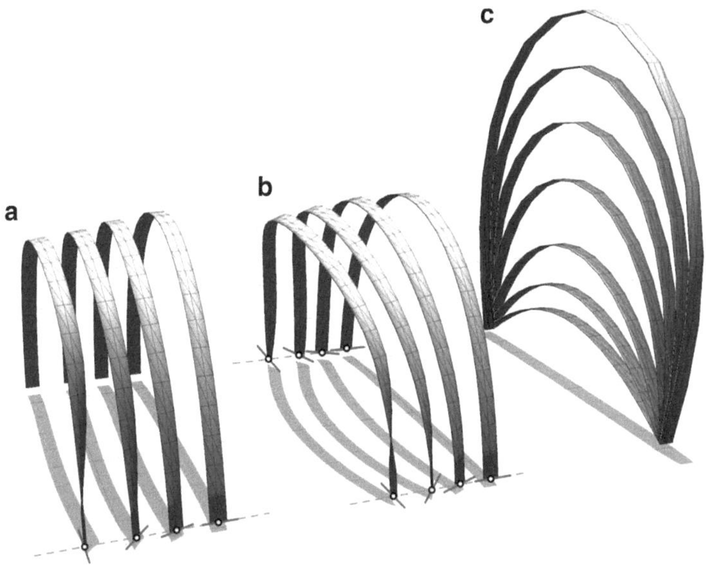

Fig. 2 Elastica variants with and without end twist for timber panel 0.24 × 0.015 m spanning over 6 m: (**a**) Twist applied on single end vertex, (**b**) symmetric twist applied on both end vertices and (**c**) the zero twist Elastica solutions for various total lengths from 6.5 up to 16 m

put in practice for some case-study concept structures to demonstrate the design potential of the procedure.

1.1 Related Work

Interlaced/form-active structures still most of applications for form-active structures made from timber panels come from research institutes, hereby four among the recent pavilions are cited: the wood pavilion of Oslo School of Architecture and Design is a modular structural web, made out of thin pine veneer naturally curved and post-tensioned by material's hygroscopic effect (Hensel 2013) (Fig. 3a). ICD/ITKE research pavilion (Fleischmann and Menges 2012) (Fig. 3b) highlights an overall interlaced geometry using actively-bent timber panels connected by clip dovetail joints and clamped at both ends. The "Radical Wood Pavilion" of Design + Analysis research group of Aalto University (Niiranen 2013) is a form-active, interlaced structure, highlighting a twisted module (Fig. 3c). Timber Fabric structures of IBOIS-EPFL, Switzerland (Hudert 2012) is about interpretation of knotted and braided patterns using flexible timber panels as strands. Panels are elastically bent and twisted into a curved arch structure (Fig. 3d) where the geometry comes from the assembly logic.

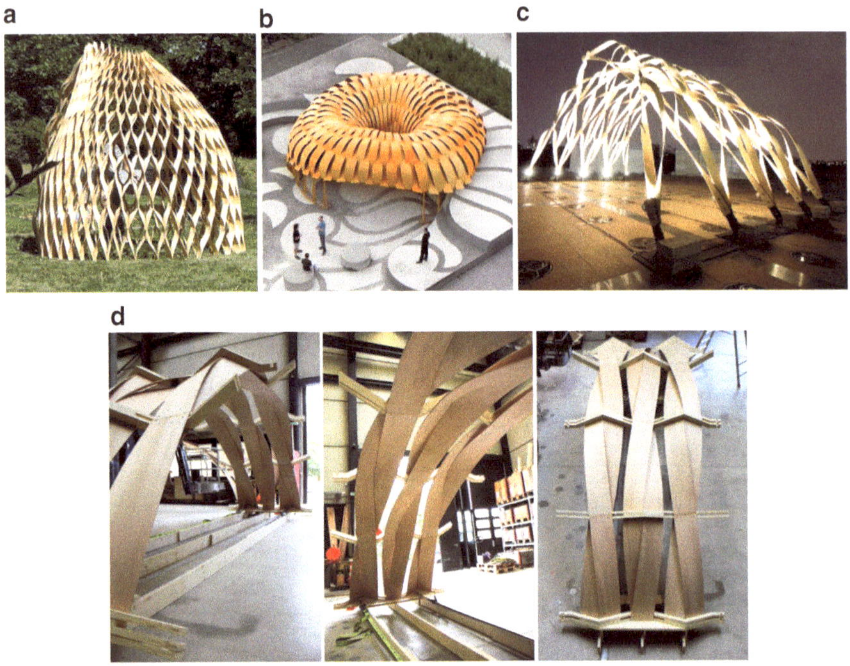

Fig. 3 Interlaced form-active structures made form flexible panels: (**a**) The wood Pavilion of Oslo School of Architecture and Design, (**b**) ICD/ITKE research pavilion 2010, (**c**) the "Radical Wood Pavilion" and (**d**) Timber Fabric Structures (Photo ©Markus Hudert)

Elastic rods Elastic rods have been topic of intense ongoing research. Covered subjects go from computational mechanics (Dias and Audoly 2014), DNA super-coiling Simulation (Olson et al. 2013) and robotics (Lock et al. 2010) up to the physically based computer animation and virtual surgery simulators (Chentanez et al. 2009; Miller et al. 2014).

Thin shell physics-based models Thin shell deformable modeling has been intensively researched through diverse disciplines. Mass spring systems (Choi and Ko 2005) have been among most trivial approaches for physically simulation of cloth and deformable skins while Bridson et al. (2003) and Grinspun et al. (2003) propose a dihedral angle difference as a measure to define the thin shell out of plane behavior. An alternative approach based on discrete isometric bending model is also presented in Bergou et al. (2006). Most recently Iso Geometric analysis has also been applied to simulate cloth deformation (Lu and Zheng 2014).

1.2 *Goals and Contributions*

Our main motivation through this study is thus to formulate a physics-based tool for interlaced configurations to enable an interactive design experience.

Rich rod and shell models for structural architecture applications we offer implementations of two efficient physics-based models from the computer graphics body of knowledge (Bergou et al. 2008; Grinspun et al. 2003) to use for architectural geometry applications. The fact of using these models bring some key features into the existing particle-based form-finding tools such as (Piker 2013): (i) the possibility to deal with general rod and shell cross sections with usual elastic stiffness terms, instead of simplified spring stiffness. Spring values usually need interpretation, depends on the spatial discretization and require special care to create out-of-plane resistance for shell simulations (ii) enhancing twist degrees of freedom for rod deformation (iii) an improved out-of-plane bending behavior for thin shell physically based simulation.

Constraint formulation of an interlaced design pattern we give elements on how to interpret a desired interlacing pattern as equality constraints for a nonlinear optimization problem. Our formulation integrates also some useful architectural local connection features such as imposing a desired offset distance or the possibility to impose the alignment of panels at the overlap.

Discrete Elastic rod model employed in a static framework we reformulate the initially proposed dynamically integrated rod model of Bergou et al. (2008) into an offline constrained optimization problem with displacements and twists as variables of the optimization problem defined all at once.

2 Twisted Interlaced Structure Simulation Pipeline

The complexity of interlaced configuration form-finding can be reduced by breaking it down into pertinent sub-problems based on the flexible body model used to represent panels and the physical phenomenon to simulate. Buckling is the key curving agent in these form-active structures and supposing a static framework, it can be readily formulated as a constrained elastic energy minimization with the only constraints as total panel length and span. A flexible rod model with vertex positions and cross section orientation of edge segments as variables, can be efficiently used to get elastic energy expressions. Having end vertices twisted and panels which have to pass underneath or above of another panel of the assembly at a particular position (interlacing) are simply additional constraints on either rod vertices or cross section orientation angles (twists), which have to be injected to the initial elastic energy minimization problem of buckling. The notion of *above* and *below* have to be discussed (is it with respect to global vertical axis or the local one?) as well as the clearance at overlap nodes, the minimum required offset. Up to here if the

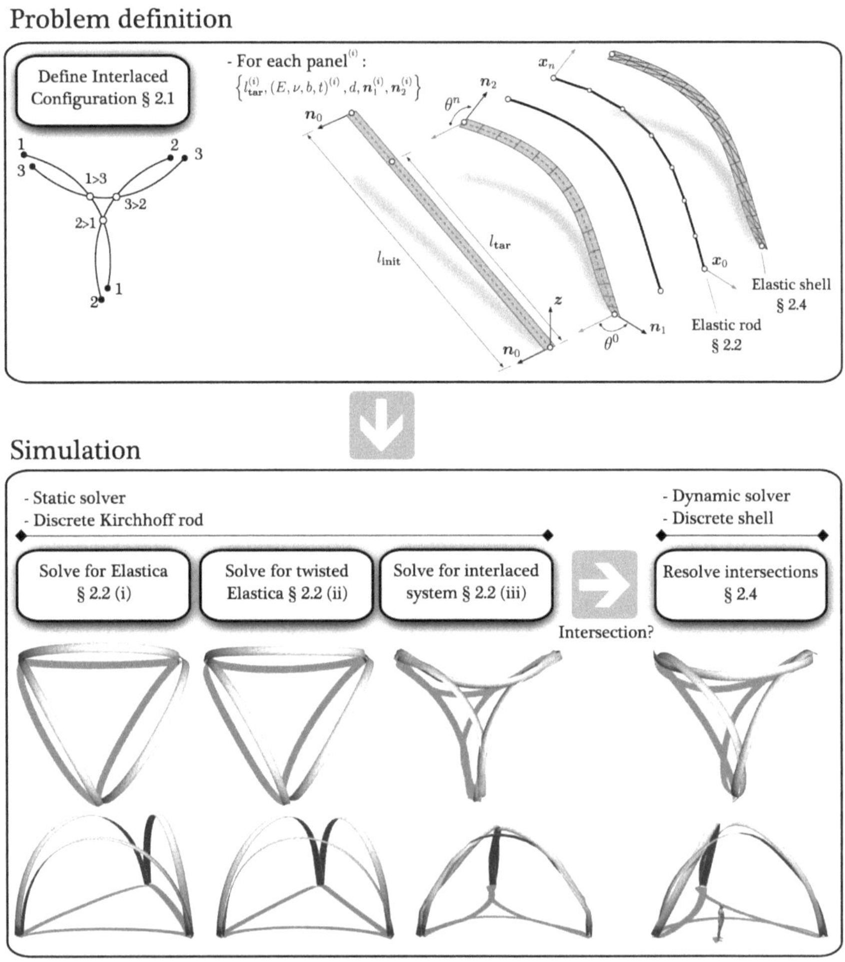

Fig. 4 Twisted interlaced structure form-finding pipeline

minimum elastic energy geometry does not contain colliding panels, all of above mentioned can be solved using constrained optimization formulated with a rod kinematics. Collision detection/resolving in its turn seems to us easier to handle with a shell kinematics in a dynamic simulation, which motivates our hybrid approach. This brief go-through also provides indications on the inputs to be specified for each problem, the design parameters. For each panel, the initial length and target span, the imposed end twist and material properties have to be specified by user. At each overlapped node the offset distance to respect, is also an input data. These design parameters are symbolically represented on Fig. 4 at problem definition level. While investigating an interlaced configuration, it is also convenient to represent and discuss the pattern in a connectivity graph. The interlaced diagram is introduced for this purpose. Since the location of overlap node is also a degree of freedom for our problem, user has also to specify where (on which rod vertices) he would like to create the overlap coupling and his desired element order at this particular overlap.

We elaborate the rest of this section as follows: the *interlaced diagram* is first introduced and technical details for the elastic rod model used to simulate the twisted Elasticas and for the thin-shell model used for intersection resolving are presented briefly afterward. We show how to configure the optimization problems to solve for the classical planar Elastica (without end twists), for the twisted one and for the coupled system of twisted Elasticas with overlap order constraints. The thin-shell model is presented next, followed by our intersection resolving implementation.

2.1 Interlaced Diagram

A graph representation inspired form projected knot diagrams is introduced in order to define and distinguish interlaced morphologies. Lines numbered at both ends represent panels and for intermediate interlace nodes, the relative overlap order of panels is noted using the > notation. The graph representation enables to effectively sweep over interlaced configurations in an abstract topological form regardless of the dual 3D shape while there might be several of those duals corresponding to the same graph depending on panel's initial lengths and span, their elastic properties, precise location of overlaps and also the crossing's linking number.

2.2 Static Simulation with Elastic Kirchhoff Rod Model

We assume the panel(s) to be represented by their centerline curve, have a cross section of $w \times t$, a flat initial state (zero initial curvature, zero initial twist) of length $l_{\mathbf{init}}$ and initial orientation defined by vector $\mathbf{n}_0$. Our goal is to solve the buckled twisted state of the panel when it spans over $l_{\mathbf{tar}}$ and end twists of θ^0, θ^n are applied on its end vertices, giving the new cross section orientation of end vertices as $\mathbf{n}_1$ and $\mathbf{n}_2$ (see Fig. 4 top). We consider the energy functional for elastic Kirchhoff rod proposed by Bergou et al. (2008) and reproduced in Eq. 1 for a rod with n linear edge segments. Basically the smooth centerline curve of the rod is divided into a polyline with linear edge segments. We follow the same notation for scalar and vector quantities as Bergou et al. (2008) in Eq. 1, over-lined terms correspond to the quantities at their reference state, ω_i^j measures the contribution of bi-normal curvature vector of vertex i on edge segment j and m_k takes the relative twist of vertex k into account. We refer interested readers to Bergou et al. (2008) for further details on the curve-angle model. The following stiffness constants are used for rectangular cross section of panel $w \times t$ with usual notation for elastic constants: the bending stiffness $\mathbf{B} = \begin{pmatrix} Etw^3/12 & 0 \\ 0 & Ewt^3/12 \end{pmatrix}$ and the torsional stiffness $\beta = \frac{Gwt^3}{3}$ where $G = \frac{E}{2(1+\nu)}$ is the material shear modulus.

$$E_{bend}(\Gamma) = \sum_{i=1}^{n-1} \frac{1}{2\bar{l}_i} \Big(\sum_{j=i-1}^{i} (\omega_i^j - \bar{\omega}_i^j)^T \mathbf{B} (\omega_i^j - \bar{\omega}_i^j) \Big) \quad E_{twist}(\Gamma) = \sum_{k=1}^{n-1} \frac{\beta\,(m_k - \bar{m}_k)^2}{\bar{l}_k} \tag{1}$$

The above mentioned energies remains the same for a general cross section, only the stiffness terms are required to be updated. Bergou et al. (2008) also gives simplified expressions for a rod with an isotropic cross section and deformed from a flat initial state. The original method in Bergou et al. (2008) is a dynamic simulation and at each time step, before integrating for positions, a quasi-static framework for twist angles has to be satisfied. The inextensibility of edge segments and is particularly imposed using the manifold projection method (Hairer et al. 2006). While the proposed curve-angle kinematic fits well to our problematics, we only require the final relaxed deformed state and thus decide to proceed instead with a static framework for all degrees of freedom and propose to reformulate it as a constrained optimization problem. The coupled system of twisted Elastica is supposed to have of a given number of rods, with ith panel noted as $\Gamma^{(i)}$. Instead of imposing all constraints at once to the flat initial state to get the buckled – twisted – interlaced configuration, we instead proceed with a three stage solution procedure formulated as constrained optimization problems detailed bellow.

I. *System of planar Elasticas of imposed total length and span* The initial optimization is to solve the Elastica problem with zero twist contribution, which turns out to be the classical bi-hinged Elastica curve. Panel$^{(i)}$ with initial length of $l_{\mathbf{init}}^{(i)}$ buckles to span $l_{\mathbf{tar}}^{(i)}$, with $n^{(i)}$ equal edges, fixed end vertices and under an additional constraint to keep the deformed configuration in the initial vertical plane. Variables to solve for are unfixed vertex positions (vector quantity $\mathbf{x}$) and edge segment twist angle (scalar quantity θ). The initial state for this first stage is the flat panel state (zero initial curvature, zero initial twist).

$$\min_{\{\theta^{(i)},\mathbf{x}^{(i)}\}} \quad E(\theta,\mathbf{x}) = \sum_{(i)=0}^{\Gamma\text{count}} E_{bend}(\Gamma^{(i)}) + E_{twist}(\Gamma^{(i)})$$

$$\text{s.t.} \quad (i)\ \left\| \mathbf{x}_{j+1}^{(i)} - \mathbf{x}_j^{(i)} \right\|^2 = \left(l_{\mathbf{init}}^{(i)}\right)^2 / \left(n^{(i)}\right)^2 : \quad j \in \{0^{(i)}, \ldots, n^{(i)} - 1\}$$

$$(ii)\ \mathbf{x}_j^{(i)} = \bar{\mathbf{x}}_j^{(i)} \ : \quad j = 0^{(i)}, n^{(i)}$$

$$(iii)\ \mathbf{n}_0^{(i)} \cdot \left(\mathbf{x}_1^{(i)} - \mathbf{x}_0^{(i)}\right) = 0$$

$$(iv)\ \mathbf{n}_0^{(i)} \cdot \left(\mathbf{x}_{n^{(i)}-1}^{(i)} - \mathbf{x}_{n^{(i)}}^{(i)}\right) = 0$$

II. *System of twisted Elasticas:* as the next step we solve for the twisted Elasticas with both end edges supposed to be clamped. The initial state for this analysis are deformed Elastica retrieved by solving problem I described above. The cross section orientation at both ends for each panel are obtained by rotating the corresponding initial normal vector $\mathbf{n}_0^{(i)}$ respectively by $\theta^{0^{(i)}}$ and $\theta^{n^{(i)}}$.

$$\min_{\{\theta^{(i)},\mathbf{x}^{(i)}\}} \quad E(\theta,\mathbf{x}) = \sum_{(i)=0}^{\Gamma\text{count}} E_{bend}(\Gamma^{(i)}) + E_{twist}(\Gamma^{(i)})$$

$$\text{s.t.} \quad (i) \left\|\mathbf{x}_{j+1}^{(i)} - \mathbf{x}_j^{(i)}\right\|^2 = \left(l_{\textbf{init}}^{(i)}\right)^2 / \left(n^{(i)}\right)^2 : \quad j \in \left\{1^{(i)},\ldots,n^{(i)}-2\right\}$$

$$(ii)\ \mathbf{x}_j^{(i)} = \bar{\mathbf{x}}_j^{(i)} : \quad j = 0^{(i)}, 1^{(i)}, n-1^{(i)}, n^{(i)}$$

(iii) Twist angle of material frames for first and last edges of panel$^{(i)}$ respectively $(\theta^0)^{(i)}$, $(\theta^{n^{(i)}-1})^{(i)}$ aligned w.r.t $\mathbf{n}_1^{(i)}$ and $\mathbf{n}_2^{(i)}$

III. *Interlacing twisted Elasticas:* The final simulation step is dedicated to couple twisted panels according to the schematic interlace pattern. The initial state for this analysis is the result of problem II. For each overlap panel$^{(i)}$ > panel$^{(k)}$, a handle reference point have to be defined by the user to determine where exactly the overlap has to be passed. The vertices of the panels involved in the overlap (the ones closest to this handle point), noted as $\mathbf{x}_j^{(i)}$ and $\mathbf{x}_l^{(k)}$, will be the ones to hold the constraints. The distance between corresponding nodes of the overlapping panels, $\left\|\mathbf{x}_j^{(i)} - \mathbf{x}_l^{(k)}\right\|$, is set to be equal to the given offset distance d and the top/bottom in the overlap is imposed using the projected distance of the involved nodes with respect to the material frame vector $\mathbf{m}_2^{(i)}$ of the panel$^{(i)}$. (The out of plane direction of the panel supposed to pass on top of the other one.) The former axis gives a local order depending on panel orientation. The global z axis should be used instead of $(\mathbf{m}_2^j)^{(i)}$ for a vertically oriented overlap.

$$\min_{\{\theta^{(i)},\mathbf{x}^{(i)}\}} \quad E(\theta,\mathbf{x}) = \sum_{(i)=0}^{\Gamma\text{count}} E_{bend}(\Gamma^{(i)}) + E_{twist}(\Gamma^{(i)})$$

$$\text{s.t.} \quad (i) \left\|\mathbf{x}_{j+1}^{(i)} - \mathbf{x}_j^{(i)}\right\|^2 - \left(l_{\textbf{init}}^{(i)}/n^{(i)}\right)^2 = 0 : \quad j \in \left\{1^{(i)},\ldots,n^{(i)}-2\right\}$$

$$(ii)\ \mathbf{x}_j^{(i)} - \bar{\mathbf{x}}_j^{(i)} = 0 : \quad j = 0^{(i)}, 1^{(i)}, n-1^{(i)}, n^{(i)}$$

(iii) Twist angle of material frames for first and last edges of panel$^{(i)}$ respectively $(\theta^0)^{(i)}$, $(\theta^{n^{(i)}-1})^{(i)}$ aligned w.r.t $\mathbf{n}_1^{(i)}$ and $\mathbf{n}_2^{(i)}$

(iv) **foreach** $\left(\mathbf{x}_j^{(i)} > \mathbf{x}_l^{(k)} \text{order constraint}\right)$

$$\left\|\mathbf{x}_j^{(i)} - \mathbf{x}_l^{(k)}\right\|^2 - d^2 = 0$$

$$\left(\mathbf{x}_j^{(i)} - \mathbf{x}_l^{(k)}\right) \cdot (\mathbf{m}_2^j)^{(i)} - d = 0$$

end foreach

The above cited problems are solved using a Quasi-Newton procedure with BFGS approximation of the Hessians using a primal-dual interior-point algorithm with a filter line-search method (Wächter and Biegler 2006; HSL 2013). Notice that only energy gradients with respect to the vertex positions and twist angles are required (and not the Hessians, see Bergou et al. (2008) for forces).

2.3 *Pseudo-dynamic Simulation with Elastic Shell Model*

In some interlaced configurations with repeatedly inversed overlaps (like overlaps in a braid $\{1 > 2,\ 2 > 1,\ 1 > 2\ldots\}$) or when the overlap offset distance is small with respect to the panel width, the minimum energy configuration obtained from the coupled Elastica optimization will not be intersection free. We propose to dynamically integrate an elastic shell model of the discretized mid-surface of panels for intersection detection and resolving. Among existing physically based thin shell model we use Grinspun et al. (2003) for triangle meshes with flexural and membrane energy functionals in their discrete form. The flexural energy comes form a discrete analogy of the integral mean curvature of the surface mesh and is measured as a function of the difference between complements of the dihedral angle of mesh edges on deformed and reference state, the former quantities are marked with an over line. Principals of discrete elastic shell settings and energies are briefly reminded in Fig. 5. Around a given vertex $\mathbf{v}$ we identify a set of direct dihedrals and corresponding edges noted as $e_d(\mathbf{v})$ contributing to membrane forces (in plane tension or compression) and a set of indirect dihedrals and corresponding edges, noted as $e_i(\mathbf{v})$ which apply the flexural forces according to the stiff hinge effect around the corresponding edge e_i. The term $\bar{h}_{e_i}$ is the one third of the sum of the height of the two triangles sharing the edge $e_i(\mathbf{v})$ at the reference configuration. See also the index of Grinspun et al. (2003) for another possible definition for $\bar{h}_{e_i}$. We apply corresponding values of k_L, k_A, k_B and derivate these energies to get the forces acting on vertices through vector calculus manipulation which we do not reproduce here for conciseness (Grinspun et al. (2003) also proposes automatic derivation). The Newmark integration scheme is used for position update with $\gamma = 0.5, \beta = 0.25$ with the predictor-corrector approach (Hughes et al. 1979). Mesh intersection detection is performed using the fast triangle-triangle intersection test (Möller 1997) and resolved by applying the corrective penalty forces in direction which minimizes the intersection line as detailed in Volino and Magnenat-Thalmann (2006). This method is particularly adopted to our problematic because it does not necessarily require the mesh intersection to be a closed contour (unlike Baraff et al. (2003)) and moreover, it is history-free.

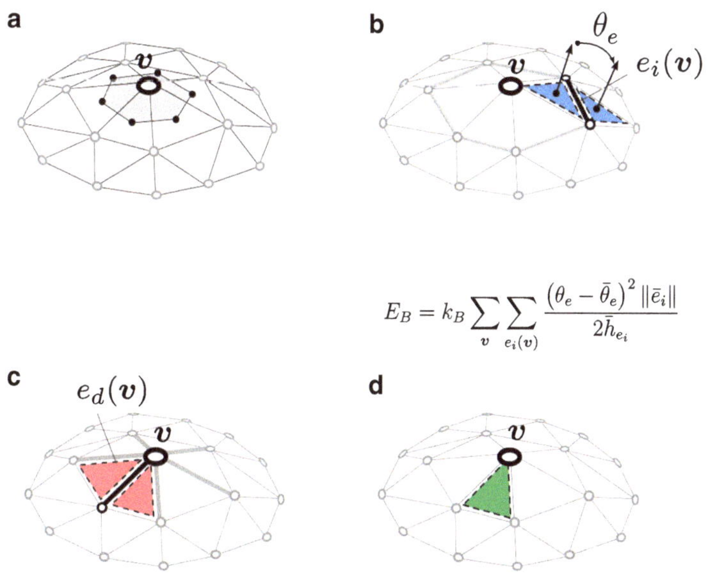

$$E_B = k_B \sum_{v} \sum_{e_i(v)} \frac{(\theta_e - \bar{\theta}_e)^2 \|\bar{e}_i\|}{2\bar{h}_{e_i}}$$

$$E_L = k_L \sum_{v} \sum_{e_d(v)} \left(1 - \frac{\|e_d\|}{\|\bar{e}_d\|}\right)^2 \|e_d\| \qquad E_A = k_A \sum_{f} \left(1 - \frac{\|A\|}{\|\bar{A}\|}\right)^2 \|A\|$$

Fig. 5 Discrete elastic shell setting and energies, contributions acting on vertex **v**: (**a**) The area (mass) related to vertex, (**b**) flexural energy contributions of indirect dihedrals around vertex, (**c**) membrane energy contributions of direct dihedrals around vertex and (**d**) in plane shear energy contributions of faces connected to vertex

3 Case Studies

Four case studies are discussed in this section to demonstrate our form-finding procedure: a gridshell like structure built with twisted panels, a trefoil arch, a pentafoil roof and a twisted truss-beam module. In all these examples panels are supposed to be made from timber and mean elastic properties of wood $E = 8{,}000$ Mpa, $\nu = 0.3$, $\rho = 500\,\text{kg/m}^3$ are used.

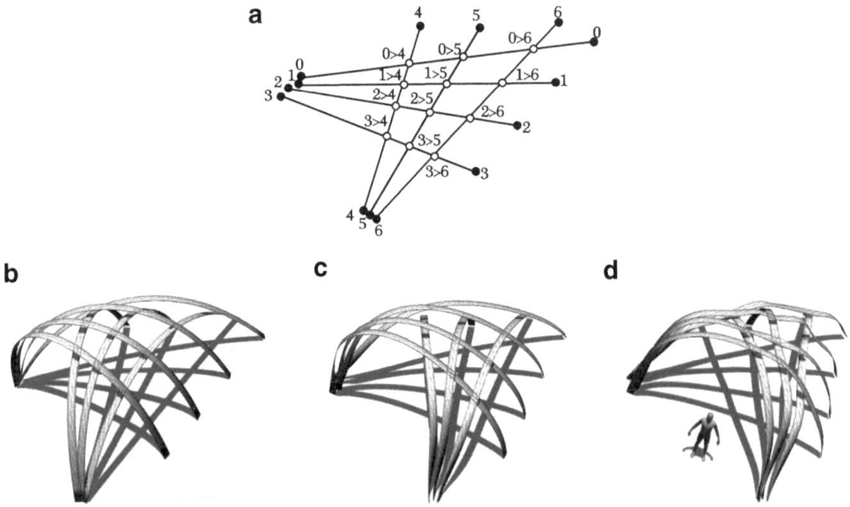

Fig. 6 A gridshell structure with twisted panels: (**a**) The graph representation, (**b**) the solution to Elastica problem, (**c**) twisted Elastica configuration and (**d**) interlaced configuration

3.1 A Gridshell Structure with Panels

We provide an example of a gridshell type structure (see Fig. 6) this time built with panels instead of regular circular cross-section beams and we additionally impose a uniform twist of 90° on start vertex of all panels. The beams are all 10 m long buckled to span various lengths: 6.25 m $< l <$ 8.46m. The offset distance is 0.2 m and alignment of panels is imposed at overlaps. Panels are all 0.24 × 0.015 m. The structure is at most 3.2 m high at the level of entrance.

3.2 An Interlaced Trefoil

The concept structure illustrated in Fig. 7 is an example of interlacing which requires an intersection handling step. The trefoil arch structure is supposed to be made with timber panels 0.24 × 0.015 m, all panels are 14 m long spanning over 8.5 m and respective twist of (−15°, −30°) is applied on start and end vertices of each panel. The arch becomes 4.8 m high at middle-vertex at its intersection-free relaxed state.

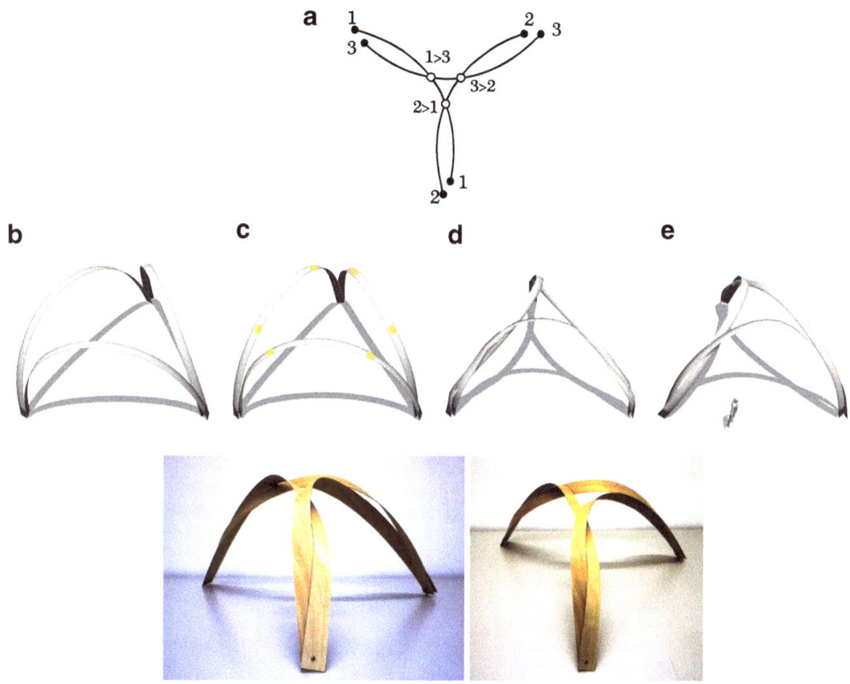

Fig. 7 A trefoil interlaced arch: (**a**) The graph representation, (**b**) the solution to Elastica problem, (**c**) twisted Elastica configuration with overlapping vertices highlighted, (**d**) interlaced configuration (contains intersection) and (**e**) intersections handled

3.3 An Interlaced Pentafoil

The concept structure illustrated in Fig. 8 is a complex structure with multiple overlap nodes and colliding components. The pentafoil roof structure is supposed to be made with timber panels 0.65×0.03 m, all 14 m long spanning over 11.8 m with support points equally spaced on a circle of 6.3 m radius. Respective twist of $(34°, 34°)$ is applied on start and end vertices of each panel. As a result of interlacing the arch lifts up to clear 1.98 m vertically with respect to the support point at its intersection-free relaxed state.

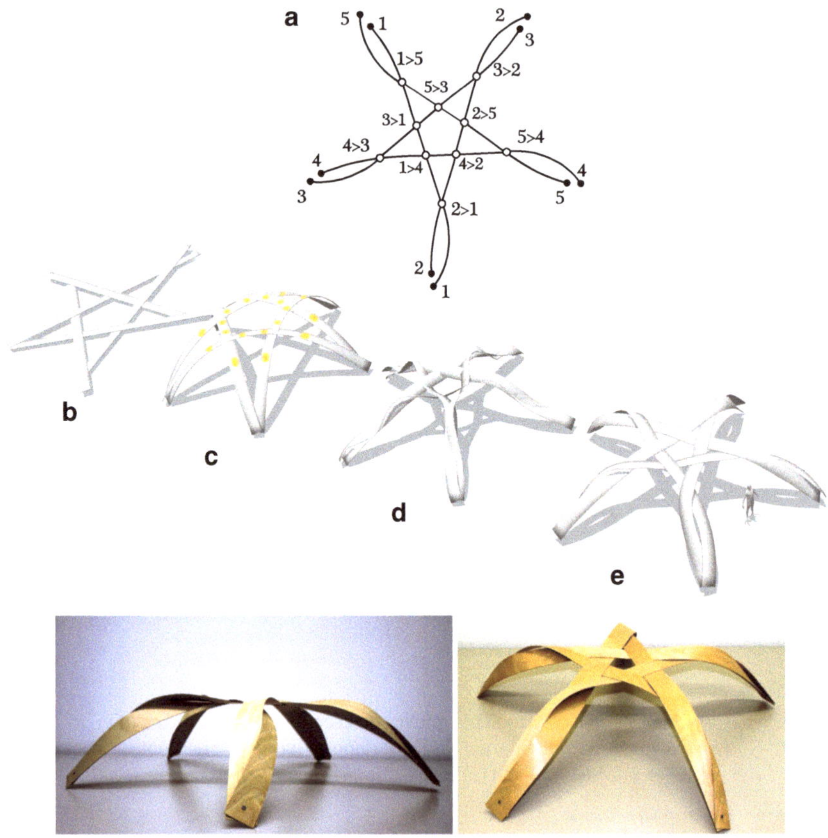

Fig. 8 A pentafoil interlaced roof: (**a**) The graph representation, (**b**) flat initial state, (**c**) solution to the twisted Elastica problem with overlapping vertices highlighted to be interlaced at next step, (**d**) interlaced configuration (contains intersection) and (**e**) intersections handled

3.4 Twisted Truss Module

Figure 9 demonstrates an example where the coupling constraint is only a proximity one and no alignment is imposed. Exterior beams are 6.8 m long and interior ones 6.3 m and the span is 6 m. The width of the beam (axis to axis between exterior beams) is 0.7 m, the middle proximity constraint is set also to 0.7 m and the two top proximity constraints are 0.4 m. (constrained vertices are highlighted in Fig. 9c) All panels are all 0.2×0.01 m. In concept structure Fig 9d, the rebars are positioned to represent the imposed proximity while solving the constrained system and make the entire module behave as an active prestressed truss beam.

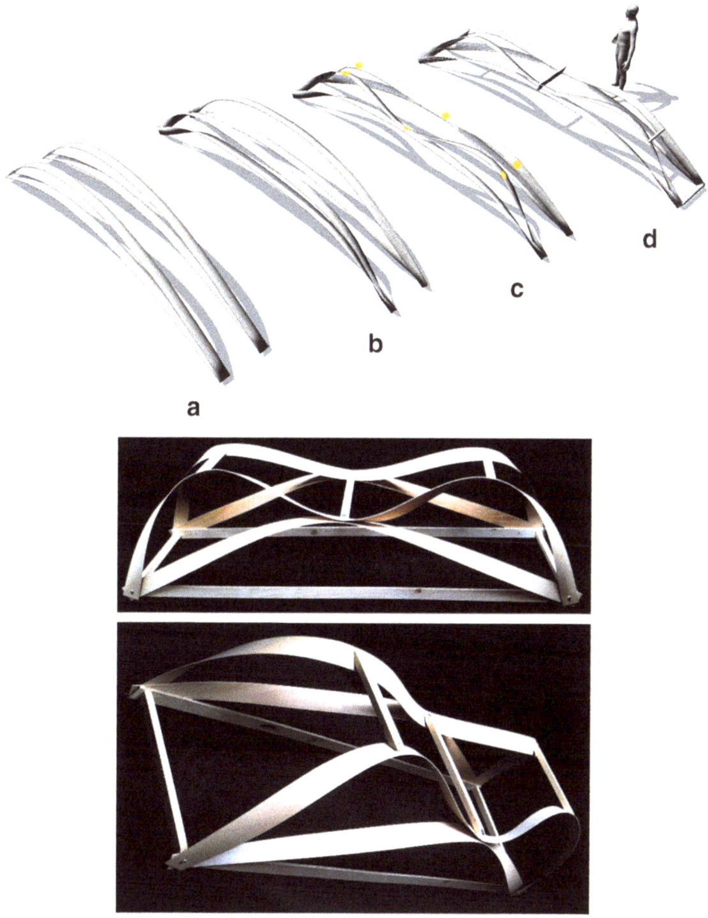

Fig. 9 A twisted truss beam active module: (**a**) The solution for Elastica problem, (**b**) twisted non-coupled beams, (**c**) coupled system of twisted Elasticas with constrained vertices highlighted and (**d**) a concept structure with connectors maintaining the enforced constraints

Conclusion and Outlook

We borrowed two elegant models and techniques employed for physically based simulation of flexible bodies from computer graphics to use in structural form-finding and combined them as a simulation pipeline with a particular focus to explore interlaced structures and modules and twisted configurations currently out of reach of non-engineers. The framework is subdivided into optimization subproblems to handle the Elastica problem in its zero-twist and twisted form and also in case of coupled system of panels with or without

(continued)

overlap constraints. Intersecting panels are cleared using a pseudo-dynamic shell model and an intersection contour minimization algorithm. Both rod and shell frameworks have been implemented as Grasshopper plug-ins and are concerted as described to run presented case studies.

Acknowledgements This work was sponsored by the Swiss National Science Foundation (SNSF) under grant No. 200021_137884/1. This support is gratefully acknowledged. Authors would like also to thank Basile Audoly and Eitan Grinspun for their constructive remarks on problem formulation and implementation.

References

Baraff, D., Witkin, A., Kass, M.: Untangling cloth. ACM Trans. Graph. **22**(3), 862–870 (2003)

Bergou, M., Wardetzky, M., Harmon, D., Zorin, D., Grinspun, E.: Discrete quadratic curvature energies. In: ACM SIGGRAPH 2006 Courses, pp. 20–29. ACM, Boston (2006)

Bergou, M., Wardetzky, M., Robinson, S., Audoly, B., Grinspun, E.: Discrete elastic rods. ACM Trans. Graph. (TOG) **27**(3), 63 (2008)

Bridson, R., Marino, S., Fedkiw, R.: Simulation of clothing with folds and wrinkles. In: Proceedings of the 2003 ACM SIGGRAPH/Eurographics Symposium on Computer Animation, pp. 28–36. Eurographics Association, San Diego (2003)

Chentanez, N., Alterovitz, R., Ritchie, D., Cho, L., Hauser, K. K., Goldberg, K., Shewchuk, J. R., O'Brien, J. F.: Interactive simulation of surgical needle insertion and steering. ACM Trans. Graph. **28**(3), 1–10 (2009). ACM

Choi, K.-J., Ko, H.-S.: Stable but responsive cloth. In: ACM SIGGRAPH 2005 Courses, p. 1. ACM (2005)

Dias, M.A., Audoly, B.: A non-linear rod model for folded elastic strips. J. Mech. Phys. Solids **62**, 57–80 (2014)

Fleischmann, M., Menges, A.: Icd/itke research pavilion: A case study of multi-disciplinary collaborative computational design. In: Gengnagel, C., Kilian, A., Palz, N., Scheurer, F. (eds.) Computational Design Modelling, pp. 239–248. Springer, Vienna (2012)

Grinspun, E., Hirani, A.N., Desbrun, M., Schröder, P.: Discrete shells. In: Proceedings of the 2003 ACM SIGGRAPH/Eurographics Symposium on Computer Animation, 62–67. Eurographics Association, San Diego (2003)

Hairer, E., Lubich, C., Wanner, G.: Geometric Numerical Integration: Structure-Preserving Algorithms for Ordinary Differential Equations, vol. 31. Springer, Berlin/Heidelberg (2006)

Hensel, M.: Performance-Oriented Architecture: Rethinking Architectural Design and the Built Environment. Wiley, Chichester/West Sussex (2013)

HSL: A collection of fortran codes for large scale scientific computation (2013). http://www.hsl.rl.ac.uk.

Hudert, M.: Structural timber fabric: applying textile principals in building scale. PhD thesis, IBOIS, EPFL (2012)

Hughes, T.J., Pister, K.S., Taylor, R.L.: Implicit-explicit finite elements in nonlinear transient analysis. Comput. Methods Appl. Mech. Eng. **17**, 159–182 (1979)

Lock, J., Laing, G., Mahvash, M., Dupont, P.E.: Quasistatic modeling of concentric tube robots with external loads. In: 2010 IEEE/RSJ International Conference on Intelligent Robots and Systems (IROS), pp. 2325–2332. IEEE, Taipei (2010)

Love, A.E.H.: A Treatise on the Mathematical Theory of Elasticity. Cambridge University Press, Cambridge (2013)
Lu, J., Zheng, C.: Dynamic cloth simulation by isogeometric analysis. Comput. Methods Appl. Mech. Eng. **268**, 475–493 (2014)
Miller, J., Lazarus, A., Audoly, B., Reis, P.: Shapes of a suspended curly hair. Phys. Rev. Lett. **112**(6), 068103 (2014)
Möller, T.: A fast triangle-triangle intersection test. J. Graph. Tools **2**(2), 25–30 (1997)
Niiranen, J.: kierretty ja kaareva: Warped and twisted. Puu — Wood — Holz — Bois **1**, 22–27 (2013)
Olson, S.D., Lim, S., Cortez, R.: Modeling the dynamics of an elastic rod with intrinsic curvature and twist using a regularized stokes formulation. J. Comput. Phys. **238**, 169–187 (2013)
Piker, D.: Kangaroo: Form finding with computational physics. Archit. Des. **83**(2), 136–137 (2013)
Volino, P., Magnenat-Thalmann, N.: Resolving surface collisions through intersection contour minimization. ACM Trans. Graph. (TOG) **25**, 1154–1159 (2006). ACM
Wächter, A., Biegler, L.T.: On the implementation of an interior-point filter line-search algorithm for large-scale nonlinear programming. Math. Program. **106**(1), 25–57 (2006)

A Graph-Based Approach for Discovery of Stable Deconstruction Sequences

Lukas Beyeler, Jean-Charles Bazin, and Emily Whiting

Abstract The aim of object pile deconstruction is to safely remove elements one by one without compromising stability. The number of combinations of removal sequences increases dramatically with the number of objects and thus testing every combination is intractable in practical scenarios. We model the deconstruction sequencing problem using a disassembly graph, and investigate and discuss search strategies for discovery of stable sequences in an architectural context. We run and compare techniques in a large-scale experiment, on various virtual scenes of architectural models composed of different shapes, sizes and number of elements.

1 Introduction

When a building collapses, many parts of the structure (wall, roof beams, water pipes, etc.) get piled up in an unstable and complicated arrangement. It is very difficult for rescuers to decide which objects can be safely removed, avoiding additional collapse that might injure survivors still blocked in the debris. A simplification of such a challenging problem is the Mikado game whose goal is to remove sticks from a pile one by one without disrupting the remaining pile. This paper is a preliminary study for pile deconstruction in the context of architecture, where the goal is to automatically find a sequence of objects which can be safely removed from the pile one by one without compromising the pile stability.

One of the main challenges in finding such a sequence is the combinatorial aspect. Consider the naive approach: for a pile of n objects, there are n possible objects to remove at the first step, $n - 1$ at the second step, and so on, which results in $n!$ possible sequences. Testing them all is intractable even for relatively small n (e.g. for Mikado where $n = 41$, there are $41! > 3 \times 10^{49}$ possible sequences).

L. Beyeler (✉) • J.-C. Bazin
ETH Zurich, Universitatstrasse 6, 8092 Zurich, Switzerland
e-mail: lukas.beyeler@bluewin.ch; jean-charles.bazin@inf.ethz.ch

E. Whiting
Dartmouth College, Sudikoff Lab, 9 Maynard St., Hanover, NH 03755, USA
e-mail: emily.whiting@dartmouth.edu

P. Block et al. (eds.), *Advances in Architectural Geometry 2014*,
DOI 10.1007/978-3-319-11418-7_10

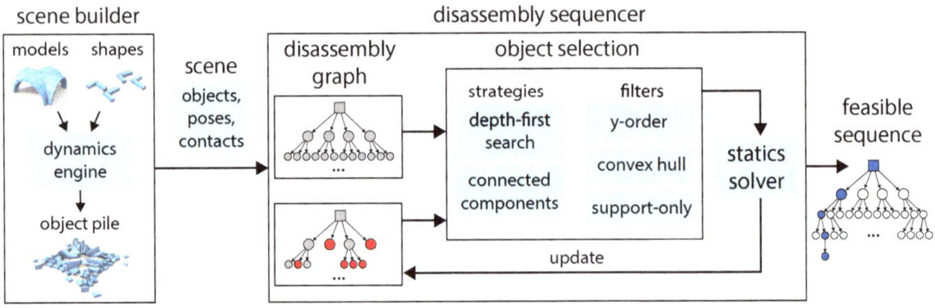

Fig. 1 System overview

While pile construction and simulation have been extensively studied (see related work section), the pile deconstruction problem has been largely ignored. The study of stable deconstruction sequence in an architectural context constitutes our first contribution. For this, we develop an entire pipeline for a deconstruction study (Fig. 1): from pile generation to disassembly sequencing searching, making use of physics simulation and static equilibrium analysis. We cast the problem as a graph search and investigate search strategies. Finally, we conduct a large-scale experiment to study a wide range of piles and experimentally compare search strategies and heuristics.

2 Related Work

Object disassembly Disassembly sequencing is a heavily studied topic, particularly within the realm of production research. Disassembly refers to the process of separating an assemblage of parts, for example, with the purpose of replacement, repair, or disposal at a product's end-of-life. For a historical survey of techniques, see Lambert (2003). Lambert and Gupta (2005) review fundamentals of sequence representation and search techniques, where the emphasis is on geometric constraints for topologically complex assemblies. Abe et al. (1999) apply stability for verification in assembly planning, while Loomis and Balkcom (2006) focus on the performance of static analysis, by re-using computations from sub-structures. Guo et al. (2013) present a method for disassembling models and visualizing the disassembly process. Stability is considered by having the user specify fixed parts of the model that are then kept unchanged through the disassembly process. In contrast, our work surveys performance for discovery of feasible sequences. Schwartzburg and Pauly (2013, xxxx) investigate the analogous problem of *assembling* objects, in the context of interlocking planar pieces. They similarly model the elements with a graph representation, but focus on geometric constraints rather than stability.

A robotic perspective A related problem in robotics is the bin-picking task, that is teaching a machine to pick up an object from a pile. This simple daily-life task for humans combines many key challenges from the fields of computer vision and robotics. Ikeuchi et al. (1983) use photometric stereo to identify objects and avoid collision while removing one object from the pile. Katz et al. (2013) use a machine learning approach to estimate object affordances and select a suitable action. These methods are restrictive in the shapes of objects and, while trying to avoid collisions, usually do not consider the stability of the pile after removing the object. Wang et al. (2009) developed a robot to remove blocks from a Jenga tower but follow a greedy approach for the removable block identification. In contrast, we consider the deconstruction problem at the global scale, i.e. we do not aim to find just one next block to remove, but an entire deconstruction sequence.

Object piles Numerous systems exist for simulating piles of objects (Guendelman 2003; Kaufman et al. 2005; Erleben 2007). Hsu and Keyser (2010) directly model cone shaped piles using a heuristic to freeze objects temporarily and remove them from dynamic simulation. Hsu and Keyser (2012) use automated constraint placement allowing for more artistically guided piles and to achieve enhanced realism. Although related in the context of stability concerns for assemblages of objects, these solutions deal with an opposite motivation of constructing piles rather than disassembling them.

Graph algorithms and game theory Identifying a sequence of actions (in our context, selecting and removing an object from a pile) to reach a certain goal also arise in game theory. The moves of a game are often represented as a graph, like in our work. Graph algorithms like α-β-pruning (a depth-first based search) and killer lookup tables (hashing), among many others, have been applied successfully for various games like checkers (Schaeffer 1997), Othello (Buro 2002) and chess (Campbell et al. 2002). They remove unprofitable plays early and allow for deeper inspection of the decision tree. Even with early abortion, the space of moves in a game is usually huge. Thus most of the existing optimization techniques rely on the rules of the games to develop specific heuristics and strategies. Unfortunately these game-specific heuristics cannot be applied in our context of pile deconstruction.

3 Background

3.1 Disassembly Graphs

The goal is to find a disassembly sequence, that is the order in which the objects can be removed from the pile without compromising its stability. Given a pile composed of n objects (see Fig. 2c), the choice of an object to remove can be modeled as a decision tree called a disassembly graph (Lambert and Gupta 2005) (see Fig. 2d). The nodes represent the pile state (that is the remaining objects) and the edges

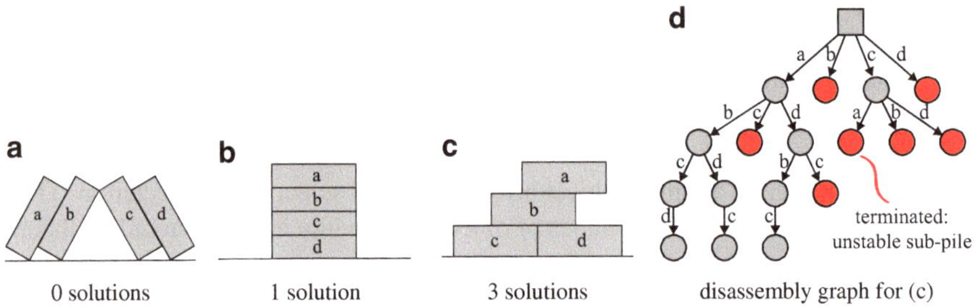

Fig. 2 Different arrangements of n objects. (**a**) No feasible solutions exist. (**b**) The one feasible sequence is (a, b, c, d). (**c**) Only 3 sequences are feasible, namely (a, b, c, d), (a, b, d, c) and (a, d, b, c). (**d**) The disassembly graph for (**c**): edges represent an element removal, *red nodes* indicate an unstable arrangement where the sequence is aborted

represent the object removed. The root of the decision tree is the initial pile. We call a "path" a sequence of removed objects from the top node to one leaf, that is of length n. A "solution path" (also called "feasible sequence") is a path which retains stability at every stage of disassembly. The depth of the disassembly graph is n, with $n!$ leaves (i.e. paths to explore). Note that some paths can be aborted early as soon as one object along the path leads to an unstable sub-pile (red nodes in Fig. 2d). Exhausting the decision tree is infeasible even for small numbers of objects as the number of paths in the graph for arbitrary piles grows exponentially. Note that there might be zero, one or many solution paths depending on the arrangement of the objects (Fig. 2a–c).

3.2 *Static Equilibrium Analysis*

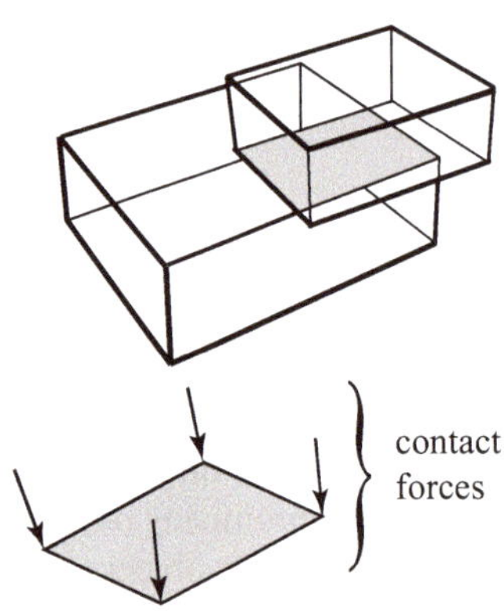

Static equilibrium analysis identifies whether a certain configuration of objects is stable or not. Every object requires zero net force and net torque for the configuration to meet equilibrium conditions. As in Livesley (1978), we assume a polygonal contact interface between adjacent objects. The forces between two objects are distributed as discrete vectors at the vertices of the contact polygon (see inset figure). It has been shown in Livesley (1978) that stability analysis of rigid block assemblies can be formulated as a linear program. The unknown is the vector of contact forces **f** (consisting of normal and shear components), and the constraints encode static equilibrium conditions, friction cone constraints, and compression-only constraints. If a feasible **f** exists, then the configuration of objects is stable.

4 The Picking Algorithm

We present the different aspects of the search strategies employed in our system. Provided a set of n objects in a pile, the goal is to find a sequence of objects of length n such that removing the objects one-by-one from the pile according to this sequence does not lead to unstable configurations. Representing the initial pile P as a set of objects, we define the pile obtained by removing objects from P according to the sequence S_i as a set difference $P \setminus S_i$. We can formally express the problem:

$$\begin{aligned}&\text{given } P = \{1, \ldots, n\}\,, \text{ a set of } n \text{ objects in a pile}\\&\text{find a sequence of objects } S = (o_1, \ldots, o_n) \text{ with } o_i \in P,\ \forall i = 1 \ldots n\\&\text{such that } \textsc{isStable}(P \setminus S_i) = 1, \forall i = 1 \ldots n,\ S_i = (o_1, \ldots, o_i)\end{aligned}$$

where $\textsc{isStable}(P') = 1$ when the pile P' is stable, and returns 0 otherwise. We see that the sequence S belongs to the set of permutations of P. To decide if a subset of P is stable or not, we apply static analysis as described in Sect. 3.2. Static equilibrium analysis determines feasibility of an arrangement of objects, and thus whether it is possible to remove an object safely.

5 Search Strategies

As discussed, exhaustively searching all possible paths is intractable in practical scenarios. In the following we discuss and investigate several alternative search strategies.

5.1 Depth-First Search vs. Breadth-First Search

The two main strategies to perform a complete tree search are breadth-first and depth-first search. In a breadth-first search, visiting the k-th level of the tree requires storing $\binom{n}{k}$ nodes in the worst case, which is intractable even for small k, and thus could not be applied for our evaluations. Therefore we choose depth-first search as it is memory-conserving: only a portion of intermediate results up to the depth of the tree has to be stored. The basic implementation of depth-first search is straightforward: while traversing the decision tree, it is needed to keep track of the path from the root to the current node. Then it only needs to detect when the empty pile has been reached and then inspect the stack to return the solution path.

5.2 Connected Components

Oftentimes, a scene is composed of distinct piles that do not share objects. Piles may be considered independent if no interacting forces exist between any pair of objects. The naive approach would be to analyze the scene as a single assemblage. In contrast, since each grouping has no influence on the others, they can be processed (i) independently, which drastically reduces the number of possible combinations, and (ii) in parallel, which further speeds up the evaluation. For a scene composed of n objects in total, with K independent piles of $m < n$ objects each, the number of combinations is $K(m!)$, which is far less than the naive $n!$ combinations.

Independent piles can be quickly and easily identified by connected components on a connectivity graph, which is a well established technique in graph theory (Hopcroft and Tarjan 1973). Each node on the connectivity graph represents an object, and the edges indicate the connectivity of the objects (whether contact points exist). Examples are shown in Fig. 3.

Connected components can be easily incorporated in depth-first search. In practice, we build the connectivity graph from the scene at initialization, and the nodes of the graph (i.e. the objects) are inserted and removed in accordance to the algorithm stepping through the possible sequences. A solution path for the whole scene is simply any concatenation of the solutions paths of the independent piles.

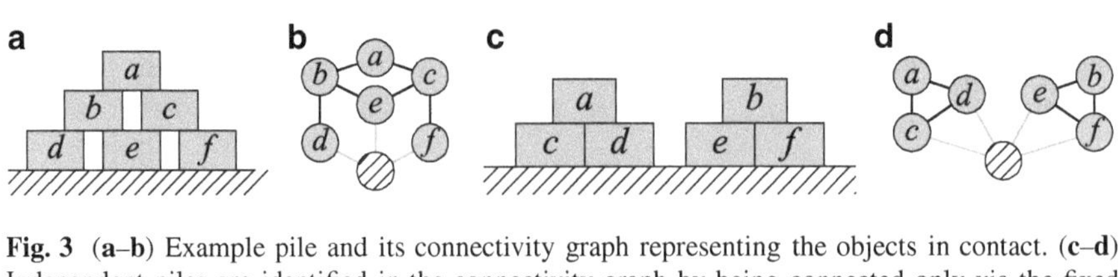

Fig. 3 (**a–b**) Example pile and its connectivity graph representing the objects in contact. (**c–d**) Independent piles are identified in the connectivity graph by being connected only via the fixed ground node. Here two independent piles are identified, namely (a,c,d) and (b,e,f)

5.3 Heuristics

In the following, we investigate additional heuristics. In the best case, depth-first search finds a successful solution path with a complexity $O(1)$, i.e. the first tested path is feasible, but this is unlikely to happen in practice. An effective heuristic can be used to efficiently drive the search: choosing objects in a clever way to greatly reduce the number of steps required in finding a solution.

Heuristics can be soft or hard. A soft heuristic performs a sorting of all the candidates, according to a criteria preference. A hard heuristic selects some candidates and discards the others. Hard heuristics can effectively remove choices at a given node in the decision tree. Note that they reduce the number of paths searched which implies that a badly designed hard heuristic might accidentally remove solution paths. Soft and hard heuristics can be combined: a hard heuristic is generally applied first to select candidates, followed by a soft heuristic for sorting. We developed and tested several approaches:

Y-order Ranks objects based on the height of their center of mass (soft heuristic). The motivation is that objects at the top of the pile (high y values) are more likely to maintain the pile stability than objects at the bottom and thus should be removed first. See Fig. 4a.

Convex hull Filter object selection to those in contact with the convex hull (hard heuristic). The motivation is that objects at the external boundary/layer of the pile are more likely to maintain the pile stability than objects inside the pile. See Fig. 4b.

Support-only Filter object selection to those where all contact interface normals point in the direction of gravity (hard heuristic). The implication is that these objects do not support other objects, and thus are more likely to maintain the pile stability when removed. The directional constraint applies to the entire half-space of the gravity vector. See Fig. 4c.

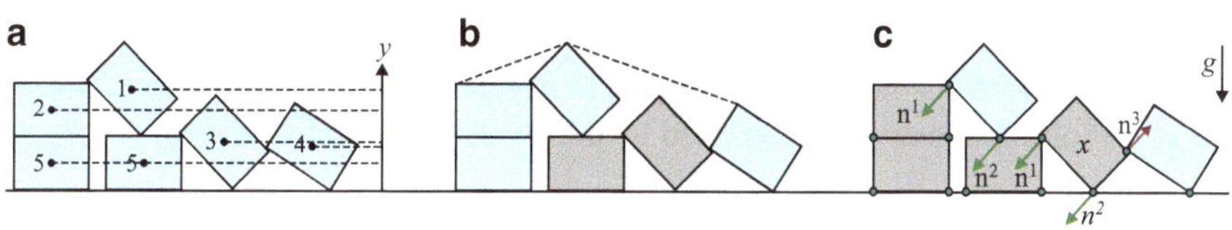

Fig. 4 Heuristics for object selection. (**a**) y-order. (**b**) Convex hull. (**c**) Support-only

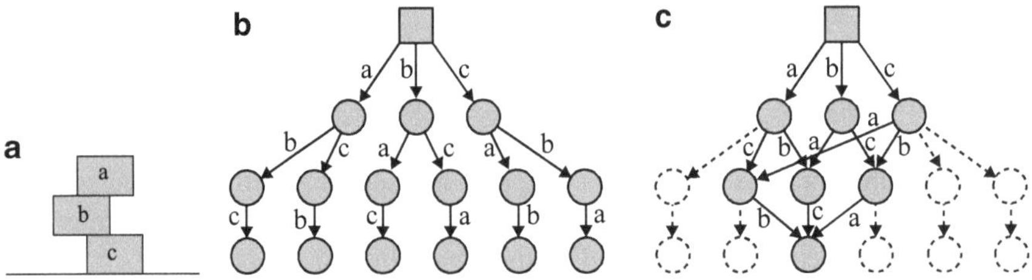

Fig. 5 Disassembly graph (**b**) of the scene pile (**a**), and disassembly graph with equivalent sub-piles merged (**c**)

5.4 *Redundant Paths*

An additional noteworthy point is the existence of redundant paths. Given a node in the disassembly graph, finding a solution path for the remaining sub-pile does not depend on the order in which the current node was reached. All the paths can be represented in a "merged" disassembly graph, where every node is the order-independent sub-pile (see Fig. 5c). Consequently, if a node leads to a dead-end, all searches reaching this node can be directly aborted. To achieve this merged representation, solvability information for each node must be stored. The total number of nodes in the merged graph are $\sum_{k=1}^{n} \binom{n}{k}$ where n is the number of objects in the pile. For example a pile of $n = 30$ object would require storing more than 1.07×10^9 nodes. While this is theoretically possible, it has clear practical limitations. Defining an appropriate data structure and dynamically removing the subsequent paths of non-solvable nodes is a key challenge in itself and beyond the scope of this paper.

6 Experiments

This section presents how data were generated, and then shows and compares the results of pile deconstruction obtained by the investigated search strategies.

6.1 *Scene Generation*

To evaluate the search strategies for disassembly, we simulated numerous virtual scenes using the Bullet open source physics engine. We considered two types of scenes: randomly generated piles (see Fig. 6) and collapsed architectural masonry models (see Fig. 7). To promote entanglement of the random piles we used L-shaped objects. The input architectural models were obtained from Panozzo et al. (2013). We extracted contact point information once the scene reached static equilibrium.

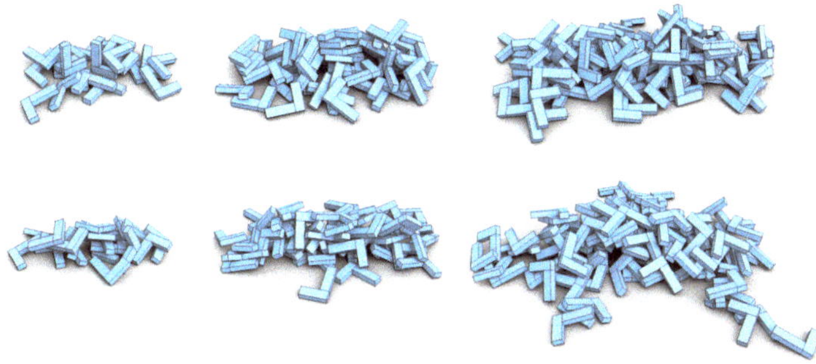

Fig. 6 Instances of randomly generated piles composed of $n = 20$ (*left*), $n = 50$ (*middle*), $n = 100$ (*right*) L-shaped objects

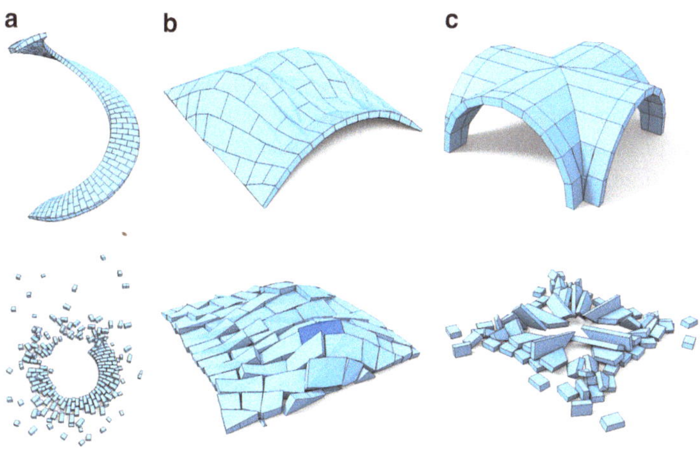

Fig. 7 Piles (*bottom*) generated from collapsed architectural models (*top*). (**a**) Staircase model. (**b**) Shell model. (**c**) Groin vault model

All materials were presumed to have the same properties of rigid bodies with zero tensile resistance and friction coefficients of 0.8.

In this paper we were interested in evaluating: "if a feasible disassembly sequence exists, can a given search strategy find a solution in a reasonable amount of time?". To determine in advance whether a pile has a solution or not, one could apply an exhaustive search, however this is not possible for large piles due to time complexity. In contrast, we generated solvable piles by *adding* objects one by one in a stable sequence. The reverse order of this assembly sequence is guaranteed as a valid disassembly sequence. In the following, we run the various strategies on these solvable piles.

We used architectural models of varying complexity. In the staircase model (Fig. 7a) the collapsed pile has high scatter of the individual blocks. This distribution can be quickly solved using the connected components strategy, as all lone blocks can be removed in parallel in a single step. In contrast, the shell model (Fig. 7b) is tightly packed and behaves as one large connected assemblage despite its shallow spread. The collapsed groin vault model (Fig. 7c) contains more stacking of individual blocks. The disassembly requires ample searching among possible combinations before arriving at a feasible sequence.

6.2 Results

Large scale evaluation We compared search strategies according to success rates. A success occurs when a disassembly solution path is found for a given pile within a time limit of one hour. The success rate is defined as the number of successfully disassembled piles relative to the total number of piles tested.

The efficiency of the search methods depends on the number of objects and also on the scene (especially the spatial arrangement of the objects). Therefore for each given number of objects, we generated 1,000 unique piles. Evaluating these numerous piles and comparing the different search strategies requires extensive processing time. We conducted the large-scale experiments in parallel on a high-end cluster with node specifications: AMD Opteron 6174 (2.2 GHz) and 1 GB of memory. In total, more than 100,000 jobs were processed on 416 cores.

Comparison We consider the exhaustive approach (also called naive in the following) with the y-ordering heuristic as the baseline. We count the number of successes for each search strategy and compute the improvement rate with respect to the baseline. Results are shown in Fig. 9(a). It can be seen that for naive search, the convex hull and support-only heuristics improve performance over the baseline strategy, where the improvement gradually increases with the number of objects. For 52 objects, the improvement rate by the support-only and convex-hull heuristics is respectively +17 and +28 %.

Overall, the best performance among all investigated strategies is obtained by connected components coupled with y-ordering. It can be seen that the y-ordering heuristic outperforms both the convex-hull and support-only heuristics. The performance gradually increases along with pile size. Compared to the baseline, the improvement rate is approximately +46 % for 52 objects.

For completeness, one may note that some improvement rates are negative for piles composed of few objects. This occurs for the support-only heuristic (both naive and with connected components). This is because the support-only heuristic discards the objects that support other objects. If the solution path goes through at least one of these supporting objects, then it will not be found. The same behavior theoretically applies for the convex-hull strategy, since it is also a hard heuristic that discards objects which do not comply. However, in practice, our experiments have not shown a negative improvement rate.

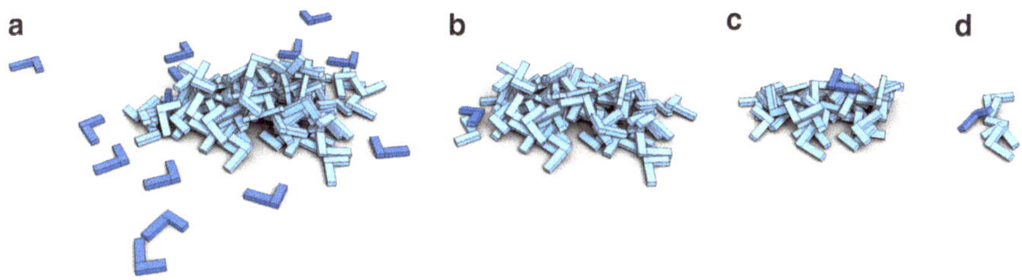

Fig. 8 Pile of randomly assembled L-shaped objects at different stages of disassembly obtained by the connected-component approach coupled with the y-ordering heuristic. *Dark blue* indicates the next object(s) selected for removal. In (**a**), note all the independent lone objects are removed in a single step. (**a**) initial pile: $n = 100$ objects. (**b**) $n = 65$. (**c**) $n = 35$. (**d**) $n = 5$

Visualization of pile deconstruction The large-scale study indicates that the best performance is obtained by the connected-component approach coupled with the y-ordering heuristic. For visualization purposes, we show results of the pile deconstruction sequence obtained by this approach. Figure 8 is a representative result of pile deconstruction, here for a random pile composed of $n = 100$ L-shaped objects. We invite the readers to see additional animated deconstruction sequences in the **supplementary video**.

Solvability Our large-scale experiments led to the following interesting observation: object piles are far more likely to be unsolvable as the number of objects increases. Unsolvability is the scenario where every possible deconstruction sequence contains an unstable sub-structure. We conducted extensive experiments with more than 48,000 piles for each tested number of objects. In each trial we applied exhaustive search to determine existence of a feasible sequence. Since the complexity of exhaustive search is $n!$, we applied it only to small piles of up to 10 objects. Nevertheless, Fig. 9(b) shows how the percentage of solvable piles drops rapidly below 50 % in the span of 1–10 objects.

Limitations and discussions Our large-scale experiments showed that heuristics can greatly improve the performance of the search. Nevertheless, it must be noted that incorrectly designed hard heuristics may inadvertently remove solution paths from the search, and in turn, prevent discovery of a feasible sequence.

We obtained deconstruction sequences for piles of up to 200 objects using the pipeline described in this paper (Fig. 1). However the current heuristics are not sufficient to deal with piles composed of thousands of highly entangled objects. Note that there might not be an approach that can solve this problem in a reasonable amount of time, in a similar way to the popular NP-hard traveling salesman problem.

In this study, an object is considered removable if it does not affect the pile stability. This is performed without considering its reachability, which is a geometric constraint, or how to grasp and extract the object. The grasping issue has been investigated for manipulator control and planning in the robotics community and is beyond the scope of this paper, which is dedicated to stability constraints.

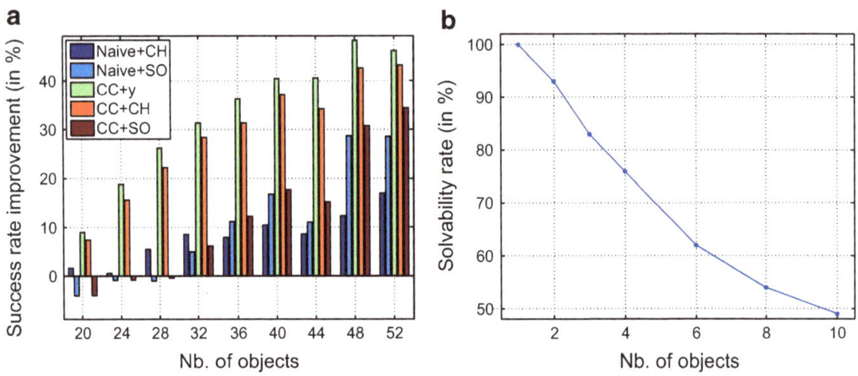

Success rate of search techniques   Solvable assemblies

Fig. 9 (**a**) Comparison of the success rate improvement obtained by various search techniques over the baseline, with a cut-off time of one hour. *CH*: convex hull, *SO*: support only, *y*: y-ordering, *CC*: connected components. (**b**) Percentage of solvable piles for randomly generated scenes using L-shaped objects, with respect to the number of objects

Conclusion

In this paper, we conducted a preliminary study of object pile deconstruction in an architectural context. For this, we developed an entire experimental pipeline: from pile generation to physical simulation and sequence optimization. Our large-scale experiments (i) allowed us to investigate and compare search strategies, (ii) indicated that the best performance is obtained by connected components coupled with the y-ordering heuristic, and (iii) showed that the number of solvable piles declines rapidly with the number of objects. We also applied these search strategies on architectural models.

We hope that our work will encourage future research in the field of deconstruction. While our experiments relied on virtual scenes, a compelling challenge is to operate on real-world data, requiring digitization of the scene. Numerous potential directions can also be investigated for search strategies. For example, identification of substructures that have no feasible deconstruction sequence could abort the dead-end paths, and thus speed up the search. It would also be worth investigating the role that randomized search techniques, like Monte Carlo tree search methods and random walks, could play in the search strategy. Finally, an exciting direction for piles unsolvable via the one-by-one object removal approach is the study of supporting structures to maintain stability during intermediate phases, in a similar way to supporting struts used for the 3D printing of complex objects.

Acknowledgements We thank Daniele Panozzo and Philippe Block for providing the architectural masonry models. Emily Whiting was supported by the ETH Zurich/Marie Curie COFUND Postdoctoral Fellowship.

References

Abe, S., Murayama, T., Oba, F., Narutaki, N.: Stability check and reorientation of subassemblies in assembly planning. IEEE Int. Conf. Syst. Man Cybern. **2**, 486–491 (1999)

Buro, M.: Improving heuristic mini-max search by supervised learning. Artif. Intell. **134**(1–2), 85–99 (2002)

Campbell, M., Joseph Hoane, Jr., A., Hsu, F.-H.: Deep blue. Artif. Intell. **134**(1–2), 57–83 (2002)

Erleben, K.: Velocity-based shock propagation for multibody dynamics animation. ACM Trans. Graph. **26**(2), 12 (2007)

Guendelman, E., Bridson, R., Fedkiw, R. Nonconvex rigid bodies with stacking. ACM Trans. Graph. **22**(3), 871–878 (2003)

Guo, J., Yan, D.-M., Li, E., Dong, W., Wonka, P., Zhang, X.: Illustrating the disassembly of 3D models. Comput. Graph. **37**(6), 574–581 (2013)

Hopcroft, J., Tarjan, R.: Efficient algorithms for graph manipulation. Commun. ACM **16**(6), 372–378 (1973)

Hsu, S.-W., Keyser, J.: Piles of objects. ACM Trans. Graph. **29**(6), 155:1–155:6 (2010)

Hsu, S.-W., Keyser, J.: Automated constraint placement to maintain pile shape. ACM Trans. Graph. **31**(6), 150:1–150:6 (2012)

Ikeuchi, K., Horn, B.K., Nagata, S., Tom, Callahan, T., Feingold, O.: Picking up an object from a pile of objects. In: Proceedings of the First International Symposium on Robotics Research, pp. 139–166. MIT, Massachusetts (1983)

Katz, D., Venkatraman, A., Kazemi, M., Bagnell, J.A., Stentz, A. Perceiving, learning, and exploiting object affordances for autonomous pile manipulation. In: Robotics Science and Systems Conference, 2013

Kaufman, D.M., Edmunds, T., Pai, D.K. Fast frictional dynamics for rigid bodies. ACM Trans. Graph. **24**(3), 946–956 (2005)

Lambert, A.F., Gupta, S.M.: Disassembly Modeling for Assembly, Maintenance, Reuse and Recycling. CRC (2005)

Lambert, A.J.D.: Disassembly sequencing: A survey. Int. J. Prod. Res. **41**, 16 (2003)

Livesley, R.K.: Limit analysis of structures formed from rigid blocks. Int. J. Numer. Methods Eng. **12**(12), 1853–1871 (1978)

Loomis, A., Balkcom, D.J.: Computation reuse for rigid-body dynamics. In: IEEE/RSJ International Conference on Intelligent Robots and Systems, Beijing, pp. 4181–4186, 2006

Panozzo, D., Block, P., Sorkine-Hornung, O.: Designing unreinforced masonry models. ACM Trans. Graph. **32**(4), 91:1–91:12 (2013)

Schaeffer, J.: One Jump Ahead: Challenging Human Supremacy in Checkers. Springer, New York (1997)

Schwartzburg, Y., Pauly, M.: Design and optimization of orthogonally intersecting planar surfaces. In: Computational Design Modelling Symposium, Berlin, pp. 191–199, 2012

Schwartzburg, Y., Pauly, M.: Fabrication-aware design with intersecting planar pieces. Comput. Graph. Forum (Proc. Eurograph.) **32**, 2 (2013)

Wang, J., Rogers, P.D., Parker, L.T., Brooks, D.A., Stilman, M.: Robot Jenga: Autonomous and strategic block extraction. In: IEEE/RSJ International Conference on Intelligent Robots and Systems, St. Louis, pp. 5248–5253, 2009

Advanced Topology Optimization Methods for Conceptual Architectural Design

Niels Aage, Oded Amir, Anders Clausen, Lior Hadar, Dana Maier, and Asbjørn Søndergaard

Abstract This paper presents a series of new, advanced topology optimization methods, developed specifically for conceptual architectural design of structures. The proposed computational procedures are implemented as components in the framework of a Grasshopper plugin, providing novel capacities in topological optimization: Interactive control and continuous visualization; embedding flexible voids within the design space; consideration of distinct tension / compression properties; and optimization of dual material systems. In extension, optimization procedures for skeletal structures such as trusses and frames are implemented. The developed procedures allow for the exploration of new territories in optimization of architectural structures, and offer new methodological strategies for bridging conceptual gaps between optimization and architectural practice.

1 Introduction

Topology optimization can be described as a family of computational methods aimed at finding optimal structural layouts and configurations. In the context of continuum structural representations, this corresponds to optimizing the distribution of one or several materials in a given design space. In the context of skeletal structures – e.g. trusses and frames – the optimization procedures aim at finding the best possible layouts of bar or beam members. This paper presents advanced topology optimization procedures that are tailored specifically for integration into the disciplines of architecture and industrial design. In this section, we first briefly present the fundamental concepts of topology optimization, and then we review

N. Aage (✉) • A. Clausen
DTU – Technical University of Denmark, Kongens Lyngby, Denmark
e-mail: naage@mek.dtu.dk; andcla@mek.dtu.dk

O. Amir • L. Hadar
Technion – Israel Institute of Technology, Haifa, Israel
e-mail: odedamir@cv.technion.ac.il; hadar07@gmail.com

D. Maier • A. Søndergaard
Aarhus School of Architecture, Aarhus, Denmark
e-mail: 3dana.m@gmail.com; asbjorn.sondergaard@aarch.dk

P. Block et al. (eds.), *Advances in Architectural Geometry 2014*,
DOI 10.1007/978-3-319-11418-7_11

various recent applications of the method in architectural design – highlighting its potential as a computational design tool for achieving innovative, efficient structures.

1.1 Overview of Topology Optimization Procedures

Following intense research since its introduction by Bendsøe and Kikuchi in the late 1980s (Bendsøe and Kikuchi 1988; Bendsøe 1989), topology optimization of continua is now considered an integral part of the design process of load-bearing structural components in the automotive and aircraft industries (Bendsøe and Sigmund 2003; Sigmund and Bendsøe 2004). Significant research effort has been invested over the past decade in further development of the method for utilization in other engineering fields. Some of the most exciting and promising examples are optimization of photonic crystal waveguides (Borel et al. 2004); microstructural design of materials with extreme properties (Sigmund and Torquato 1997); systematic design of phononic band-gap materials (Sigmund and Jensen 2003); and design of micro-actuators and micro-mechanisms (Jonsmann et al. 1999; Sardan et al. 2008) (Fig. 1).

In contrast to the wide acceptance of continuum-based computational procedures, truss- and frame-based approaches had less industrial impact so far. This is despite their earlier conception which can be traced back to the groundbreaking work of Michell (1904). In fact, most of the mathematical background and fundamental knowledge in the field of topology optimization, as well as in the more general discipline of structural optimization, has been developed based on studying optimal design problems involving trusses and frames. This is clearly reflected in various textbooks and monographs (e.g. Hemp 1973; Kirsch 1993).

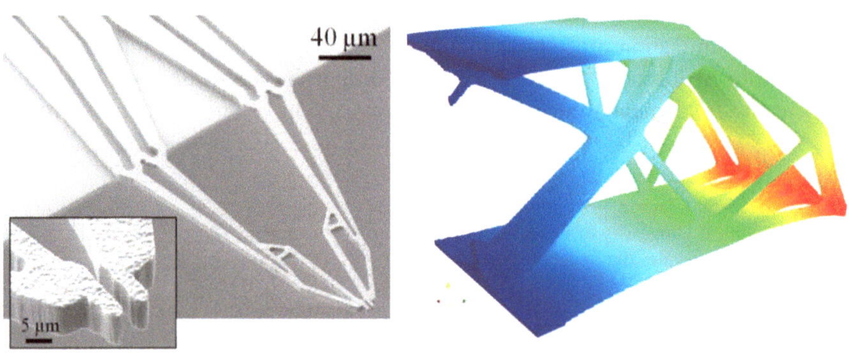

Fig. 1 A topology-optimized micro-gripper (Sardan et al. 2008) (*left*) and an optimized 3-D cantilever beam, based on a model with nearly five million finite elements (Aage and Lazarov 2013) (*right*)

Even though topology optimization was originally conceived as a particular method of structural optimization, it did not yet succeed in transforming the construction industry as reflected in the practices of architects and structural engineers. Nevertheless, over the past few years one can observe a clear trend of expanding interest within the architectural community in topology optimization as a means of generating aesthetic and efficient structural forms and configurations. Various recent examples are reviewed in the following section. Another important factor is the growing awareness for sustainable development, calling for reducing the consumption of materials and resources used by the construction industry, namely concrete and steel. It could be that in the near future, environmental considerations will be the driving force behind the integration of advanced optimization techniques in architectural and structural design, just as reducing weight is for the aeronautics industry.

1.2 Architectural Applications of Topology Optimization

Possibly the earliest real-world example of application is the façade of the Doha Convention Center, designed by Japanese architect Arata Isozaki and structural engineer Mutsuro Sasaki. Continuum topology optimization suggested a tree-like structure that replaced traditional vertical columns, leading Sasaki to the terminology "Morphogenesis of Flux Structure" (Sasaki et al. 2007).

Various recent studies have highlighted the potential for disruptive design innovation and efficiency gains by adopting topology optimization procedures as a constitutive design tool (Frattari 2011; Dombernowsky et al. 2012). Nevertheless, the full integration of such methods into the architectural design process is still in its infancy, as also is the coupling with construction technologies. Efforts led by authors of this paper have demonstrated the realization of optimized concrete structures through architectural robotic fabrication in respectively large scale robotic CNC-milling (Dombernowsky and Søndergaard 2011) and robotic hotwire cutting (Søndergaard and Feringa 2014) (Fig. 2).

Fig. 2 Unikabeton prototype structure, (Dombernowsky and Søndergaard 2011) (*left*) and proposed Opticut prototype structure (Søndergaard and Feringa 2014) (*right*)

Several examples of engagement into studying topology optimization for conceptual architectural design can be found in the work of Skidmore, Owings and Merrill LLP (SOM). Both continuum- and frame-based topology optimization inform the design process of bracing systems for high-rise buildings (Stromberg et al. 2012) and of long-span structures (Beghini et al. 2013). Furthermore, computational algorithms in general facilitate tight architect-engineer collaboration at SOM, requiring both practitioners to collaborate in defining performance goals and constraints, in parameterizing geometric forms and in evaluating the qualities of design alternatives. This opens up new possibilities of exploring theoretical paradigms and emergent formal typologies (Besserud et al. 2013).

Continuum topology optimization clearly facilitates the generation of novel structural morphologies, transcending traditional typological classification. However, the realization of optimized continuum layouts is feasible primarily through fabrication processes delivering continuum bodies, such as in-situ cast concrete or – extrapolating from current trends in additive manufacturing – 3-D printing of concrete (Le and Austin 2011; Buswell et al. 2007). In practice, a significant part of architectural structures are still assembled from discrete members such as bars, beams and façade elements, produced in formative processes and assembled on-site. This calls for a slight shift in the discussion regarding the application of topology optimization. While continuum procedures facilitate the form-finding of advanced structural design, they demand high degrees of time-demanding geometrical post-processing. Optimal design of trusses and frames offers by comparison a path for closer integration between optimization and fabrication, and it is therefore arguable that they should take a more prominent role in the investigations of the field.

One approach suggested recently is to involve continuum topology optimization in the design process while aiming at constructing discrete members, by essentially translating the free-form layout into geometries of nodes and bars (Mostafavi et al. 2013). It is shown that translational algorithms such as "skeletonization" and node finding methods can open the black box of topology optimization, hence making it more applicable by architects and designers. It was pointed out that the structural layout optimization (i.e., procedures for finding optimal arrangements of structural elements in trusses and frames) has the potential to help reduce the polarization between the visual and the technical elements of design, therefore it can be considered as an 'integrative' form generation tool (Park et al. 2012). Early investigations into the implementation of truss-based optimization and its coupling with digital fabrication techniques have been conducted recently by authors of this paper. These include for example topology optimization and robotic fabrication of space trusses, resulting in non-standard structural layouts that offer weight reductions and can only be assembled effectively by robotic techniques (Søndergaard et al. 2013) (Fig. 3).

The variety of recent contributions reveal the tremendous potential of embedding topology optimization into architectural design processes. However, the currently available computational tools are intended primarily for the automotive and aeronautical industries that were first to adopt structural optimization methodologies. Recent work by the authors has demonstrated the applicability of industrial solvers

Fig. 3 Topology optimized, robotically manufactured space-frame structure under construction at Aarhus School of Architecture. The structure was optimized with the truss module of the TopOpt-plugin, facilitating a contextualization of the structural layout to irregular site-settings through laser-scanning of the physical environment and adaption of ground structure configuration; and direct coupling between optimization and fabrication procedures through linking of the result data to rapid-code-generating modules

for the design of optimized concrete structures (Dombernowsky and Søndergaard 2011; Feringa and Søndergaard 2014), while simultaneously identifying the following conceptual and practical short-comings of this adaption:

(a) **Design process**: Topology optimization using continuum procedures results in structural morphologies of high geometrical complexity. As such, they introduce significant design impact on the architectural appearance, in case the structure is visually exposed. Topology optimization is by nature an autonomous process, that 'grows' results morphologically unpredictably and beyond direct design control. Design control can however be regained through indirect measures, by reiterating the optimization parameters, such as the design space geometry, position of supports, and fabrication constraints, in response to the emerging optimization results. This process is however complicated and vastly time-consuming with existing procedures, and hence impractical to conventional design practice.

(b) **Design prescription**: Due to the strong visual impact resulting from the application of topology optimization in architectural design, an early integration in the design phase is desirable, allowing for higher degrees of coordination with complementary design parameters that cannot be numerically prescribed. However, because the optimization process assumes an unambiguous definition of the optimization parameters, early integration requires a forestalling of the design configuration that under normal conditions is clarified only in later design phases. This can result in either (1) a pre-determination of functional or architectural parameters, such as façade-compositions, building component geometries or spatial organization structures that conflicts structural criteria, leading to later sub-optimal optimization results; or (2) the transfer of purely formal aspects of optimization results to later design stages, in which the structural congruency is compromised by late-coming parameters that contradicts early stage optimization assumptions.

(c) **Tension-compression prioritization**: In the design of architectural structures, geometric configurations that discretise building components into classes predominantly resisting tensile or compressive forces are often desirable. Similarly, a prioritization between optimization solutions that favour the emergence of either a predominantly compressive or tensile system would be strongly advantageous in conceptual structural design of material systems with disparate tensile and compressive strengths, e.g. concrete. However, currently available methods are targeted at homogenous materials with linear elastic properties, and allow for the optimization of results with uniform strength distributions only. This leads to unnecessary complexities in dealing with inhomogeneous material systems and omits the potential exploration of optimized designs based in divergent material properties.

(d) **Material representation**: Currently available continuum procedures assumes isotropic material configurations, as found in pure concrete or metal bodies. However, design of unreinforced concrete structures remain a rare exception in architectural practice. Application of available tools for isotropic continuum structures to optimization of concrete structures assumes a coarse 'hack' in which concrete bodies are optimized assuming uniformly distributed tensile properties of reinforced concrete and subsequent empirical reinforcement design by conventional engineering means. As such, available methods remain incapable of suggesting dual material configurations that optimally distributes e.g. both steel and concrete according to their individual properties.

(e) **Result rationalization**: A final complexity arising from the application of continuum topology optimization procedures is the above mentioned geometric rationalization of results, required for translating optimized meshes into buildable geometry. This rationalization process is often laborious, involving development of bespoke rationalization modules and state-of-the-art project design tools, while simultaneously requiring advanced digital manufacturing procedures for its realization. This challenge can be addressed through closer coupling between optimization and fabrication procedures

2 Advanced Topology Optimization in Grasshopper

Motivated by the conclusion that significant enhancements can be achieved in the applicability of topology optimization to architectural design by addressing the challenges identified in preceding chapters, the authors of this paper mobilized for a collaborative research effort, targeting the development of new optimization methodologies. The challenges are distinguished in two sub-aspects: (1) The availability of experimental procedures targeted specifically at conceptual architectural design; (2) the availability of such procedures within a software platform commonly applied in architectural research.

Currently available topology optimization tools counts commercial solvers such as Altair's Inspire and Optistruct and FE-Design's TOSCA and TOSCA Fluid. In addition to this, a number of freely available tools such as TopoStruct and Millipede for Grasshopper, developed by Panagiotis Michalatos and Sawako Kaijima; or Karamba developed by Clemens Preisinger in cooperation with Bollinger-Grohmann-Schneider ZTGmbH Vienna. The optimization procedures offered in these applications are based on respectively the homogenization method applied in Ole Sigmund's 99-line topology optimization code (Sigmund 2001) and the Evolutionary Structural Optimization (ESO) method (Huang 2010) and provides optimization for classical mechanical stiffness problems that does not methodologically address the specific challenges outlined in earlier chapters of applying topology optimization in architectural design.

The ongoing research efforts discussed in this paper seek to address this challenge by introducing experimental topology optimization procedures specifically developed for the related problems of architectural design identified in Sect. 2 in chapter "Simulation of Aggregate Structures in Architecture: Distinct-Element Modeling of Synthetic Non-convex Granulates" through: (a) Interactive control and continuous visualization; (b) embedding flexible voids within the design space; (c) consideration of distinct tension / compression properties; (d) consideration of dual material systems and (e) optimization procedures for skeletal structures such as trusses and frames. The methodological developments are implemented within the framework of a Grasshopper-extension, the TopOpt plugin. The Grasshopper platform is chosen for its flexibility with regard to allowing users to connect the procedures to related geometric modules, and for its relative wide-spread adoption within design research activities.

The implementation of continuum topology optimization follows the classical density-based approach, overviewed in the monograph (Bendsøe and Sigmund 2003). The design problem is essentially to find the best trade-off between stiffness and weight, i.e. maximize stiffness for a given volume of material. The basic idea of the computational method is that the distribution of material is defined by attaching a design variable ρ to each finite element in the computational model of the design space. Solid material is represented by $\rho = 1$ while void is represented by $\rho = 0$. During the iterative optimization process, the variable ρ is gradually "pushed" towards these extreme values due to the penalization of stiffness of intermediate values. For this purpose we utilize the popular SIMP rule in its modified form (Bendsøe 1989; Sigmund and Torquato 1997), defining the stiffness modulus E as follows:

$$E = E_{min} + (E_{max} - E_{min})\,\rho^p$$

Here, E_{min} and E_{max} represent the minimum (typically close to zero) and maximum stiffness of the material and p is a penalization power larger than 1. Finally, we note that in many topology optimization procedures a certain regularization is necessary in order to control the physical size of structural members appearing in the optimized design. This is most frequently achieved by a filter operation, with a single parameter R_{min} controlling the filter radius.

The most basic functionality in the TopOpt plugin for Grasshopper, capable of finding optimal distributions of elastic isotropic material, resembles earlier implementations in publicly available MATLAB codes and in the TopOpt app for mobile devices (Mathworks 2013; Sigmund 2001; Andreassen et al. 2012; Aage et al. 2013). In the following sections, we describe the more advanced features that are particularly intended for conceptual architectural design within the Rhino-Grasshopper environment.

2.1 Interactive Exploration of Design Concepts

One of the primary considerations influencing the development of the TopOpt plugin, responding to challenge (a) of Sect. 2 in chapter "Simulation of Aggregate Structures in Architecture: Distinct-Element Modeling of Synthetic Non-convex Granulates", is the aim to provide an interactive computational tool that architects and designers can utilize for design experimentation. By providing live feedback through direct interactivity, design cycles can be significantly shortened, while simultaneously, and most importantly, provide an intuitive insight to the effect of changing parameters. Several aspects of our implementation combine to serve this goal: (1) Computational speed, relying predominantly on the efficiency of the finite element analysis solver; (2) Continuous graphical display of results (topological layouts) during the optimization process; and (3) Direct representation of the optimized geometry in Rhino for further exploration. These three aspects are further elaborated in this section.

Computational Efficiency It is well-known that the computational cost involved in topology optimization procedures is typically dominated by the effort invested in repeatedly solving the finite element analysis equations, once per design cycle. The first version of the TopOpt plugin, released May 2013, relied on self-developed routines programmed in C#, exhibiting satisfactory performance. In the current version however, we have based the FEA solver routines on the highly acclaimed SuiteSparse package (Davis 2006). Furthermore, highly efficient 3-D FEA procedures have been integrated based on the recently developed Multigrid-CG framework for topology optimization (Amir et al. 2014).

Continuous Display In our view, enabling interactive "play" between the user and the software is crucial for promoting the utilization of topology optimization by architects and designers. This drives a different user interface approach in comparison to software intended for engineers: While for engineers it seems natural that an optimization process is executed in the background and results are displayed only upon completion, for architects such an approach may not suffice. The possibilities to interactively view the evolution of the optimized layout and to interfere with it by graphically by changing the input are necessary for fruitful user-computer interaction.

In the TopOpt plugin, any change in input – for example change of geometry, loads or supports in the Rhino model, or change of a parameter value in Grasshopper – automatically initiate a new optimization process. Furthermore, in the current version the ongoing optimized geometry is displayed during the execution – a rather rare functionality in Grasshopper components. This is achieved by attaching a Hoopsnake Grasshopper component (developed by Yiannis Chatzikonstantinou) that manages recursive executions of the TopOpt optimization engine. In future development, we will consider a tighter integration of the concepts behind Hoopsnake into our plugin for achieving a more effective implementation.

Direct Representation of Geometry The TopOpt plugin has two types of components for displaying optimized layouts. The {Preview.1Mat} and {Preview.2Mat} components provide solid-void (black-white) and compressive-tensile-void (red-blue-white) distribution pictures, respectively. The {Geometry.1Mat} and {Geometry.2Mat} components draw the contours of the boundaries between materials, i.e. either between solid and void (for single material optimization) or between compressive material, tensile material and void (for dual material optimization). The exact geometry of the contour depends on a user defined cut-off value between 0 and 1. The latter components facilitate the immediate export of the geometry without any need for post-processing actions (Fig. 4).

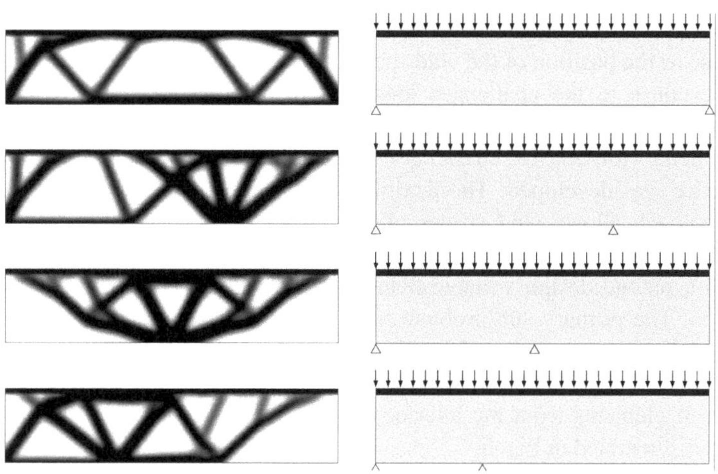

Fig. 4 Interactive control of optimization parameters. By performing gradual changes in the optimization variables – for example as a linear translation of the position of the right corner point support – insights in the influence of these parameters with regard to the non-linear transformations of the topology can be obtained. By facilitating dynamic explorations through interactivity, the procedure allow for a mapping of relationships between the influences of different variables, hereby allowing designers to understand and exploit these conditions actively in the design process

2.2 Embedding Flexible Voids

A common restriction in structural design problems is that void space should be reserved for functional, construction-specific or aesthetic reasons. Functional restrictions can include for example other structural components passing through the structure. Holes for such purposes may or may not have a rigidly prescribed geometry and position. Construction restrictions include reserving space for the construction phase or maintenance. The exact position and shape of the corresponding void spaces are usually not strictly defined. Void spaces included for aesthetic reasons may have arbitrary geometric freedom.

In an architectural setting the void space restriction manifests itself in both 2-D and 3-D applications (currently, the computational functionality is available in a 2-D version only). A typical 2-D optimization problem is the design of a façade structure. The supporting structure of a façade may be attributed a role in the overall architectural expression, thus implying significant aesthetical requirements on the design. A typical 3-D application is flexible room modeling, where the internal room design and functional organization may be defined as flexible void spaces, this way allowing for a higher degree of synthesis between constructive and functional considerations.

The usual approach to include restrictions on void space in topology optimization is by defining passive areas in the design space, which have a fixed shape and position. This method, however, has two main drawbacks. First, unless the shape of the void space is carefully defined, the resulting optimized structure may contain irregular structural members. Second, the method lacks flexibility for cases where the shape or the position of the void space do not need to be strictly defined.

In response to the challenges identified in subtopic (b), Sect. 2 in chapter "Simulation of Aggregate Structures in Architecture: Distinct-Element Modeling of Synthetic Non-convex Granulates" an alternative approach for inclusion of void space was developed: The flexible void area method (Clausen et al. 2014). This approach allows void spaces of fixed volume to be flexibly reshaped and repositioned. The flexible void spaces are introduced into the optimization problem through a second design variable field. The formulation is based on a combined approach: The primary sub-problem is to maximize stiffness for a given volume of material, with a secondary sub-problem of minimizing the disturbance from the flexible void spaces. The void area is updated by adding and removing an equal number of elements from the interface of the flexible void area. The idea of the method is illustrated in Fig. 5.

For some applications, the void area may be allowed to deform or move freely during the optimization, but for many applications the degree of flexibility is constrained by practical limitations. For this reason two different measures may be included into the optimization problem: A location measure and a deformation measure. These may be included either individually or combined. This paper presents examples where the location measure is applied. The measure works by penalizing the distance from the geometrical center of the flexible void area to

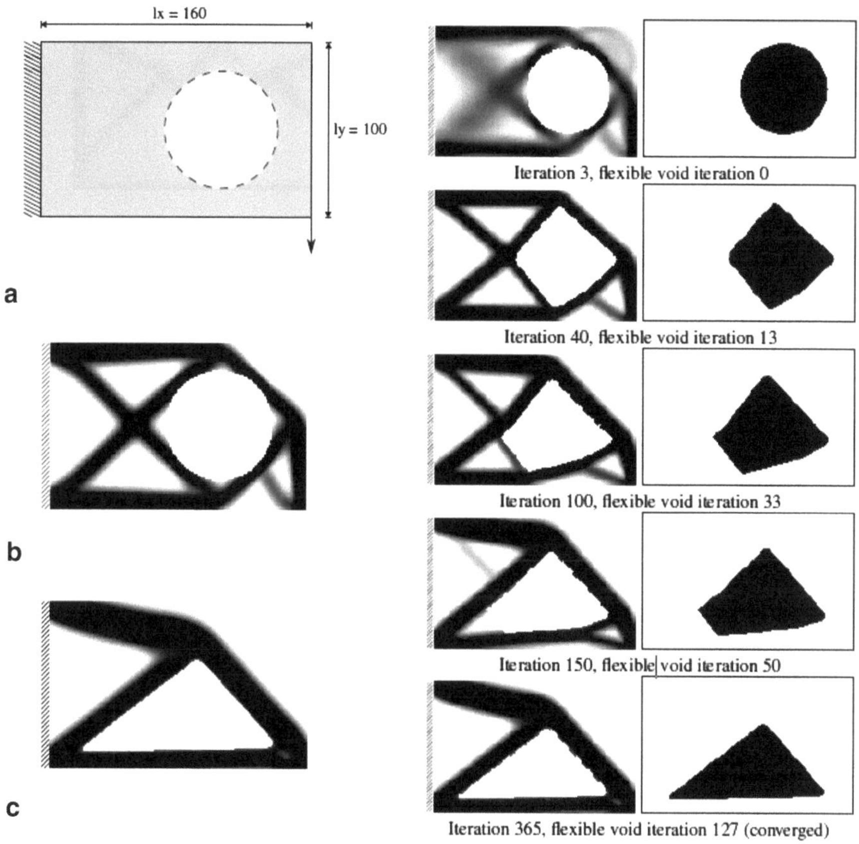

Fig. 5 An illustration of the flexible void method. Throughout the optimization process, the shape and position of the void change so that the disturbance is minimized. The stiffness with the flexible void area method is 27 % higher than with a passive area solution (Clausen et al. 2014). (**a**) Problem definition with passive void. (**b**) Optimized layout with passive void. (**c**) Optimized layout with flexible void

a given reference point. The penalization is applied with a given weight, α. The higher the value of α, the stronger the constraint is imposed on the problem. For applications of the deformation measure readers are referred to the original paper.

Example Application: Façade Design Most façade design problems are associated with void area requirements. A typical reason is that windows should be included or that the façade should include void areas to gain a lighter appearance. An example of a simple façade optimization problem is presented in Fig. 6. In the optimized design with passive voids the inclined bars carrying the central load are strongly weakened by the singularities at the corners of the void area. This artifact is avoided by applying the flexible void approach which leads to a 19 % higher stiffness. Furthermore, the resulting façade structure has a more aesthetically

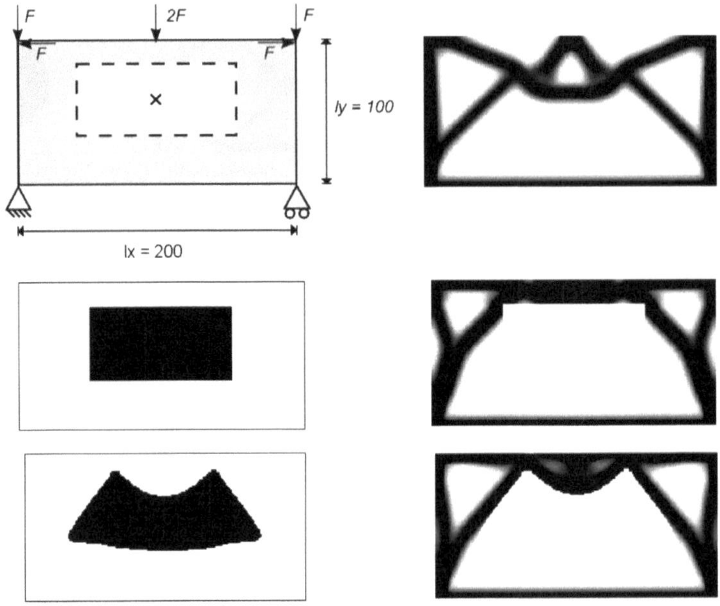

Fig. 6 The flexible void method applied to a façade design problem. *Top*: design space and solution without imposed voids; *middle*: passive voids solution; *bottom*: flexible voids solution

pleasing expression than when rigidly implying the void area. The introduction of an arch to support the center is not trivially deducted from neither the standard optimization without a void area nor the passive elements method. This way the example illustrates how the flexible void method is a useful tool in the design phase to work creatively with critical structural members without compromising unnecessarily on the structure's load-carrying capacity.

Implementation in the Grasshopper Plugin The flexible void area method will be implemented into the TopOpt plugin in the near future. At the submission deadline for this paper the method is already implemented in another interactive platform – a stand-alone application developed by members of the TopOpt research group at DTU (Aage et al. 2013). The implementation into the TopOpt plugin will be based on the learnings (e.g. in terms of interactivity) from this work.

2.3 *Optimization with Tension-Compression Prioritization*

Responding to challenge (c) of Sect. 2 in chapter "Simulation of Aggregate Structures in Architecture: Distinct-Element Modeling of Synthetic Non-convex Granulates", the TopOpt plugin in Grasshopper introduces a novel methodology, allowing to explore designs involving materials with different stiffness and strength in tension and compression. We believe this is an essential capacity in any computational framework intended for architectural / structural design, due to the widespread use of such materials, primarily concrete. The vast majority of explorations reviewed above were based on computational procedures for linear, elastic, isotropic and homogenous material – evidently because the currently available software is intended primarily for automotive and aeronautical applications where metals are dominant.

A consistent computational representation of a quasi-brittle material such as concrete requires the integration of nonlinear material laws, so that damage and cracking under tension are taken into account. Several computational procedures for optimizing the distribution of reinforcement while accounting for nonlinear material damage have been proposed recently (Kato et al. 2009; Kato and Ramm 2010; Amir and Sigmund 2013). Furthermore, unified procedures for optimizing the distribution of both concrete and reinforcement bars have been developed (Bogomolny and Amir 2012; Amir 2013). However, due to the computational effort involved in such procedures they are not yet suitable for implementation within interactive applications.

In light of these difficulties, the challenge of obtaining optimized layouts with distinct tension and compression properties is approached by assuming orthotropic material properties. This means that each material point has distinct material properties in two orthogonal axes. With an orthotropic representation of material, the optimization problem involves two separate actions: (1) Distribute material in an optimal manner, following the same principles as for isotropic materials; (2) Find the optimal rotation angle at each material point so that it is adequately aligned with the stresses and strains at that specific point – meaning material is locally tailored to effectively deal with tension or compression. We note that optimal orientation of orthotropic materials has been studied in various contexts which are beyond the scope of this article (e.g. Pedersen 1989).

A tension-compression prioritization procedure is developed and implemented in the Grasshopper plugin under the label {TenCom.1Mat} and is intended for optimizing the distribution of a single orthotropic material with one stiff material direction and one soft material direction. The ratio between these stiffnesses is defined by a single parameter, which also defines the prioritization towards tension or compression. This functionality facilitates the design of a structure with a single material where there is clear preference towards transferring forces either in tension or in compression – according to the specified material parameters. Examples of a structure suspending a load with various tension or compression prioritizations are presented in Fig. 7. The principles of {TenCom.1Mat} and the algorithmic details are presented in the following. First, the inputs required from the user are as follows:

1. Design domain properties;
2. Loading configuration;
3. Support configuration;
4. Optimization settings:

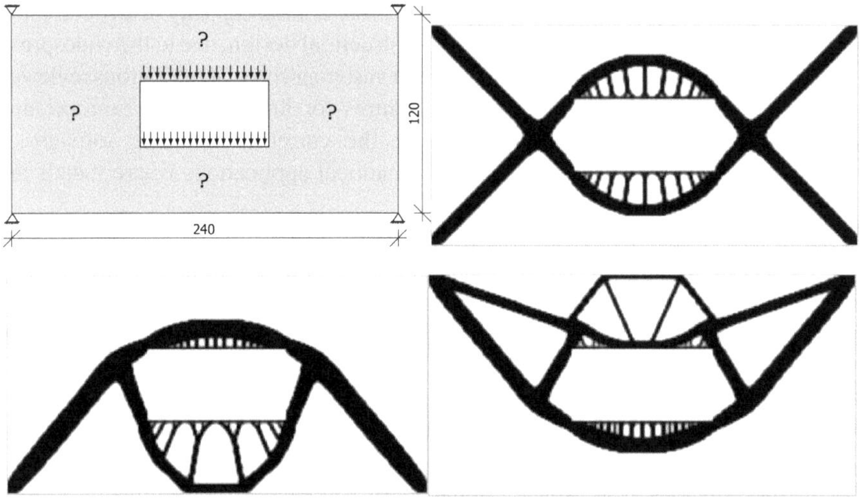

Fig. 7 Examples of a structure suspending a central load achieved by TenCom.1Mat with various prioritizations towards either tension or compression. All layouts comprise of a material volume fraction of 20 %. From *top*, *left* to *right*: problem settings; standard topology optimization with an isotropic material (*black* = solid, *white* = void); layout with compression prioritization, *ratio* = 5; layout with tension prioritization, *ratio* = 0.2

(a) Volume fraction of available material;
(b) Penalty parameter in the SIMP law;
(c) Filtering radius, regularizing the size of members in the layout.

5. Compression-to-tension ratio, where *ratio* = 1 means no prioritization; *ratio* > 1 means compression is preferred; and *ratio* < 1 means tension is preferred.

For simplifying the presentation, we focus on the case of *ratio* > 1 meaning that the user chooses to distribute a material which is stiffer in compression and softer in tension. The optimization process begins with a uniform distribution of material and a uniform alignment of the compressive stiff axis parallel to the global x-axis. Then, an iterative process is initiated where within each cycle the following steps are performed:

I. Compute the current stress state and rotate the material axes of each finite element such that the x-axis, with stiffness modulusE_x, is aligned with the compressive principal stress.
II. Compute the stiffness matrix of the orthotropic material, with the following stiffness moduli for each orthogonal axis:

$$Ex = E_{min} + \left(E_{stiff} - E_{min}\right) \rho^p$$

$$E_y = E_{min} + \left(E_{soft} - E_{min}\right) \rho^p$$

This means that the material's x-axis, which is intended to resist compressive forces, is stiffer than the y-axis.

III. Solve the equilibrium equations resulting from a finite element analysis.
IV. Perform an adjoint sensitivity analysis where for each finite element the contribution to stiffness per unit volume is found.
V. Update the distribution of material such that the available amount is fully utilized. In this phase, each finite element is assigned a value of ρ ranging from 0 to 1, corresponding to void and solid material.
VI. If material distribution did not change significantly with respect to the previous cycle – stop iterating; otherwise, return to step (I).

2.4 *Optimization of Dual Material Structures*

Responding to the challenge of representing composite material systems described in preceding chapters, is the TopOpt-plugin introduces a novel procedure for the optimization of dual-material layouts – aimed at distributing simultaneously two distinct materials with different tension-compression priorities. This can facilitate the exploration of concrete-steel layouts which are highly relevant for architects.

The suggested approach relies again on an orthotropic representation of material. Intuitively, optimal design with concrete and steel can be achieved by positioning concrete members or domains where compressive forces are dominant, while steel is positioned where tensile forces are dominant – either in the form of embedded reinforcement or as separate members. A similar approach has been followed recently in the context of generating strut-and-tie models for reinforced concrete design (Victoria et al. 2011), while procedures intended for the same goal but involving nonlinear structural analysis have also been suggested (Cai 2011; Liu and Qiao 2011).

The implementation in the TopOpt plugin is labeled {TenCom.2Mat} and is intended for optimizing the distribution of two materials. Each material has one stiff direction and one soft direction, while they differ in their prioritization towards tension or compression. This functionality facilitates the design of a structure with two materials, one of which is preferable for transferring compressive forces and the other is preferable for transferring tensile forces. Different layouts are obtained depending on the availability of each material which is specified by the user. Example designs of a simply-supported beam where the amount of tensile material is smaller than that of compressive material are presented in Fig. 8.

Despite their distinct purposes, the algorithmic principles of {TenCom.2Mat} follow directly those of {TenCom.1Mat}, whereas the few additional features are emphasized herein. First, the inputs required from the user are the same, with one exception – instead of a single volume fraction the user should supply two volume fractions corresponding to the two types of materials. Each finite element is assigned two values ρ_1 and ρ_2 corresponding to the existence of each type of material at that specific point. Steps I, III, IV and VI are the same as for {TenCom.1Mat}, otherwise:

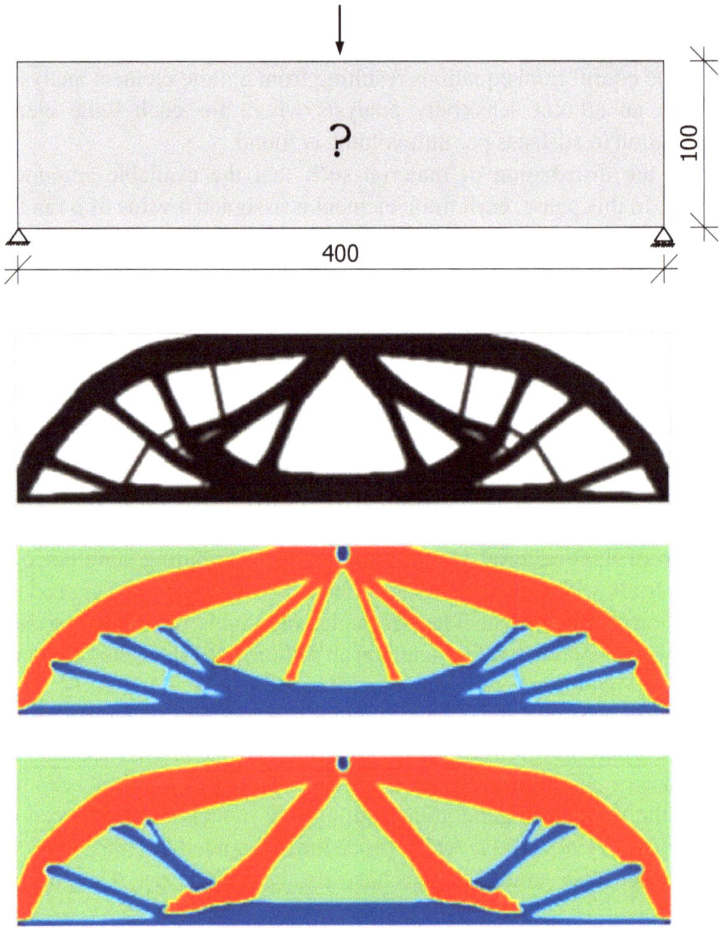

Fig. 8 Examples of a simply-supported beam designed by TenCom.2Mat. From *top*: problem settings; standard topology optimization with a volume fraction of 50 % using a single isotropic material (*black* = solid, *white* = void); layout with 30 % compressive material and 20 % tensile material; layout with 35 % compressive material and 15 % tensile material. Compressive material appears in *red* and tensile material in *blue*. *Dark blue* represents material that is stiff in both tension and compression, while *green* represents void. The ratio between stiff and soft material properties is 10

II. Compute the stiffness matrix of the orthotropic material, with the following stiffness moduli for each orthogonal axis:

$$E_x = E_{min} + \left(E_{stiff} - E_{min}\right)\rho_1^p + \left(E_{soft} - E_{min}\right)\rho_2^p + \left(E_{min} - E_{soft}\right)\rho_1^p\rho_2^p$$

$$E_y = E_{min} + \left(E_{soft} - E_{min}\right)\rho_1^p + \left(E_{stiff} - E_{min}\right)\rho_2^p + \left(E_{min} - E_{soft}\right)\rho_1^p\rho_2^p$$

This means that the following material properties can be obtained for the extreme values of ρ_1 and ρ_2:

ρ_1	ρ_2	x-axis	y-axis
0	0	Minimum stiffness	Minimum stiffness
1	0	Stiff	Soft
0	1	Soft	Stiff
1	1	Stiff	Stiff

V. In odd cycles, update the distribution of material 1 represented by the values of ρ_1, such that the available amount is fully utilized, for the given distribution of material 2; in even cycles, update the distribution of material 2 represented by the values of ρ_2, such that the available amount is fully utilized, for the given distribution of material 1.

2.5 *Truss-Based Optimization*

As mentioned in the introduction, even though the theoretical development of truss-based optimization procedures preceded continuum-based methods, the former have not seen a widespread adoption in industry. Nevertheless, for the purpose of designing architectural structures, truss-based results enjoy the advantage of facilitating a prescription of thc optimization results within a pre-defined range of components, leading to straightforward interpretation and realization processes.

The initiation of research on optimal structural layouts and configurations is frequently attributed to Michell's work on minimum-weight grid-like continua (Michell 1904). Many years later, this field evolved into the general layout theory for frames and flexural systems (e.g. Hemp 1973; Rozvany 1976). Probably the most important advancement that stimulated the development of topology optimization of truss structures was the introduction of numerical methods for automatic optimum design (Dorn et al. 1964). Until today, most of the established computational procedures follow the so-called "ground structure" approach, where the design space is discretized using a fixed set of nodal points, which are then connected by a set of potential truss bars. This is also the underlying method implemented in the current version of the TopOpt plugin, even though other attractive approaches exist, for example the "Growth Method" that integrates topology, shape and sizing optimization (Martinez et al. 2007). Early implementations of the ground structure approach show multiple benefits from integration within the Grasshopper environment: complex ground structure matrices derived from non-uniform surface geometries are easily generated and manipulated; non-uniform subdivision schemes that would be highly challenging to implement and control in existing environments are generated at ease through GH's inbuilt mapping functions; direct coupling to digital and robotic manufacturing machinery is achieved through processing of the optimization output in Rapidcode-generating modules (Fig. 9).

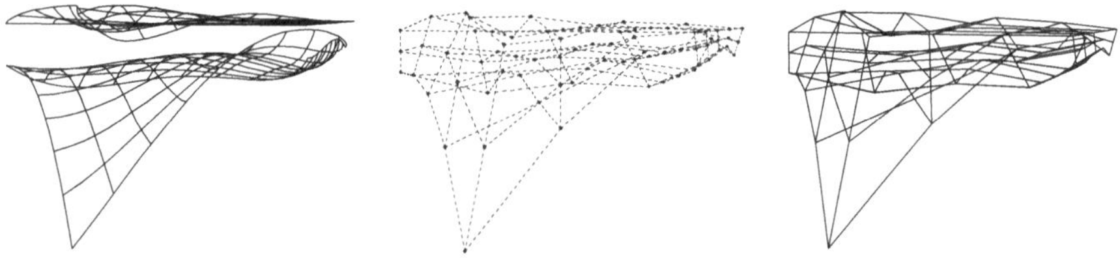

Fig. 9 Generation of a non-uniform ground structure in three, simple steps: input of NURBS surface (*left*); computing of node points (*middle*); generation of ground-structure. The geometric flexibility of the procedure allows for adapting to challenging contextual settings, facilitating the form-finding of context-specific space-frame structures

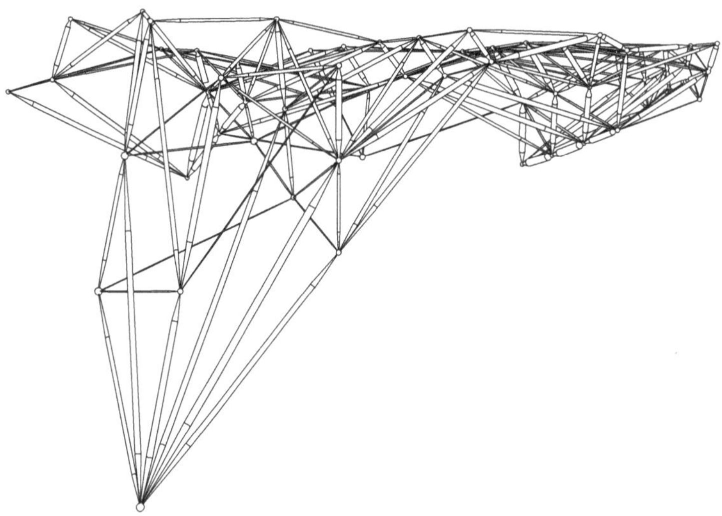

Fig. 10 Structural configurations of hybrid space frame structure. The canopy structure, a pre-study for a digitally fabricated, full scale prototype currently under construction, is adapted from laser-scanned context studies, fitted between walls in an irregular corner of two adjacent buildings. The optimization result drives robotic procedures directly within the GH environment through translation of bar lengths, orientations and node positions to the corresponding rapid-code instructions

One of the attractive aspects of truss-based optimization is that in its most basic form, a linear programming problem is obtained that can be solved very efficiently and results in a globally optimal solution. This formulation corresponds to the minimization of the structure's weight while constraining the allowable stresses in the bars. In the context of the ground structure approach, the resulting layout will indicate the optimal cross-section areas of all potential bars, including eliminating unnecessary bars by assigning them a zero cross-section area. From an engineering perspective, the optimal structural configuration provides fundamental understanding regarding the desirable flow of forces, thus providing the designer with valuable feedback regarding the physical sensibility of the design. This is demonstrated via an example involving 3-D truss optimization of a hybrid space structure as presented in Fig. 10.

Even though the globally optimal layouts are often not directly suitable for practical construction, they can serve as an ideal starting point for advanced investigations. In particular, we are currently focusing our efforts on implementing the following functionalities: (1) Optimizing node locations; (2) Fitting the globally-optimal cross-sections to a prescribed set of bars; and (3) Imposing constructability limitations on the number of bars to be connected at each node. We note that these computational procedures have already been developed and tested in MATLAB and only need to be adapted to Grasshopper.

Conclusions

In this paper, a variety of advanced topology optimization methods, intended exclusively for conceptual architectural design of structures, have been presented. The purpose is to highlight computational procedures that can be integrated directly into the architect's desktop tools thus facilitating the design of innovative, efficient structural forms.

Throughout the paper we provide details regarding the implementation of the computational procedures in the framework of the TopOpt plugin for Grasshopper. The plugin offers several unique features, namely: (1) Interactive control and continuous visualization; (2) Embedding flexible voids within the design space; (3) Consideration of distinct tension / compression properties; and (4) Optimization procedures for skeletal structures such as trusses and frames. We believe that by developing advanced computational tools and user interfaces that are tailored specifically for the architectural community, the potential of embedding topology optimization into architectural design processes could be fully explored.

Future work will focus primarily on improving the optimization procedures for skeletal structures – e.g. space trusses and frames. This is due to the possibility of directly representing computational results as real-world constructable configurations. Various methodological challenges arise, such as the consideration of buckling and manufacturing restrictions. This leaves much room for continuing research and development efforts, with the aim to provide architects and engineers with consistent, efficient and reliable computational design tools.

Acknowledgements The authors would like to thank Ole Sigmund for plenty of fruitful discussions and suggestions, and for his active support in the development of the TopOpt plugin. Financial support for the second author, received from the European Commission Research Executive Agency, grant agreement PCIG12-GA-2012-333647, is gratefully acknowledged.

References

Aage, N., Lazarov, B.S.: Parallel framework for topology optimization using the method of moving asymptotes. Struct. Multidiscip. Optim. **47**(4), 493–505 (2013)

Aage, N., Nobel-Jørgensen, M., Andreasen, C.S., Sigmund, O.: Interactive topology optimization on hand-held devices. Struct. Multidiscip. Optim. **47**(1), 1–6 (2013)

Amir, O.: A topology optimization procedure for reinforced concrete structures. Comput. Struct. **114**, 46–58 (2013)

Amir, O., Sigmund, O.: Reinforcement layout design for concrete structures based on continuum damage and truss topology optimization. Struct. Multidiscip. Optim. **47**, 157–174 (2013)

Amir, O., Aage, N., Lazarov, B.S.: On multigrid-CG for efficient topology optimization. Struct. Multidiscip. Optim. **48**, 815–829 (2014)

Andreassen, E., Clausen, A., Schevenels, M., Lazarov, B.S., Sigmund, O.: Efficient topology optimization in MATLAB using 88 lines of code. Struct. Multidiscip. Optim. **43**(1), 1–16 (2011)

Beghini, A., Beghini, L., Baker, W.: Applications of structural optimization in architectural design. Struct. Congress **2013**, 2499–2507 (2013)

Bendsøe, M.P.: Optimal shape design as a material distribution problem. Struct. Optim. **1**, 193–202 (1989)

Bendsøe, M.P., Kikuchi, N.: Generating optimal topologies in structural design using a homogenization method. Comput. Meth. Appl. Mech. Eng. **71**, 197–224 (1988)

Bendsøe, M.P., Sigmund, O.: Topology Optimization – Theory, Methods and Applications. Springer, Berlin/New York (2003)

Besserud, K., Katz, N., Beghini, A.: Structural emergence: architectural and structural design collaboration at SOM. Archit. Des. **83**(2), 48–55 (2013)

Bogomolny, M., Amir, O.: Conceptual design of reinforced concrete structures using topology optimization with elasto-plastic material modeling. Int. J. Numer. Meth. Eng. **90**, 1578–1597 (2012)

Borel, P.I., Harpøth, A., Frandsen, L., Kristensen, M., Shi, P., Jensen, J., Sigmund, O.: Topology optimization and fabrication of photonic crystal structures. Opt. Express **12**(9), 1996–2001 (2004)

Buswell, R.A., Soar, R.C., Gibb, A.G.F., Thorpe, A.: Freeform construction: mega-scalerapid manufacturing for construction. Autom. Constr. **16**, 224–231 (2007)

Cai, K.: A simple approach to find optimal topology of a continuum with tension-only or compression-only material. Struct. Multidiscip. Optim. **43**(6), 827–835 (2011)

Clausen, A., Aage, N., Sigmund, O.: Topology optimization with flexible void area. Struct. Multidiscip. Optim. (2014, in press)

Davis, T.A.: Direct Methods for Sparse Linear Systems. SIAM, Philadelphia (2006)

Dombernowsky, P., Søndergaard, A.: Unikabeton prototype. In: Fabricate: Making Digital Architecture. Riverside Architectural Press, Toronto (2011)

Dombernowsky, P., Søndergaard, A.: Design, analysis and realization of topology optimized concrete structures. Int. Assoc. Shell Spat. Struct. **53**, 209–216 (2012)

Dorn, W.S., Gomory, R.E., Greenberg, H.J.: Automatic design of optimal structures. J. Mecanique **3**(1), 25–52 (1964)

Frattari, L.: The structural from: topology optimization in architecture and industrial design. Doctoral dissertation, School of Advanced Studies, University of Camerino (2011)

Hemp, W.S.: Optimum structures. Clarendon, Oxford (1973)

Huang, X., Xie, Y.M.: Evolutionary Topology Optimization of Continuum Structures: Methods and Applications. Wiley, Chichester (2010). 235 pp. ISBN 9780470746530

Jonsmann, J., Sigmund, O., Bouwstra, S.: Compliant electro-thermal microactuators. In: Micro Electro Mechanical Systems, pp. 588–593. IEEE (1999)

Kato, J., Ramm, E.: Optimization of fiber geometry for fiber reinforced composites considering damage. Finite Elem. Anal. Des. **46**, 401–415 (2010)

Kato, J., Lipka, A., Ramm, E.: Multiphase material optimization for fiber reinforced composites with strain softening. Struct. Multidiscip. Optim. **39**, 63–81 (2009)
Kirsch, U.: Structural Optimization. Springer, Berlin/New York (1993)
Le, T., Austin, S.: High-performance printing concrete for freeform building components. In: Proceedings of the FIB Symposium, PRAGUE 2011 (2011)
Liu, S., Qiao, H.: Topology optimization of continuum structures with different tensile and compressive properties in bridge layout design. Struct. Multidiscip. Optim. **43**(3), 369–380 (2011)
Martinez, P., Marti, P., Querin, O.M.: Growth method for size, topology, and geometry optimization of truss structures. Struct. Multidiscip. Optim. **33**(1), 13–26 (2007)
Michell, A.G.M.: The limits of economy of material in frame-structures. Lond. Edinb. Dublin Philos. Mag. J. Sci. **8**(47), 589–597 (1904)
Mostafavi, S., Beltran, M. M., Biloria, N.: Performance Driven Design and Design Information Exchange (2013)
Park, P., Gilbert, M., Tyas, A., Popovic-Larsen, O.: Potential use of structural layout optimization at the conceptual design stage. Int. J. Archit. Comput. **10**, 13–32 (2012)
Pedersen, P.: On optimal orientation of orthotropic materials. Struct. Optim. **1**(2), 101–106 (1989)
Rozvany, G.I.: Optimal Design of Flexural Systems: Beams, Grillages, Slabs, Plates, and Shells. Pergamon Press, Oxford/New York (1976)
Sardan, O., Eichhorn, V., Petersen, D.H., Fatikow, S., Sigmund, O., Bøggild, P.: Rapid prototyping of nanotube-based devices using topology-optimized microgrippers. Nanotechnology **19**(49), 495–503 (2008)
Sasaki, M., Itō, T., Isozaki, A.: Morphogenesis of Flux Structure. AA Publications, London (2007)
Sigmund, O.: A 99 line topology optimization code written in Matlab. Struct. Multidiscip. Optim. **21**(2), 120–127 (2001)
Sigmund, O., Bendsøe, M.P.: Topology optimization: from airplanes to nano-optics. In: Stubkjær, K., Kortenbach, T. (eds.) Bridging From Technology to Society. Technical University of Denmark, Lyngby (2004)
Sigmund, O., Jensen, J.S.: Systematic design of phononic band–gap materials and structures by topology optimization. Philos. Trans. R. Soc. Lond. A Math. Phys. Eng. Sci. **361**(1806), 1001–1019 (2003)
Sigmund, O., Torquato, S.: Design of materials with extreme thermal expansion using a three-phase topology optimization method. J. Mech. Phys. Solids **45**(6), 1037–1067 (1997)
Søndergaard, A., Feringa, J.: An integral approach to structural optimization and fabrication. In: Acadia, 491–497 (2014)
Søndergaard, A., Amir, O., Knauss, M.: Topology optimization and digital assembly of advanced space-frame structures. In: 2013 ACADIA Conference (2013)
Stromberg, L.L., Beghini, A., Baker, W.F., Paulino, G.H.: Topology optimization for braced frames: combining continuum and beam/column elements. Eng. Struct. **37**, 106–124 (2012)
The Mathworks.: MATLAB version 8.1.0.604 (R2013a) (2013)
Victoria, M., Querin, O.M., Martí, P.: Generation of strut-and-tie models by topology optimization using different material properties in tension and compression. Struct. Multidiscip. Optim. **44**, 247–258 (2011)

Computational Design and Construction of Notch-Free Reciprocal Frame Structures

Nicolas Mellado, Peng Song, Xiaoqi Yan, Chi-Wing Fu, and Niloy J. Mitra

Abstract A reciprocal frame (RF) is a self-standing 3D structure typically formed by a complex grillage created as an assembly of simple atomic RF-units, which are in turn made up of three or more sloping rods forming individual units. While RF-structures are attractive given their simplicity, beauty, and ease of deployment; creating such structures, however, is difficult and cumbersome. In this work, we present an interactive computational framework for designing and assembling RF-structures around a 3D reference surface. Targeting notch-free assemblies, wherein individual rods or sticks are simply tied together, we focus on simplifying both the process of exploring the space of aesthetic designs and also the actual assembly process. By providing computational support to simplify the design and assembly process, our tool enables novice users to interactivity explore a range of design variations, and assists them to construct the final RF-structure design. We use the proposed framework to design a range of RF-structures of varying complexity and also physically construct a selection of the models.

Project page: http://geometry.cs.ucl.ac.uk/projects/2014/rf-aag

1 Introduction

Reciprocal frame (RF) is a 3D assembly of mutually-supported rods without any central support pillars. Such an assembly structure usually consists of simple atomic units called RF-units, which are three or more slanted rods supporting one another

N. Mellado (✉) • N.J. Mitra
University College London, London, UK
e-mail: N.Mellado@cs.ucl.ac.uk; n.mitra@cs.ucl.ac.uk

P. Song
University of Science and Technology of China, Anhui, China

X. Yan • C.-W. Fu
Nanyang Technological University, Singapore, Singapore
e-mail: CWFU@ntu.edu.sg

P. Block et al. (eds.), *Advances in Architectural Geometry 2014*,
DOI 10.1007/978-3-319-11418-7_12

Fig. 1 Physical RF-structures designed and fabricated by architects: a simple roof structure (*top-left*), a design by Michael Clarke (*top-right*), a design from Spiro-ETH (*bottom-left*), and a design from Wan Shu and Kengo Kuma (*bottom-right*)

in closed circuits. The inner end of each rod rests on and is supported by its adjacent rod, and a closed circuit is obtained as the last rod is placed over the first one in a mutually-supporting manner. At the boundary, such structures are supported by pillars, walls, or ring beams.

Beyond structural considerations, RF assemblies are simply beautiful and elegant. For centuries they have been used as roof or floor patterns in churches, or in construction of traditional houses in Japan and China (cf., Chilton 1995; Larsen 2008). Such designs, however, were largely restricted to flat grillage patterns (e.g., sketched by Honnecourt, Leonardo da Vinci) or circular plans, i.e., the rods meeting at one point in the center (e.g., an Indian tent). In the recent decades, RF-structures have inspired many architects and structural engineers, rekindling interest in age-old principles in structural systems. Moreover, emerging applications of computational optimization and CAD tools have inspired researchers to enhance, enrich, and scale up RF-structure construction (Pugnale et al. 2011). Figure 1 shows a few examples.

In this work we focus on *notch-free* RF-structures, which involve rods (e.g., bamboo sticks) that are simply tied together with strings, as opposed to wooden or metal beams joined at complementary grooves. The notch-free structures being light-weight, the binding threads are sufficient to balance the frictional forces to keep the rods in place. Further, being made of simple rods tied together, these RF-structures are highly cost-effective deployable systems, particularly suitable for rapid constructions of temporary structures (see Larsen (2008) for more details).

Creating such structures consists of two main stages: a *design phase* to specify the RF-structure spanning a given reference surface, and a *construction phase* to actually realize the designed structure tying together the sticks in an appropriate assembly sequence. The design phase comprises of selecting and laying out RF-units, and deciding their inter-connections. In classical tent-like roof design, the process is quite simple: a single RF-unit with only design freedom being the number of rods spanning the inner circle (see Fig. 1 top-left). The key technical challenge is to design for more general reference surfaces, while adopting a wide variety of RF-units. Although the design space is large, manually exploring such a space is quite cumbersome, and demands expertise in both aesthetics and engineering. Classical examples are the various architectural designs by Leonardo Da Vinci during the Renaissance period.

Furthermore, the construction phase imposes certain assembly constraints: (i) the rods or sticks should be straight and appropriately overlap at the junctions (i.e., avoid intersections or gaps); and (ii) the structure should be stable and should not fall over in the absence of suitable supports. More importantly, during the assembly process, the partial structure should also remain stable under self-weight, or appropriate intermediate support pillars, which are referred to as Charlie sticks, should be added to restore stability. Such constraints are difficult to manually account for in case of medium- or large-span RF-structures, as in the focus of this work.

We provide an interactive tool to aid the design phase of RF-structures. Based on the observation that RF-structures are self-similar and form highly symmetric patterns spanning vast architectural spaces, we establish a duality between RF-structure patterns and planar tiling theory, which is mathematically well understood and characterized. Based on the recent work by Song* et al. (2013), we split the design phase as: a planar grillage design using RF-units arranged according to various tiling patterns; and a conformal lifting phase to morph the planar grillage to the reference surface subject to various fabrication constraints.

In order to simplify the assembly process, we propose multiple strategies to determine an effective assembly sequence for the designed RF-structures, while also proposing respective insertion and removal sequence of supporting pillars. The strategies are formulated by computationally evaluating the feasibility of the intermediate RF-structures based on physical stability tests and scoring the validity of the intermediate structures. Specifically, the system proposes both the sequence of rods to be added and the placement of intermediate support pillars (see the supplementary video and Fig. 10). This vastly simplifies the assembly process since our system automatically proposes different assembly sequences, while accounting for different fabrication constraints.

Using our interactive framework, users can quickly sketch and formulate designs with extended number of RF-units, allowing them to easily manipulate, test, and preview a wide variety of coherent RF-patterns with feasible geometric parameters. We used our tool for designing various RF-structures by varying the grillage patterns and the reference surfaces. We physically constructed a selection of designs by assembling them using the guidance provided by our computational tool.

1.1 Related Work

Background For centuries reciprocal frames have been used in design and construction, e.g., the classical bridge sketches by Leonardo da Vinci, the roof of Nagasaki Castle in Japan, as well as Eskimo tents. However, we generally lack computational support to design and construct RF-unit-based structures. Hence, most existing realizations are restricted to small structures involving only a few RF-units. Current practice cannot be generalized to handle larger structures due to various challenges involving where to place the RF-units, how to interconnect them, and how to realize a meaningful aesthetic design.

RF-structures in practice Architects often manually experiment with different ways of assembling RF-units by testing physical mock-ups created using rods (e.g., in a scale of 1:5 (Chilton 2009; Gelez and Saby 2011)). Although such an approach gives full control over the design, form finding remains challenging while ensuring a valid arrangement of the RF-units. As a result even relatively simple designs can be tedious and very time-consuming to mock-up.

Pugnale et al. (2011) stressed the need for computational tools for RF-based designs. Existing attention, however, is focused on handling the engineering issues such as force analysis on the structural stability (Douthe and Baverel 2009; Kohlhammer and Kotnik 2010) and the fitting of rods to form a connected RF (Baverel et al. 2004; Parigi et al. 2012). Brocato and Mondardini (2010) proposed a geometric method to design stone domes with extended number of RF-units, but their method supports only one class of RF-patterns and offers a few parameters for user control. Thönnissen and Werenfels (2011) employed a Rhino-script to aid students to design RF-structures and arranged the RF-units over the cells obtained as the Delaunay triangulation of points on the input reference surface. However, since the point set can have arbitrary distribution, the resulting RF-structures can be rather irregular. Further, the users have little control on the RF-patterns, and have no support to interactively preview and refine their designs.

Fabrication-aware form finding In a more general context of architectural geometry, different geometric modeling methods have been proposed for different fabrication constraints. Whiting et al. (2009) explored structural feasibility in the context of modeling masonry buildings. They proposed a gradient-based optimization method to search the parameter space in procedural models to generate stable buildings, and recently (Whiting et al. 2012) introduced a stability metric and optimized building geometry to achieve stability. Concurrent efforts (Fu et al. 2010; Eigensatz et al. 2010; Singh and Schaefer 2010) introduced cost-driven methods to analyze and optimize panel types and shapes towards cost-effective constructions of free-form surfaces. Yang et al. (2011) devised a computational framework to characterize, parameterize, and navigate non-linearly constrained shape spaces to access feasible designs that satisfy a given set of constraints.

Computational design tools With growing focus on physically-manufacturable objects, e.g., papercraft models (Mitani and Suzuki 2004; Li et al. 2010), or interlocking 3D puzzles (Xin et al. 2011), there has been increased effort in developing computational design tools. Umetani et al. (2012) proposed a design suggestion interface that employs a force-analysis model to guide users to design valid shapes of furniture models under geometric and physical constraints.

Construction sequences Despite the growing interest in architecture geometry, the optimization of construction sequences is today mostly unexplored. As far as we know, only two recent works study this topic, in the context of planar-pieces assemblies (Schwartzburg and Pauly 2013; Cignoni et al. 2014). These approaches are, however, tailored for slit assemblies, and cannot be directly applied to RF-structures.

In this work, we develop a computational tool to support design and assembly of large RF-structures, which are difficult to conceive using physical mockup-based experimentation. In our tool, one can quickly sketch up an RF-structure by combining different RF-units, flexibly modify its appearance and pattern, as well as interactively experiment with different design parameters while the underlying optimization ensures connectivity and structural coherence. Assembly sequences are then interactively explored by combining criteria specific to RF-structures (e.g., rods are added by units), self-supporting structures (e.g., balance is ensured at each step) or any other specific construction criteria (e.g., aesthetic).

2 Overview

Our interactive computational tool considers two aspects: *design* and *assembly*. Figure 3 shows a preview of the design/assembly tool.

For the design aspect, our tool allows the user to create RF-patterns with self-similarities by tiling with basic construction units. A basic set of building blocks (see Fig. 3 left), which we call RF-units, are offered to the user, but the user can further design and customize them if she likes. The units are used to tessellate a plane (see Fig. 2b), before (conformally) lifting them to 3D around a prescribed reference surface (see Figs. 2c, d and 3 right). In order to guarantee that the RF-tessellation is coherent, the system first validates that a 2D tessellation can be generated from the RF-units specified by the user; then, it optimizes the 3D RF-structure to further ensure that the resulting grillage is coherently-connected in 3D, that is, the original angles between the rods of the RF-units are preserved, and the rods are properly touch in 3D space without self-intersections or gaps. Thus, our system by providing interactive feedback, readily allows the user to design and visualize large RF-structures for a given guiding surface.

The generated RF-structure is a parametric model. The user can interactively modify its parameters, preview its appearance on a guiding surface, and easily experiment with different design variations. Further, the tool assists in the assembly planning phase. Beyond enumerating the stick lengths, the tool also provides guidance pertaining to the sequence of assembling the rods and also prescribes

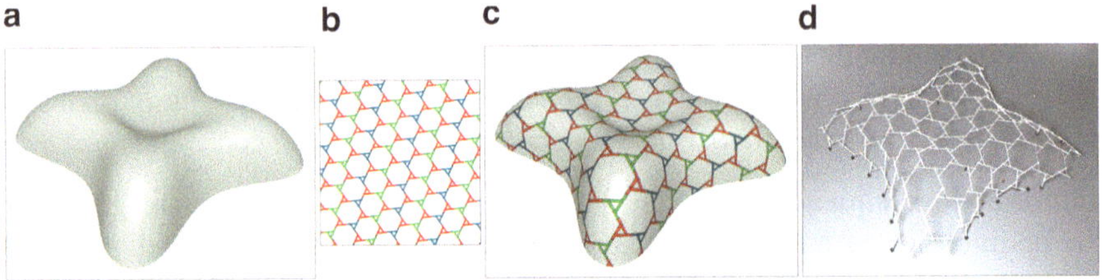

Fig. 2 Overview of the design phase: (**a**) reference surface; (**b**) planar grillage (RF tessellation pattern); (**c**) conformal mapping; and (**d**) final optimization (non-intersecting)

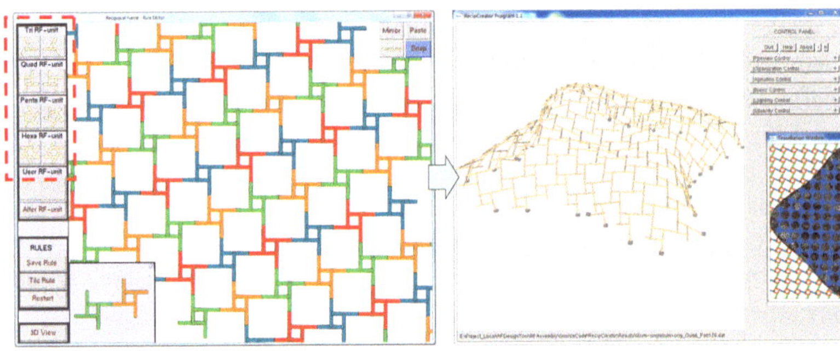

Fig. 3 User interface of the proposed RF-structure design tool. *Left*: the RF Pattern Editor for designing 2D RF-tessellation. *Right*: the RF Creator for designing the 3D RF-structure; the inset reveals the conformal map

additional (intermediate) support pillars, if necessary. The user can select among different strategies ranging from greedy ones (suitable for simple designs) to more advanced ones optimized for functional validity (i.e., basic stability test in this work).

3 RF-Structure Design

In this section, we briefly summarize three key components of the RF-structure design phase: from the design of RF-patterns to the optimization of rods arrangement in 3D (see also Song* et al. (2013)).

3.1 Duality Between RF-Tessellation and Plane Tiling

Our first contribution is on connecting RF-structures with plane tiling (Grünbaum and Shephard 1986). We found that *an RF-tessellation with rotationally-symmetric RF-units is structurally-equivalent (dual) to an edge-to-edge tiling by congruent*

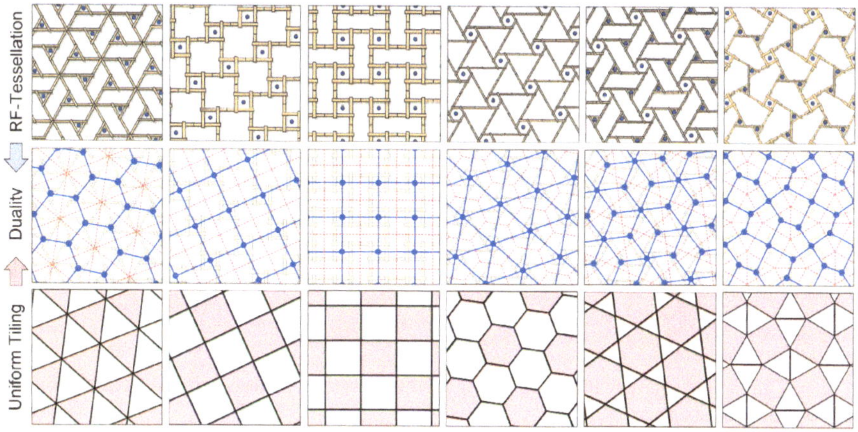

Fig. 4 Duality: RF-tessellation pattern (*top*) is a dual of uniform tiling (*bottom*)

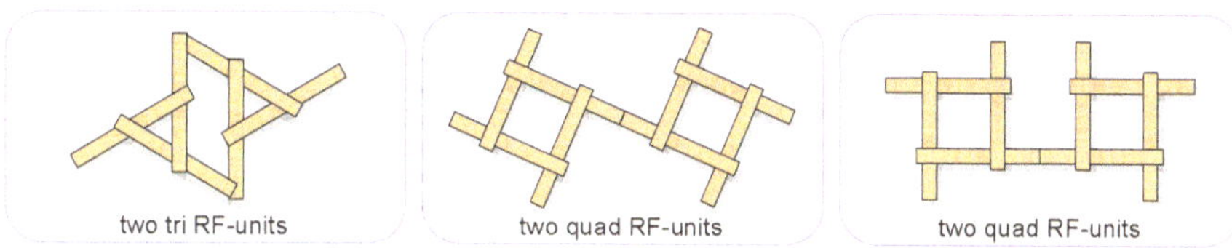

Fig. 5 The grammar rules that define the three left-most RF-structures in Fig. 4

regular polygons. From Fig. 4, we can see that if we replace faces and edges in a uniform tiling by RF-units (individual fans of reciprocal frames) and their connections, respectively, we can obtain the large RF-structures shown on top of the figure. By drawing such an analogy, we can achieve the followings:

- First, we can define grammar rules (see Fig. 5) to procedurally and effortlessly generate RF-structures according to the related tiling pattern;
- Second, we can efficiently testify the validity of a given grammar rule, and predict whether the rule can generate coherent RF-tessellations; and
- Lastly, we can manipulate a grammar rule and re-position RF-units in a complex RF-structure, thus enabling interactive exploration of RF designs (see Fig. 6).

3.2 *Lift RF-Tessellation to 3D*

Our second contribution is a fast method that lifts a 2D RF-tessellation over a 3D reference surface (see Fig. 2a–c). Here we first compute a conformal map of the given reference surface by the ABF++ method (Sheffer et al. 2005), so that we can maintain the structural symmetry during the lifting operation. Then, we construct

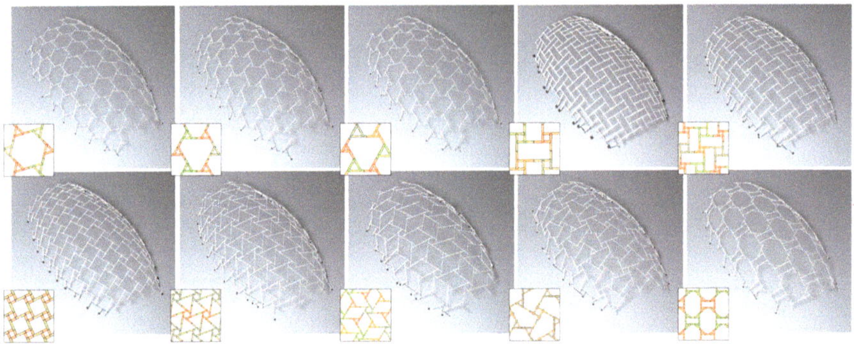

Fig. 6 Different designs realized on the CANOPY reference surface using different choice of RF-units, shown as insets

an approximate RF-structure in 3D by inversing the conformal map and lifting up each rod from the 2D RF-tessellation into the 3D space (see Fig. 3 right). Though this method may not produce coherent RF-structures with proper rods contacts, it enables high-level RF designs, where we can interactively explore different RF parameters while seeing the changes in the approximate RF-structure.

3.3 Optimize the Rods Arrangement in 3D

Since the lifted rods may penetrate or slightly float above one another in the approximate RF-structure, we use an optimization to solve for intersection-free rods placements in 3D, while adhering to the initial design. This is a two-level optimization scheme. First, we relax the RF-units centers on the reference surface to secure a better initial rods arrangement in 3D. Then, we iteratively optimize the rods positions in 3D by formulating an objective function with the rods contacts constraint (F_1), surface constraint (F_2), and conformality constraint (F_3 and F_4):

$$\min_{\text{all rods}} \alpha \sum_{i,j} F_1^2(P_j^i, P_i^j) + \beta \sum_{\{V_i\}} F_2^2(V_1 V_2 \cdots V_m) \\ + \gamma \sum_i F_3^2(V_{i-1}, V_i, V_{i+1}) \zeta \sum_i F_4^2(V_i, V_{i+1}) + \lambda \sum_i (P_i - \bar{P}_i)^2 ,$$

and then linearizing and solving this minimization problem by a least-squares method. Please refer to Song* et al. (2013) for symbol notations and details.

4 RF-Structure Assembly Sequence

4.1 Problem Formulation

Like any other structure, the assembly and construction of an RF-structure is constrained by physical laws, inducing both global and local constraints, e.g., respectively structure balance or material tensile strength. Simulations can be ran on entire structures, however they do not take into account the temporary constraints that could be applied during the construction stage. In contrary to self masonry structures that may need a supporting shape to be constructed (Panozzo et al. 2013), RF-structures are made of multiple self-supported components, that can be assembled in various order to produce the final RF-assembly. Depending on the order chosen to assemble the RF-units, the building can suffer during construction from various issues, e.g. unbalanced assembly or local constraints beyond material resistance.

In this section we propose a generic formulation to produce construction sequences and avoid invalid intermediate assemblies. Our approach takes as input a valid RF-structure, and output a list of constructions steps, including intermediate supports to stabilize the structure when it is unstable.

Our formulation is elaborated around construction strategies, which define how to split an input structure as construction elements, and in which order they should be assembled (see Sect. 4.2). We model the physical constraints that must be respected during the construction as cost functions. They are used by strategies to disambiguate between different solutions and ensure the feasibility of the construction sequence, eventually by suggesting intermediate supporting elements like pillars (see Sect. 4.3).

4.2 Construction Strategies

The aim of a construction strategy is to decide how to split an RF-structure into construction elements, and explore the different assembling alternatives with respect to a given cost function. We define in this paper two ways to split and assemble RF-units: gradually or by group, leading the names GRAD and GROUP (see Fig. 7). With GRAD strategies, fans are added one by one, for instance, by following a counter-clockwise order in each successive n-ring surrounding the starting point.

Instructions to build a structure are elaborated as follow: a user select a starting RF-unit to erect, a strategy that must be used (GRAD or GROUP), and a cost function. Starting from the user selection, other RF-units are iteratively marked as constructed and associated instructions generated. Invalid configurations are also detected at this step to avoided when possible, or fixed using pillars.

A crucial question is here to define in which order to pick unselected RF-units surrounding those already erected, and construct them. We propose two solutions,

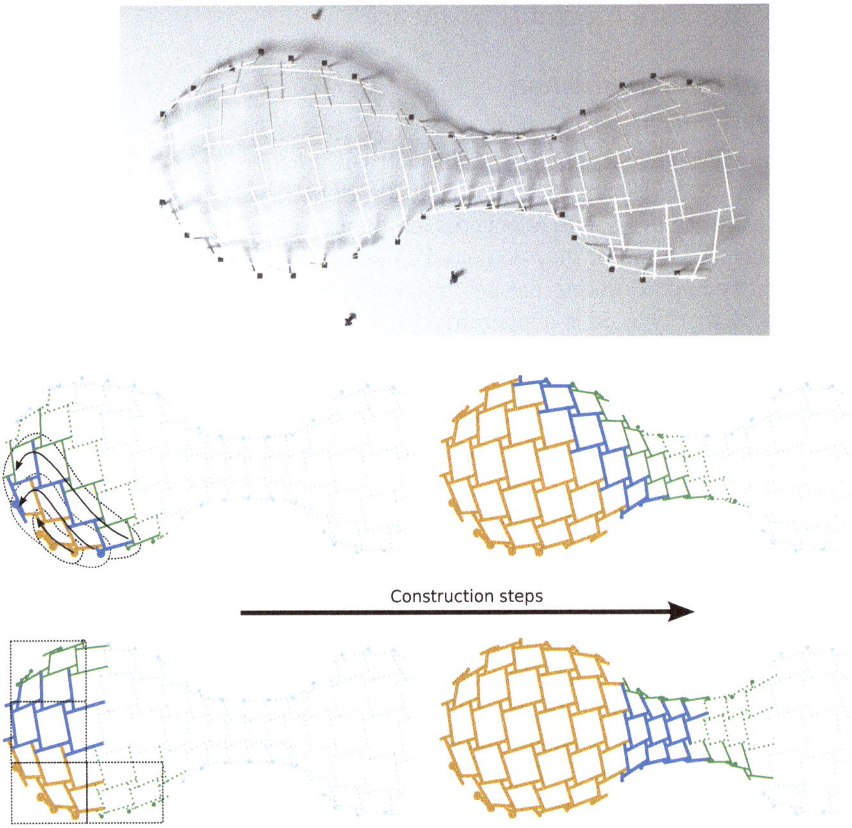

Fig. 7 (*Top*) Top view of the PEANUT model; (*bottom*) two different construction strategies, where RF-units are added from a starting point either gradually (GRAD strategy, following n-rings and counter-clockwise order), or by groups (GROUP strategy). The model on the top uses orthographic projection, and successive steps are visualized in *gold*, *blue* and *green*

that can be applied with both GRAD and GROUP approaches. A first solution is simply to add RF-units using a predefined regular order, either along the selected region for GRADREG or using coordinates in the slicing grid for GROUPREG (see Fig. 8a–c). Here, the cost function is used only to check the configuration validity and suggest pillars.

A second solution is to create a candidate configuration for each of the RF-units that could be added, analyze it using the cost function, and greedily select the configuration introducing the smaller cost. These approaches GRADGREEDY and GROUPGREEDY are illustrated in Fig. 8b–d. Potentially invalid configurations can still be added only when fixed using temporary pillars suggested by the cost function.

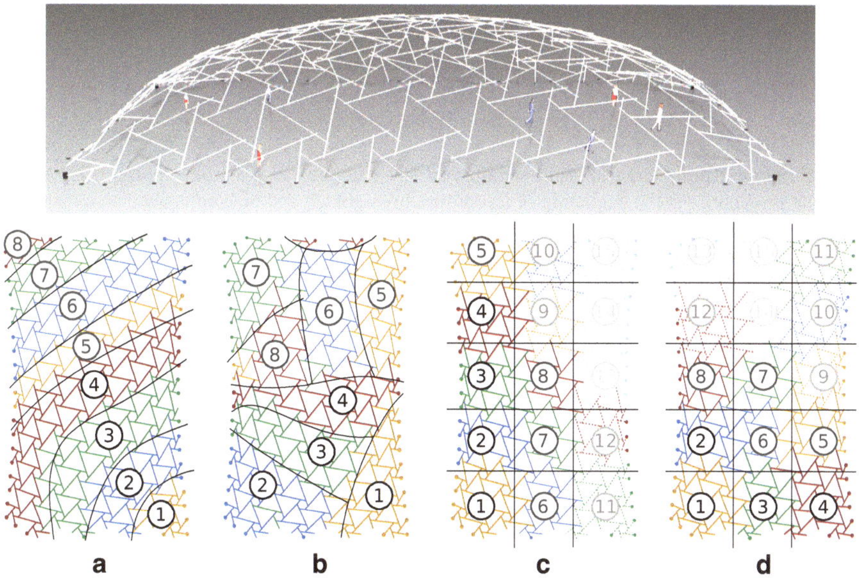

Fig. 8 Proposed building strategies. Approaches (**a**) and (**c**) use the cost function for verification and to suggest support pillars; while greedy approaches (**b**) and (**d**) use the cost function to also determine the assembly order

4.3 Cost Functions and Supporting Elements

As described in the previous section, designing cost functions is a critical step in order to obtain suitable construction sequences: they are used to detect invalid configurations, chose between valid ones, and suggest temporary pillars.

In our system, a cost function takes as input an RF-structure and offers to: (a) compute its cost and detecting invalid configurations, (b) suggest pillars positions. Thanks to this generic formulation, we can plug any analysis in our system, ranging from simple balance check to accurate physical simulations. However, we demonstrate the validity of our approach with two simple cost functions, sufficient in our case to construct real models (see Sect. 5).

Our first cost function aims at favoring RF-units close to the ground, by returning the height of potential candidates as a cost. Neither pillars nor invalid configurations are used here, so its use is limited to simple RF-structures with few elements.

Our second cost function analyzes the balance of an RF-structure to be sure it will not fall down at some point. To detect such situations, we extract the convex 2D polygon defined by the position of the supporting rods on the floor. Then, we compute the gravity center of all the rods of the structure, and check if it is vertically projected inside the convex polygon or not.

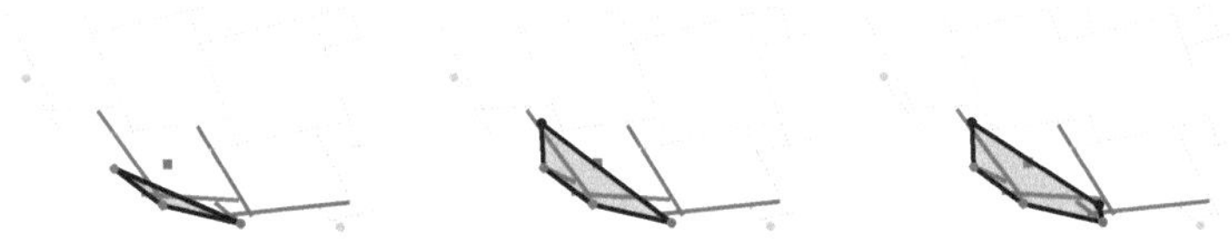

Fig. 9 Support pillar suggestion to stabilize an RF-structure (shown from the top using orthographic projection) at the beginning of its construction: golden and blue units form an unbalanced RF-structure, where their gravity center (*golden square*) is out of the polygon (*gray*) formed by its support points. Two successive pillars insertion are necessary to restore stability. Next, RF-units are shown as *dashed lines*

In the first case, the RF-structure is stable, and the cost is measured as the distance from the gravity center to the convex polygon, normalized by the area of the later. This normalization is critical to avoid favoring configurations with elongated configurations.

In the second case, illustrated in Fig. 9, the RF-structure is unbalanced. Supporting pillars are then suggested below the farthest RF-unit, until a stable configuration is obtained.

Our system also supports combining cost functions. We used this functionality to generate LILIUM and INNER COURT BUBBLE in Fig. 12 in order to add RF-units close to the floor first and avoid unbalanced configurations thanks to temporary pillars.

5 Results

Our tools are implemented in C++ and OpenGL. The main tool is dedicated to the design and assembly planning of RF-structures and it consists of two main panels (see Fig. 3): (i) *RF Pattern Editor*, for composing and editing RF-units and grammar rules; and (ii) *RF Creator*, for making RF designs with interactive editing and for performing the optimization. A second command line tool takes as input the files describing an RF-structure and computes the construction steps with needed (auxiliary) pillars, and producing the instruction list for fabrication. Both tools ran at interactive rates for all the tested models.

RF-design Our tools support a wide variety of RF-patterns, and their variations. Our interactive design tool allows users to apply RF-patterns on a 3D reference shape and to test various parameter configurations (see supplementary video[1]). Figure 6 shows ten RF-structures generated with different RF-patterns on the CANOPY reference surface. Our tool can be used to design RF-structures over guiding surfaces of many different shapes. Figure 12 shows final results using

[1]The supplementary video can be accessed on the project page:
http://geometry.cs.ucl.ac.uk/projects/2014/rf-aag/

Table 1 Statistics of RF-structures shown in Fig. 12

Reference surface	RF pattern	# fans	# rods	# pillars
HYPERBOLIC	Triangular fans	32	55	0
HYPERBOLIC	Quad fans	12	31	4
INNER COURT BUBBLE	Hex fans	84	288	2
LILIUM	Quad fans	126	280	1
PEANUT	Quad fans	80	184	2

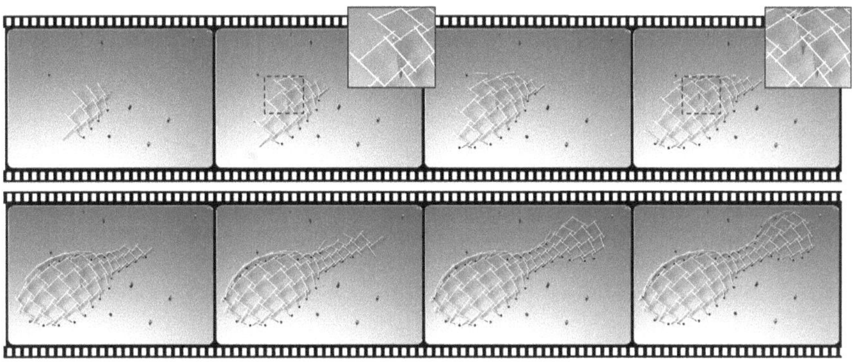

Fig. 10 Snapshots of the assembly sequence for the PEANUT model. The proposed support pillars are in *blue*

HYPERBOLIC, INNER COURT BUBBLE, LILIUM, and PEANUT, while Table 1 lists the associated statistics.

RF-construction strategies Our second tool allows users to elaborate the construction instructions for any kind of RF-structure, independently of its RF-pattern and guiding shape, as illustrated in Fig. 12 (see complete sequences in supplementary video). Regarding to its needs and constraints, one can choose to prioritize the RF-structure construction with respect to the RF-units height (HYPERBOLIC, triangular fans), or to run at each step a stability check, use this criterion to sort the units and suggest intermediate pillars (HYPERBOLIC, quad fans). Both criteria are combined to generate the construction sequences for LILIUM and INNER COURT BUBBLE, permitting us to first assemble the supporting RF-units, add pillars when necessary, and then build the roof. Note that when used in isolation, the stability cost can produce unstable configurations. In such cases, the system proposes addition of temporary pillars to restore stability (see Fig. 10). The temporary pillars can later be removed as later rods are added to restore stability as the later structures extend the building.

In addition to cost functions, our tools allows different strategies, such as regular slicing or growing strategies (see Fig. 7). Impacts on the construction instructions are shown for PEANUT in Fig. 8 and in the supplementary video.

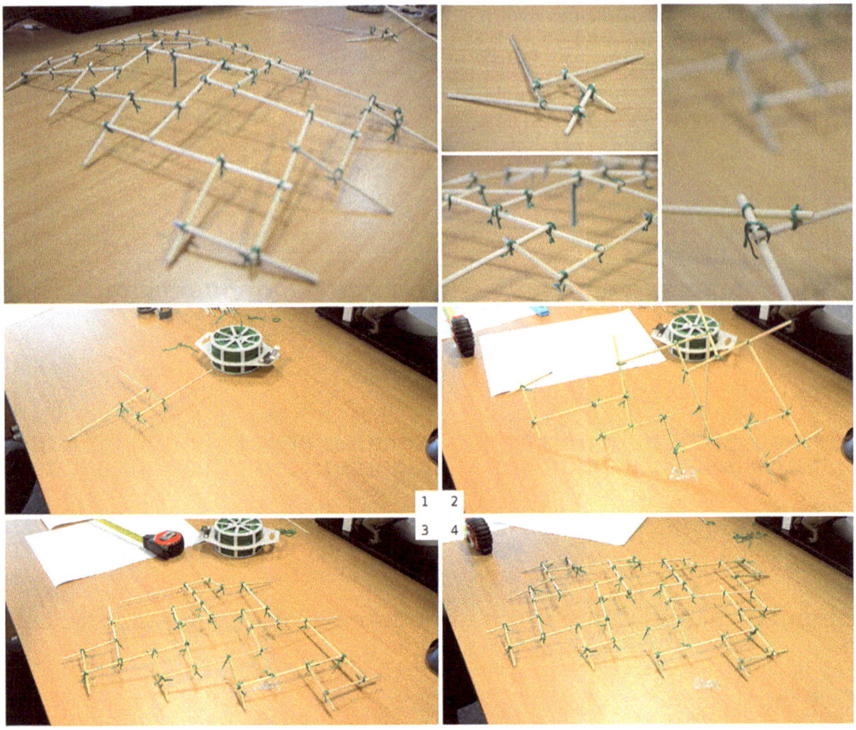

Fig. 11 Physical constructions with bamboo sticks. *Top-left*: a complete structure, with intermediate steps (*bottom*). *Top-right*: close-up view of a triangle, quad, and penta RF-unit

By adapting the cost functions and construction strategies, ones can experiment with various and practical instruction sets to build RF-structures. In our experiments, computation work in real time, making strategy exploration feasible even for complex models comprising of hundreds of units. Figure 11 shows an example of physical construction and the intermediate steps. The physical assembly time was three hours for a completely novice user, including rod cutting.

We observed a few failure cases during our experiments, where our system fails to produce a construction sequence respecting the input constraints. Each time, it was due to unbalanced configurations that could not be stabilized even by adding pillars. Such situations arise rarely, always during the early assembly steps, where there are only a few RF-units, and we can usually correct it by selecting a new starting point. A more general solution would be to allow adding pillars below rods extremities instead of units centers only. We leave this as future work.

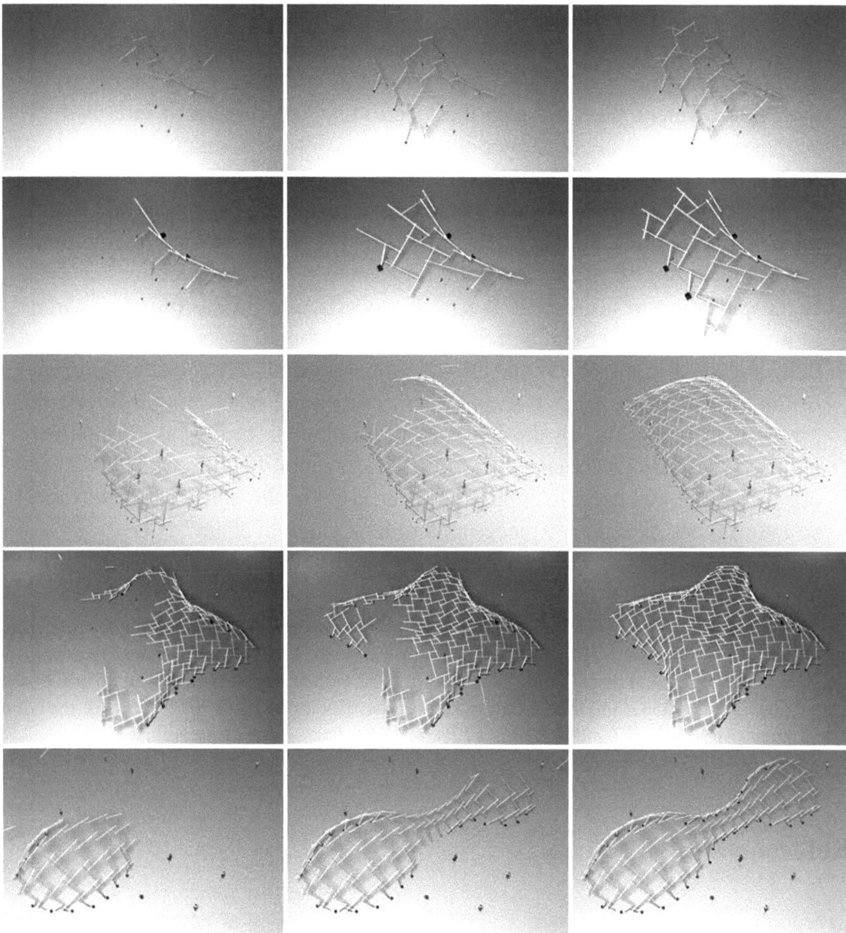

Fig. 12 Assembly process of five different RF-structures, corresponding to the five data rows in Table 1 (from top to bottom)

Conclusion

We presented an interactive tool to facilitate the design and assembly of mid- to large-scale RF-structures. In the design phase, by establishing a connection between RF-structure layout and the classical tiling theory of the plane, we propose an interactive design framework to allow users to bring together simple RF-units to form intricate design patterns over 3D reference surfaces. In the background, the computational tool automatically enforces basic fabrication constraints arising from notch-free constructions.

(continued)

Subsequently, we proposed and investigated several assembly strategies to aid in the construction of such large-scale structures, assisting in determining both the assembly sequence and also the location and placement of intermediate support pillars. We demonstrated the usefulness of the computational tool on several case studies on reference surfaces of varying geometric complexity.

In this work, we do not take into consideration the effect of force and torque on the RF-structures, neither on the final structure, nor during the intermediate construction stages. Since, a common problem with RF-structures involves sagging leading to ultimate collapse of the structure, in the future, we plan to investigate this problem and develop computational tools to appropriately reinforce the designs. We would also like to algorithmically consider the effect of the dimensional of the working environment and take into consideration the ease of manipulating the RF rods that could be heavy and difficult to maneuver, especially in narrow and restricted spaces. Finally, we believe that our formulation is generic enough to be adapted to the construction of other structure types, like planar-pieces assemblies (Cignoni et al. 2014).

Acknowledgements We thank the reviewers for their comments and suggestions for improving the paper. We thank Moos Hueting, James Hennessey and Aron Monszpart for their help on physical construction and discussions. This work was supported in part by the Marie Curie Career Integration Grant 303541, the ERC Starting Grant SmartGeometry (StG-2013-335373), the MOE Tier-2 grant (MOE2011-T2-2-041 (ARC 5/12)), and gifts from Adobe Research.

References

Baverel, O., Nooshin, H., Kuroiwa, Y.: Configuration processing of nexorades using genetic algorithms. J. Int. Assoc. Shell Spat. Struct. **45**(2), 99–108 (2004)

Brocato, M., Mondardini, L.: Geometric methods and computational mechanics for the design of stone domes based on Abeille's bond. In: *Advances in Architectural Geometry*, pp. 149–162. Springer, Wien/New York (2010)

Chilton, J.: History of timber structures, lecture E1. In STEP 2, timber engineering. STEP 2, Timber Eng. **2**, E1–E13 (1995)

Chilton, J.: Development of timber reciprocal frame structures in the UK. In: Proceedings of IASS Symposium 2009: Evolution and Trends in Design, Analysis and Construction of Shell and Spatial Structures, pp. 1877–1884, 2009

Cignoni, P., Pietroni, N., Malomo, L., Scopigno, R.: Field-aligned mesh joinery. ACM Trans. Graph. **33**(1), art.11 (2014)

Douthe, C., Baverel, O.: Design of nexorades or reciprocal frame systems with the dynamic relaxation method. Comput. Struct. **87**(21), 1296–1307 (2009)

Eigensatz, M., Kilian, M., Schiftner, A., Mitra, N., Pottmann, H., Pauly, M.: Paneling architectural freeform surfaces. ACM Trans Graph. (SIGGRAPH) **29**(4), 45:1–45:10 (2010)

Fu, C.-W., Lai, C.-F., He, Y., Cohen-Or, D.: K-set tileable surfaces. ACM Trans. Graph. (SIGGRAPH) **29**(4), 44:1–44:6 (2010).

Gelez, S., Saby, V.: Nexorades, facing an emergency situation. Int. J. Space Struct. **26**(4), 359–362 (2011)

Grünbaum, B., Shephard, G.C.: Tilings and Patterns. W. H. Freeman, New York (1986)
Kohlhammer, T., Kotnik, T.: Systemic behaviour of plane reciprocal frame structures. Struct. Eng. Int. **21**(1), 80–86 (2010)
Larsen, O.P.: Reciprocal Frame Structures. Elsevier Science and Technology (2008)
Li, X.-Y., Shen, C.-H., Huang, S.-S., Ju, T., Hu, S.-M.: Popup: automatic paper architectures from 3D models. ACM Trans. Graph. (SIGGRAPH) **29**(4), 111:1–111:9 (2010)
Mitani, J., Suzuki, H.: Making papercraft toys from meshes using strip-based approximate unfolding. ACM Trans. Graph. (SIGGRAPH) **23**(3), 259–263 (2004)
Panozzo, D., Block, P., Sorkine-Hornung, O.: Designing unreinforced masonry models. ACM Trans. Graph. (Proc. ACM SIGGRAPH) **32**(4), 91:1–91:12 (2013)
Parigi, D., KirkeGaard, P.H., Sassone, M.: Hybrid optimization in the design of reciprocal structures. In: Proceedings of the IASS Symposium 2012: From Spatial Structures to Spaces Structures, 8p, 2012
Pugnale, A., Parigi, D., Kirkegaard, P.H., Sassone, M.: The principle of structural reciprocity: history, properties and design issues. In: IASS: Internationl Conference on Space Structures, pp. 414–421, 2011
Schwartzburg, Y., Pauly, M.: Fabrication-aware design with intersecting planar pieces. Comput. Graph. Forum (Proc. Eurograph. 2013) **32**(2) (2013)
Sheffer, A., Lévy, B., Mogilnitsky, M., Bogomyakov, A.: ABF++: Fast and robust angle based flattening. ACM Trans. Graph. **24**(2), 311–330 (2005)
Singh, M., Schaefer, S.: Triangle surfaces with discrete equivalence classes. ACM Trans. Graph. (SIGGRAPH) **29**, 46:1–46:7 (2010)
Song*, P., Fu*, C.-W., Goswami, P., Zheng, J., Mitra, N.J., Cohen-Or, D.: Reciprocal frame structures made easy. ACM Trans. Graph. (SIGGRAPH) 29(4), Article 94 (2013). (* joint first authors)
Thönnissen, U., Werenfels, N.: Reciprocal frames - teaching experiences. Int. J. Space Struct. **26**(4), 369–372 (2011). (Rhino-script developed by Prof. Annette Spiro)
Umetani, N., Igarashi, T., Mitra, N.: Guided exploration of physically valid shapes for furniture design. ACM Trans. Graph. (SIGGRAPH) **31**(4), 86:1–86:11 (2012)
Whiting, E., Ochsendorf, J., Durand, F.: Procedural modeling of structurally-sound masonry buildings. ACM Trans. Graph. (SIGGRAPH ASIA) **28**(5), 112:1–112:9 (2009)
Whiting, E., Shin, H., Wang, R., Ochsendorf, J., Durand, F.: Structural optimization of 3D masonry buildings. ACM Trans. Graph. (SIGGRAPH ASIA) **31**(6), 159:1–159:11 (2012)
Xin, S.-Q., Lai, C.-F., Fu, C.-W., Wong, T.-T., He, Y., Cohen-Or, D.: Making burr puzzles from 3D models. ACM Trans. Graph. (SIGGRAPH) **30**(4), 97:1–97:8 (2011)
Yang, Y.-L., Yang, Y.-J., Pottmann, H., Mitra, N.J.: Shape space exploration of constrained meshes. ACM Trans. Graph. (SIGGRAPH ASIA) **30**(6), 124:1–124:10 (2011)

Surface Panelization Using Periodic Conformal Maps

Thilo Rörig, Stefan Sechelmann, Agata Kycia, and Moritz Fleischmann

Abstract We present a new method to obtain periodic conformal parameterizations of surfaces with cylinder topology and describe applications to architectural design and rationalization of surfaces. The method is based on discrete conformal maps from the surface mesh to a cylinder or cone of revolution. It accounts for a number of degrees of freedom on the boundary that can be used to obtain a variety of alternative panelizations. We illustrate different choices of parameters for NURBS surface designs. Further, we describe how our parameterization can be used to get a periodic boundary aligned hex-mesh on a doubly-curved surface and show the potential on an architectural facade case study. Here we optimize an initial mesh in various ways to consist of a limited number of planar regular hexagons that panel a given surface.

1 Introduction

The panelization of surfaces remains a challenge in architectural design. CAD software such as Rhino, delivers powerful NURBS surface modelling to the designer. Their ease of use have made them a de facto standard for the design of freeform (and other) shapes in architectural design – especially building envelopes and facades (Fig. 1).

The scale of buildings introduces challenges to surface-based modelling strategies: The large scale of building elements demands that they are divided into smaller elements. The cost of material and labor, standardized production lines, green building concepts, availability and redundancy during construction periods demand for a high degree of similarity of these elements. Yet, the inherent UV subdivision of NURBS surfaces offers limited control over the layout, shape and configuration of the panels. While strategies for the controlled and careful creation of freeform

T. Rörig (✉) • S. Sechelmann
Institut für Mathematik, Technische Universität Berlin, Berlin, Germany
e-mail: roerig@math.tu-berlin.de

A. Kycia • M. Fleischmann
HENN Research, HENN Architekten, Munich, Germany

P. Block et al. (eds.), *Advances in Architectural Geometry 2014*,
DOI 10.1007/978-3-319-11418-7_13

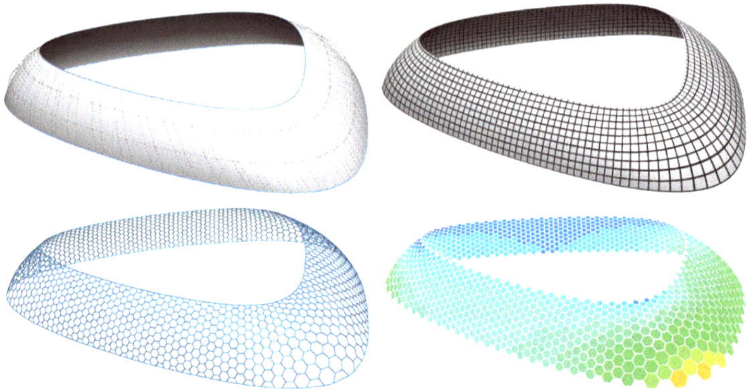

Fig. 1 Outline of the method: For a cylindrical NURBS surface (*top-left*) we create a seamless periodic conformal parameterization (*top-right*). A new mesh (*bottom-left*) is then rationalized and the panels are optimized for quantized regular hexagons (*bottom-right*)

surfaces have been presented (see Glymph et al. 2004) and realized, the tiling of true freeform surfaces through alternative algorithms is still a challenge.

The quality of a surface panelization solution can be defined in various ways. From an aesthetic point of view the shape of individual elements is important. Further there are global conditions such as alignment with surface boundaries and smooth transition of element shape. From the standpoint of fabrication, the elements should be repetitive. The contributions of this paper are:

- **Periodic discrete conformal parameterization**: We present an algorithm that maps a triangle mesh with cylinder topology conformally to a cylinder or a cone of revolution. This allows us to obtain seamless patterns on surfaces. It is a generalized version of the discrete conformal parameterization scheme described by Springborn et al. (2008).
- **Regular elements**: We show how the periodic discrete parameterization can be used to construct a panelization of the given periodic surface into a small number of repeating regular elements.
- **Applications to architectural design**: We present a case study that initiated the development and where the described methods have successfully been applied in the architectural design context.

The article is organized as follows: Sect. 2 describes various parameterization schemes available to architectural geometers. We briefly discuss the shortcomings of current methods and explain how we adress them with our method. Section 3 introduces the concept of periodic discrete conformal parameterizations. See Fig. 2 for a schematic view of the proposed method. In Sect. 4 we present a case study that led to the development of this work. Section 5 deals with the optimization of panels to meet requirements arising during the case study. We give implementational details in Sect. 6 and close with an outlook on further research.

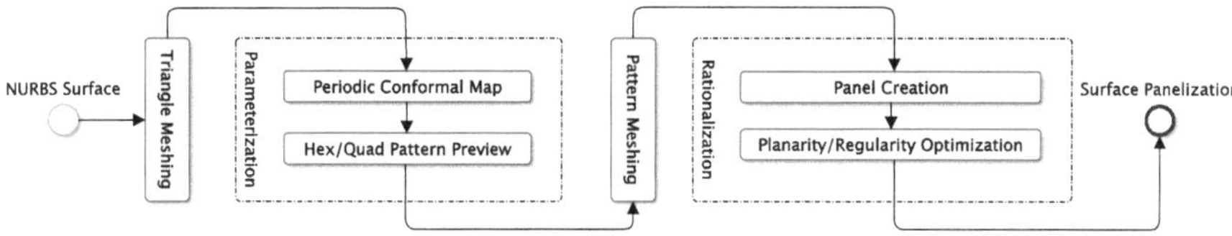

Fig. 2 Flow diagram of the algorithms described in this paper. From a NURBS surface a triangle mesh is created. The parameterization part is described in Sect. 3. A pattern-mesh is created using this parameterization. The creation and optimization of panels is described in Sect. 5

2 Related Work

In this section we will review different methods used to unroll/parameterize surfaces that are accessible in the architectural design process on the example of Rhinoceros3D and 3rd party plugins. All available tools lack at least one of the key features of the proposed method:

- *Conformality* of the mapping avoids non-uniformly stretched panels
- *Periodicity* of the parameterization is needed to layout panels seamlessly on the surface
- *Boundary alignment* avoids irregular cutting of panels at the boundary of the surface

Rhinoceros CreateUVCrv/ApplyCrv A NURBS surface is naturally equipped with a parameterization, i.e., a UV mapping from a rectangle domain to the surface. For the surface of Fig. 1 such a map can be used to project a pattern from this rectangle to the surface (`ApplyCrv` in Rhinoceros). The pattern can be constructed periodically as it is defined in the UV domain of the surface. However, in general this method does not produce satisfactory results in terms of quality of elements for complex freeform surfaces. The UV parameterization is not conformal and thus introduces non-isotropic stretch and shear preventing the elements to be regular on the surface, see Fig. 3, left. This limitation exists even for developable surfaces.

Rhinoceros Squish/SquishBack The `Squish/SquishBack` command of Rhino maps a surface to the plane minimizing the amount of stretch. While this is geometrically not a conformal map it produces acceptable patterns on the surface. It is however not capable of calculating periodic maps to the surface. Thus it is not applicable in our situation, see Fig. 3, middle.

PanelingTools for Rhino The paneling tools of Rhinoceros use the UV parameterization of the underlying surface to populate grid-points over the surface. With the help of such a grid, panels are placed onto the surface, see McNeel. The shapes of the panels depend heavily on the NURBS-parameterization. An example is shown in Fig. 3, left.

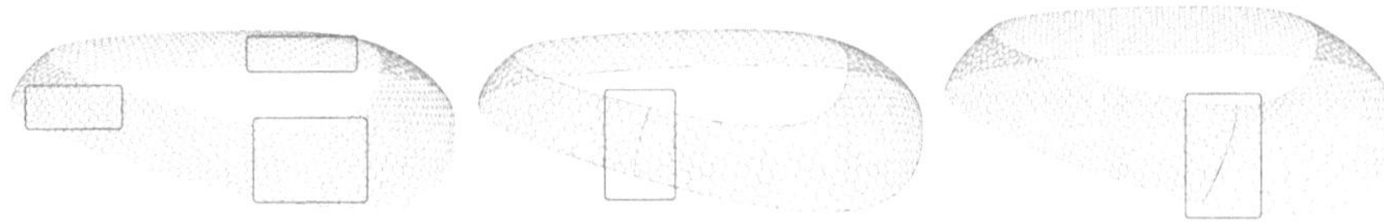

Fig. 3 State of the art unroll methods can create patterns on a closed surface. The `ApplyCrv` command of Rhinoceros produces boundary alinged periodic patterns but introduces unacceptable non-isotropic stretch (*left*). The `SquishBack` method creates sufficiently regular elements but does not respect the periodicity of the surface (*middle*). Non-periodic conformal maps align with the boundary of a cut surface. Along the cut the map is not continuous (*right*)

Hexagonal tilings In the architectural context, hexagonal panelizations have been studied by Zimmer et al. (2013), Troche (2008) and Schiftner et al. (2099). These aproaches however do not include regularity or special boundary alignment as introduced in the current work. In Schiftner et al. (2099) the result of the panelization depends on the choice of an initial triangle mesh that is optimized towards touching incircles. This allows for a torsion free support structure of a non planar hex-mesh. Hexagonal tilings for triangulated surfaces also have been studied by Nieser et al. (2012). They focus on regularity but not on boundary conditions.

Mesh parameterizations There are a vast number of parameterization schemes for meshes. To elaborate on all methods is beyond the scope of this section and we describe only the most relevant results here. General purpose parameterization methods for triangle meshes produce high quality quad or hex meshes for unstructured input data, see for instance (Bommes et al. 2009; Alexa et al. 2000; Springborn et al. 2008). They have been used with success in the architectural context, e.g., by Bo et al. (2011) and Sechelmann et al. (2013).

The basis of our method are conformally equivalent triangle meshes as described by Springborn et al. (2008). The straight forward method to map a surface with this approach is to cut it open and map it to a rectangle domain. This method yields boundary aligned conformal maps that however do not match along the introduced cut, see Fig. 3, right. How to generalize this method to overcome this limitation is the content of the following section.

3 Periodic Conformal Parametrization

In this section we describe our algorithm for the creation of periodic conformal maps for cylindrical meshes/surfaces, i.e., surfaces with the same topology as a cylinder. First we will review the discrete conformal maps of Springborn et al. (2008). Then we show how it can be generalized to yield periodic maps to cylinders or cones.

A smooth *conformal map* between two surfaces is a map that preserves angles. Intuitively, one can think of a conformal map as a map that preserves the shape but not the scale of small figures. For conformal surface parameterization, one looks for conformal maps from the plane to a surface and vice-versa. These can be used to map different patterns onto surfaces in a way that only isotropic stretch/uniform scaling is applied to the pattern elements. The method described in Springborn et al. (2008) is a triangle mesh based discretization of conformal maps. For each vertex v of the surface – interior or boundary – we may prescribe an angle θ_v that corresponds to the angle sum of adjacent triangles in the target mesh. Starting from an input mesh and target angles θ_v, the method calculates new edge lengths for the triangles of the target mesh such that the angle sums at the target vertices are as prescribed. This goal is achieved by minimizing a convex functional. The prescribed angles have to satisfy a Gauss-Bonnet type condition, i.e., the angles at interior vertices have to match the angles at the boundary vertices. We will state the condition for the special cases treated later in the article, see Eq. (1).

For the parameterization problem, we want to construct a map from a surface to the plane. To get a planar target mesh, the target angles have to be set to 2π for all interior vertices, i.e., the angles of the triangles adjacent to every interior vertex sum up to 2π. Thus the computed target triangles can be laid out in the plane. At the boundaries there is still a certain degree of freedom, which allows to map the surface to different shapes, e.g., a rectangle or a more general polygon with prescribed angles. An alternative choice of boundary conditions yields a target mesh whose boundary edges have the same lengths as the original mesh. Then the control over the boundary angles is no longer possible.

This method for the parameterization of triangle meshes can be generalized to triangle meshes with cylinder topology, see Fig. 4. Instead of constructing a discrete conformal map from the surface to the plane, we construct a map to a cylinder or cone, whose image is isometric to a polygonal region in the plane, see Fig. 5. This works with an approach very similar to the previous one. We start with the definition of a periodic parameterization (Fig. 6).

Definition 1 Let $M = (V, E, F)$ be a mesh with cylinder topology and vertices V, edges E, and triangles F. Let $D \subset C$ be a region on a cone/cylinder of revolution.

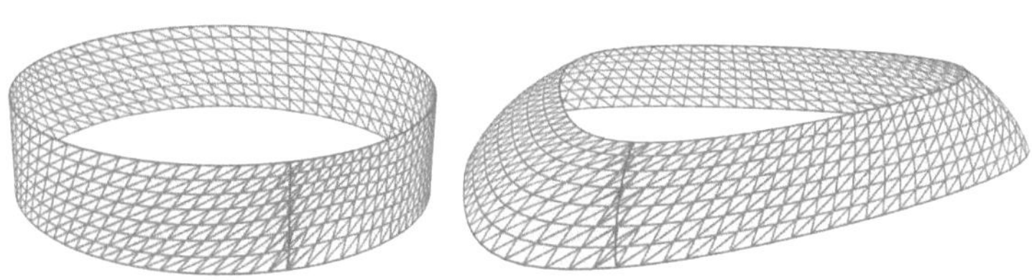

Fig. 4 A discrete periodic map from a cylinder to a triangulated surface. On the cylinder all edges of the triangulation are geodesic arcs. If the cylinder is cut at the vertical orange path, then it can be unrolled to the plane creating a rectangular domain

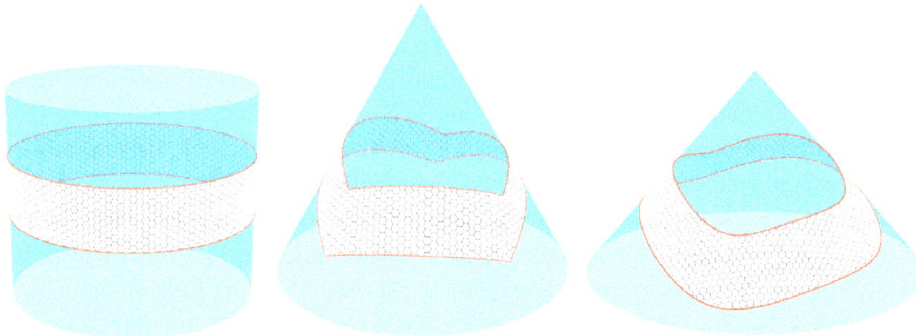

Fig. 5 Periodic domains of parameterization of the surfaces shown in Fig. 6. *Left*: Map to a cylinder with geodesic boundary curves. *Middle*: Map to a cone of revolution with hex-pattern-adapted angle. The domain is a polygon with quantized angles. *Right*: Isometric boundary on a cone with hex-pattern-adapted angle. Panelizations created with the help of these maps are shown in Fig. 6

A continuous bijection $\Phi : D \to M$ is called a *discrete periodic parameterization*. D is called the *domain of parameterization*.

In the latter we always assume that the preimages of edges of M are geodesic arcs on the cone/cylinder C. For panelization of periodic surfaces we need to make sure, that different patterns match around the cone or cylinder. This yields certain restrictions on the cone that serves as domain of parameterization.

Definition 2 Let C be a cone with aperture φ and $\Phi : C \supset D \to M$ be a discrete periodic parameterization of a triangle mesh M with domain D. The map Φ is called *triangle adapted* if the cone angle α is a multiple of $\frac{\pi}{3}$ and *quad-pattern adapted* if it is a multiple of $\frac{\pi}{2}$.

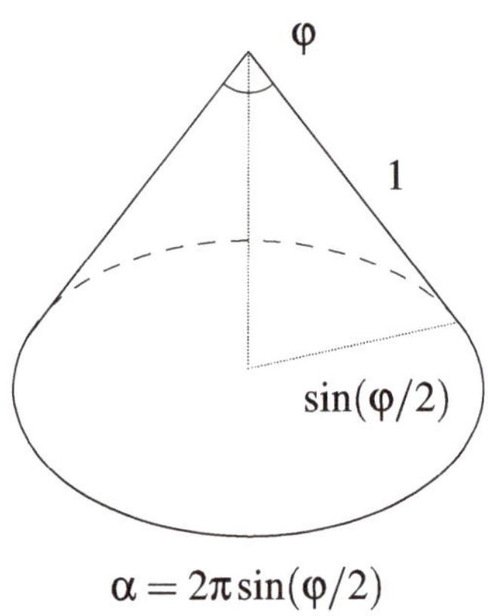

This definition ensures that either a quad-, triangle-, or hex-pattern fits seamlessly onto the surface after parameterization.

Fig. 6 Quantized periodic hexagonal panelizations. Boundary conditions affect the amount of stretch in the interior of the surface. *Top*: Hexagonal pattern aligns with the boundary, a strong condition that produces large deviation of panel sizes. Middle: Map to a pattern-adapted polygon on a cone of revolution. The pattern contains exceptional points at the boundary. The stretch is minimized while at the same time the pattern alignes with the boundary. *Bottom*: Conformal map with the least stretch in the interior, pattern can be optimized to consist of congruent hexagons alone. In each image, panels with the same colors are congruent. The corresponding domains of parameterization are shown in Fig. 5

3.1 Periodic Boundary Conditions

If we want to construct periodic conformal maps we are allowed to specify angle sums θ_v at boundary vertices. The condition for the sums of boundary angles differs from the simply-connected case in the following way: The curvature at a boundary vertex v is given by $\kappa_v = \pi - \theta_v$, where θ_v is the angle sum of the adjacent triangles in the target mesh. For the two boundary loops $(v_1, \ldots, v_n)$ and $(w_1, \ldots, w_m)$ we have:

$$\sum_{i=1}^{n} \kappa_{v_i} + \sum_{j=1}^{m} \kappa_{w_j} = 0. \tag{1}$$

This condition makes sure that the two boundary curves "bend" the same amount and can hence be wrapped around a cone. We will now show how boundary conditions can be used to construct periodic patterns on the studied models. We start with a discrete conformal map of the doubly-curved model from Fig. 1 to a standard cylinder.

3.1.1 Straight Cylinder

The simplest way to generate a map to the cylinder is to set the target angles for all boundary vertices to π. Hence the curvatures at the boundary vertices are zero and the two boundary loops are mapped to "straight" curves. In this case both angle sums of Eq. (1) vanish and the target mesh can be wrapped around a cylinder, see Fig. 5, left. The new edge lengths computed with the variational principle correspond to the lengths on a cylinder. This cylinder can be unrolled into the plane preserving angles and lengths. Therefore the two boundary polygons are mapped to straight lines in the plane. These two straight lines have to be parallel and of equal lengths. If the lengths of the boundary curves in the original model differ a lot, then a map to a cylinder induces a lot of conformal stretch. This stretch can be reduced by specifying special boundary conditions for a parameterization on a cone of revolution.

3.1.2 Cone of Revolution

As long as Eq. (1) is satisfied we obtain a map to a general cone of revolution. In our case, we require that the periodic parameterization is adapted to the target pattern. This means that the two sums of Eq. (1) need to be (the same) multiples of $\frac{\pi}{3}$ (triangle or hex) or multiples of $\frac{\pi}{2}$ (quad). We present two methods to achieve this requirement: a uniform distribution and a concentration of curvature.

If the boundary of the mesh should align with the pattern, then boundary angles need to be quantized, i.e., multiples of $\frac{\pi}{3}$ or $\frac{\pi}{2}$ need to be chosen as target angles. In Fig. 5, middle, three vertices of the top and bottom boundary curve were manually assigned to $\frac{4}{3}\pi$ and $\frac{2}{3}\pi$, respectively. All other boundary angles are set to π, i.e., straight. Such a map can be used as a starting point to obtain a tesselation with quantized hexagons as described in Sect. 5.

It is known that a discrete confomal map that does not change the lengths of the boundary edges exhibits the least stretch in the interior of the surface. To obtain such a parameterization we first construct a periodic conformal mapping onto an arbitrary cone such that the lengths of the boundary edges are equal. The resulting angle sums at boundary vertices of the target mesh determine the cone angle φ of the map. The cone angle of a pattern adapted periodic parameterization is the closest multiple of the desired quantization. We distribute the difference to the closest quantized angle uniformly to the individual boundary vertices and recompute the map with these angle conditions. The obtained map is periodic and exhibits the lowest stretch of all periodic conformal maps (see Fig. 5, right).

3.1.3 Design and Structural Opportunities

Opposed to achieving a homogenous pattern distribution, as described previously, it is also possible to use special boundary conditions to support structural purposes or design requirements. If one aims for a panelization with boundary aligned patterns, then the target boundary angles must be quantized.

To include entrances in a facade it is possible to incorporate special boundary conditions. An example with special boundary vertices with domain angles $\frac{4}{3}\pi$ and $\frac{\pi}{3}$ is shown in Fig. 7, left. In the remeshed surface, the lower boundary curve bends inside at the vertices with angle $\frac{4}{3}\pi$ around the vertex with angle $\frac{\pi}{3}$. This incorporates a natural entrance into the facade.

Another effect of such angle conditions is a densification of the pattern at the vertices with small angles. Such a concentration of elements can be used to enforce structural properties of a geometry. An example of a diagrid generated with such boundary conditions is shown in Fig. 7, right.

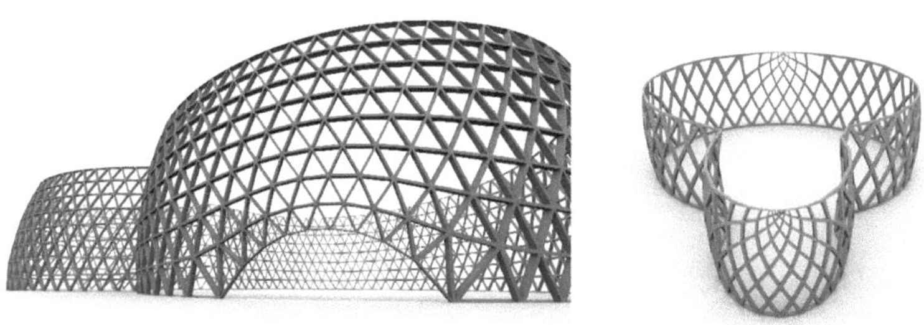

Fig. 7 A periodic conformal map onto a cylinder with special vertices creates the opportunity to incorporate entrances (*left*) or concentration of support structure (*right*)

4 Case Study: Hexagonal Surface Panelization

Here we present the findings of a first case study, where the method of discrete conformal mappings was utilized for a real-world project. In this case study, a facade design, important questions concerning panel layout, similarity and therewith constructability had to be addressed at the early design stage.

Discrete conformal maps were used, as they allow the designer to explore alternative surface textures and surface panelizations with great design flexibility. This distinguishes the method from more constrained modelling techniques (Glymph et al. 2004). Through the method of conformal mapping, opposed to naive UV mapping, the density of the surface panelization varies across the entire surface, yet the shape of elements does not. This can be used for structural purposes, such as diagrid layouts, or design driven, such as window distribution, see Fig. 7. The optimization of the surface panelization towards multiple criteria such as edge length and planarity was consequential.

For a commissioned competition entry we tested and developed the method of periodic conformal mappings. The project, which served as a case study, was highly constrained, as the architects were asked to propose an alternative facade design for an existing design proposal of a multifunctional exhibition center in China, see Fig. 8. The massing was fixed, but there were 2 alternative massing options (1 single curved and 1 doubly-curved envelope) to be explored. Also, the client wanted a hexagonal tiling on the facade but only had a very limited budget of approximately 200 Euro/sqm for the entire facade including sub structure in mind.

These limitations, in combination with the very short timeframe of 2 weeks for the entire redevelopment of the facade including a feasibility study drove the development of the conformal mapping method. Especially, since existing solutions such as UV mapping led to unsatisfactory results producing anisotropic stretch and

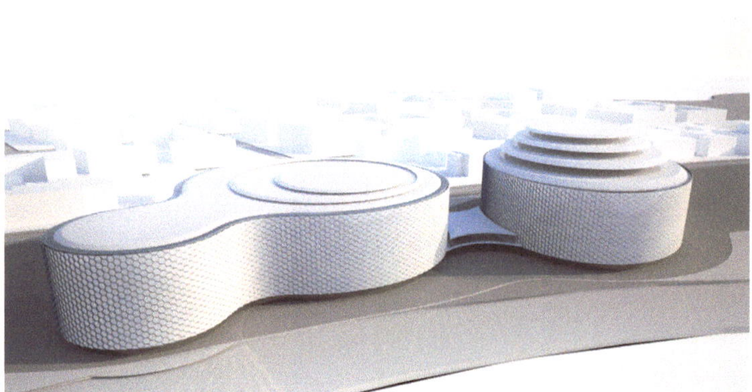

Fig. 8 Rendering of Case Study – A project where the method of conformal mapping was utilized for the facade design

shear of some regions in the master surfaces. Some specific questions that had to be addressed for each massing option were:

- How many (different) panels would we need?
- Can we clad the entire surface with planar tiles?
- Can we equalize the edge lengths of each hexagon?
- Can we control the orientation of the panels?
- Can we achieve a regular pattern with a homogeneous visual appearance?

In the end, all the above questions were answered/solved.

The first step of development focused on achieving periodicity across the surface and alignment with the boundary. While the issue of periodicity directly addressed the last question, it is strongly related to the others as they could be achieved by successive optimization steps.

Already during the design phase a fully periodic tiling was achieved. In a following step the panels were planarized, grouped by dimension and their edge lengths were equalized. Finally, a control for the panel orientation based on the tangents of the NURBS master surface was implemented. This hexagonal pattern served as a base for the facade engineering team. Due to the high cost demands, a simple component system that served as a sub-structure for each panel was developed, see Fig. 9.

Unfortunately the given massing options for the building were not very challenging in terms of geometry. One massing option was a simple extrusion and the other had very little distortion. After the successful submission of the project, we decided to continue the development and test the method of discrete conformal mappings on more extreme base geometries.

During these tests a Grasshopper plug-in for VaryLab, see Sechelmann and Rörig (2013) has been developed and refined. We focused on tiling surfaces with a large distortion/stretch and double curvature. The main aim was to tile these

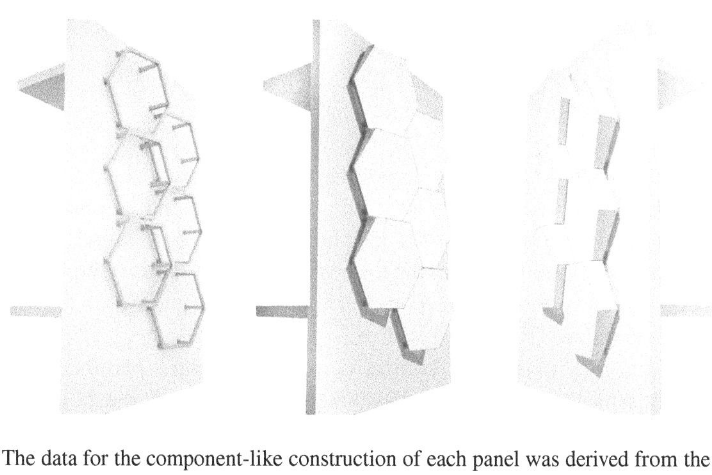

Fig. 9 The data for the component-like construction of each panel was derived from the mesh

surfaces without distorsion. This led to focus on the boundary conditions. The designer is now able to choose between an aligned mapping where the tile-pattern aligns with the underlying surface boundary. The trade-off being, that panels need to vary in sizes. Or one chooses a "homogeneous tiling", where all the tiles are the same, but do not align with the boundary. VARYLAB's numerous optimization algorithms can be applied and combined with either of the two approaches, see Fig. 6. During the development, we realized that through singularities and special boundary conditions, one is able to control the density and distribution of the pattern on the surface and along its boundaries, see Fig. 7.

5 Rationalization: Hexagon Optimization

Starting from the conformal parameterization we optimize the obtained hex-mesh to have identical regular hexagons. We use a global optimization approach and define energies to achieve *planarity*, *regularity*, and *equality*.

Planarity The planarity function is a simple adaptation of the usual energy used to planarize quad-meshes: A quadrilateral $\{A, B, C, D\}$ is planar if the volume of the tetrahedron $\{A, B, C, D\}$ is zero. So if we require the volume of all tetrahedra spanned by the vertices of a polygon to be zero we obtain a planar polygon. The planarity energy E_{pl} can easily be expressed in terms of determinants.

Regularity A regular planar polygon is characterized by having equal edge lengths and equal angles at all vertices. As for planarity we define an energy that is minimized in case of regular polygons. The interior angle at a vertex of a regular p-gon is $\frac{p-2}{p}\pi$. So an energy E_{reg} that is minimized for a regular p-gon with vertices $\{v_1, \ldots, v_p\}$ and corresponding angles $\{\alpha_1, \ldots, \alpha_p\}$ is

$$E_{reg}(P) = \lambda_P\, E_\alpha(P) + \mu_P\, E_\ell(P) \quad \text{with}$$

$$E_\alpha(P) = \sum_{i=1}^{p}(\alpha_i - \frac{p-2}{p}\pi)^2 \qquad E_\ell(P) = \sum_{(v_i, v_{i+1})} (\|v_i - v_{i+1}\| - \ell_P)^2,$$

where ℓ_P is the desired target edge length for the polygon and λ_P and μ_P weights for the different energies. In a first step, the target length can be chosen to be the average edge length of the polygon or the shortest edge length among the edges to avoid overlap. Note that, the normalization of the angles already implies planarity of the polygons. Nevertheless, we consider the planarity energy since it increases the rate of convergence (Fig. 10).

Starting from a cylindrical or conical periodic conformal parameterization we construct a hex-mesh that may or may not be aligned with the boundary. As a consequence of the conformality of the parameterization the angles of the hexagons are almost $\frac{2\pi}{3}$. In the following we do not work with a water tight mesh any more

Fig. 10 Close-up rendering of facade

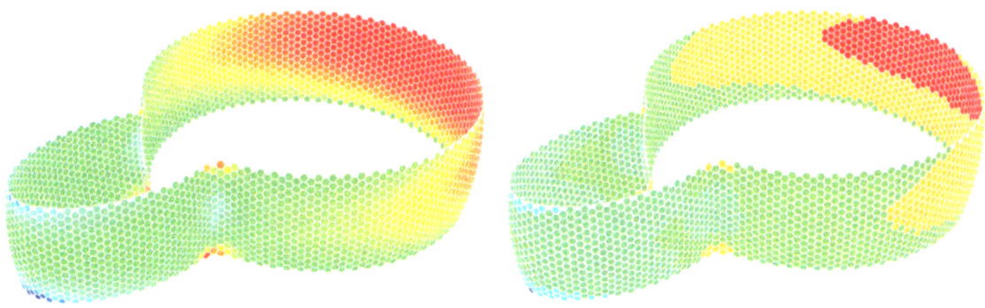

Fig. 11 Panelization of a doubly curved design alternative of the case study shown in Fig. 8. *Left*: Unquantized panelization. *Right*: Quantization to 11 panel sizes with edge lengths varying from 1.5 to 2 m.

but split the surface into individual hexagonal panels. We optimize the edge lengths of the hexagons to be constant per face using E_ℓ. To avoid overlap we choose the length of the shortest edge of each face as target length. We add the planarity and angle regularity functionals E_{pl} and E_α to the optimization and obtain planar and regular hexagons. Each of the hexagons has its own constant edge length. Finally, we can rationalize the panelization further, by choosing a discrete set of edge lengths as target lengths for the polygons in the edge length functional E_ℓ. Due to the symmetry of the edge length functional for regular hexagons edge length optimization will not destroy the planarity and regularity of the hexagons. So it is possible to adjust the edge lengths using E_ℓ only. This quantization process is illustrated in Fig. 11. The used method allows to include additional functionals into the optimization, e.g., functionals minimizing the distance to a reference surface or the distribution of gaps tangent and normal to the surface. We have not yet added these functionals due to the early phase of the presented project.

The range of lengths obtained depends on the initial hex-mesh constructed on the chosen target geometry. The effect of the different periodic conformal parametrizations on the quantization is shown in Figs. 5 and 6.

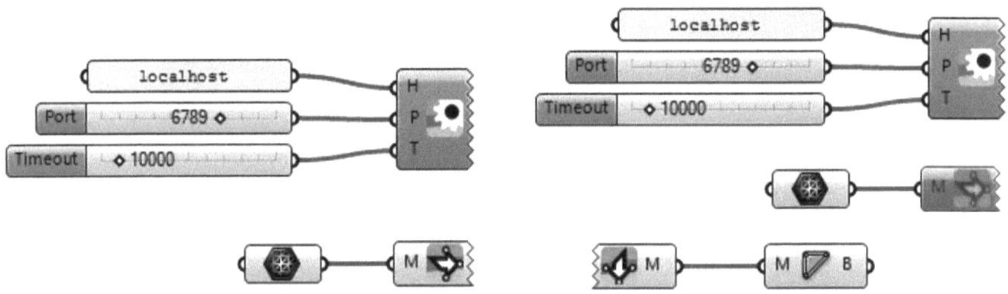

Fig. 12 Grasshopper networks connecting Rhino and VaryLab. In the first step we send mesh data to VaryLab running on the same machine at `localhost:6789` (*left*). In the second step we collect the result from this VaryLab instance and create polygons from the mesh's faces (*right*)

6 Implementation

We use the software package VaryLab (Sechelmann and Rörig 2013), in combination with Rhino's Grasshopper to calculate discrete conformal maps to the cylinder or cone as shown in Fig. 5. Figure 12 shows the Grasshopper network used to connect Rhino and VaryLab. VaryLab uses the optimization package TAO/PETSc, see Benson et al. (2007), Balay et al. (2011), and Sommer (2010), to perform energy minimization. We calculated the exact gradients of the functionals for the optimization and used the conjugate gradient method of TAO.

Conclusion

The collaboration between architect and math department proved to be very satisfactory for both parties: The architects did provide specific questions related to real world projects whilst the mathematicians were able to translate these questions into mathematical formulas and provided meaningful results that could not have been achieved alternatively. A common design framework such as Rhino and the basic knowledge of NURBS geometry and modelling techniques proved to be of essential importance for the successful collaboration between the teams.

As a result of this collaboration, we presented a method for homogeneous periodic panelization of NURBS surface geometry of cylinder type. A natural future development is the design of suitable support structures possibly with torsion free nodes. This can be derived from the panel layout by intersecting the panel planes. Furthermore the quantization of edge lengths of regular hexagonal panels is not yet fully explored. A denser distribution of quantized edge lengths in regions with great edge length variance can possibly improve the layout and number of different panels. For a later stage of the project one could add optimization for the gaps between the panels, which has not

(continued)

been incorporated yet. It would also be interesting to look at more extreme examples, even if those might not fit the architectural context.

We also started to look into how patterns can be applied across multiple surface patches, such that the pattern aligns at the crease where the patches meet. This would necessitate to develop methods for more general mappings and the applications of multiple boundary conditions; A field that is yet to be developed.

While the method presented does require a certain level of expertise from the user, we see the potential of this post-panelization strategy of freeform surfaces in the architectural design process: Opposed to carefully constrained (parametric) modelling approaches this method allows to realize even distributed, homogenous panelizations on arbitrary surfaces with cylinder topology at early design stages - something a lot of designers are interested in.

Acknowledgements We would like to thank Boris Springborn for sharing his knowledge on discrete conformal maps and the anonymous referees for their comments. Thilo Rörig and Stefan Sechelmann are supported by SFB/TR 109: Discretization in Geometry and Dynamics and DFG Research Center MATHEON.

References

Alexa, M., Cohen-Or, D., Levin, D.: As-rigid-as-possible shape interpolation. In: Proceedings of the 27th Annual Conference on Computer Graphics and Interactive Techniques (SIGGRAPH '00), New Orleans, pp. 157–164. ACM/Addison-Wesley, New York (2000)

Balay, S., Brown, J., Buschelman, K., Gropp, W.D., Kaushik, D., Knepley, M.G., McInnes, L.C., Smith, B.F., Zhang, H.: PETSc Web page (2011). http://www.mcs.anl.gov/petsc

Benson, S., McInnes, L. C., Moré, J., Munson, T., Sarich, J.: TAO user manual (revision 1.9) (2007). http://www.mcs.anl.gov/tao

Bo, P., Pottmann, H., Kilian, M., Wang, W., Wallner, J.: Circular arc structures. ACM Trans Graph. **30**(#101), 1–11 (2011). Proc. SIGGRAPH

Bommes, D., Zimmer, H., Kobbelt, L.: Mixed-integer quadrangulation. In ACM SIGGRAPH 2009 Papers (SIGGRAPH '09), New Orleans, pp. 77:1–77:10. ACM, New York (2009).

Glymph, J., Shelden, D., Ceccato, C., Mussel, J., Schober, H.: A parametric strategy for free-form glass structures using quadrilateral planar facets. Autom. Constr. **13**(2), 187–202 (2004). Conference of the Association for Computer Aided Design in Architecture

McNeel. Paneling tool documentation (2014). http://wiki.mcneel.com/labs/panelingtools.

Nieser, M., Palacios, J., Polthier, K., Zhang, E.: Hexagonal global parameterization of arbitrary surfaces. IEEE Trans. Vis. Comput. Graph. **18**(6), 865–878 (2012)

Schiftner, A., Höbinger, M., Wallner, J., Pottmann, H.: Packing circles and spheres on surfaces. ACM Trans. Graph. 28(5), #139,1–8 (2009). Proc. SIGGRAPH Asia

Sechelmann, S., Rörig, T.: VARYLAB web page (2013). http://www.varylab.com

Sechelmann, S., Rörig, T., Bobenko, A.I.: Quasiisothermic mesh layout. In Hesselgren, L., Sharma, S., Wallner, J., Baldassini, N., Bompas, P., Raynaud, J. (eds.) Advances in Architectural Geometry 2012, pp. 243–258. Springer, Vienna (2013)

Sommer, H.: JPETSCTAO, JNI wrapper (2010). http://jpetsctao.zwoggel.net/
Springborn, B., Schröder, P., Pinkall, U.: Conformal equivalence of triangle meshes. ACM Trans. Graph. **27**(3), 77:1–77:11 (2008)
Troche, C.: Planar hexagonal meshes by tangent plane intersection. In: Advances in Architectural Geometry 2008, pp. 57–64. Springer, Vienna (2008)
Zimmer, H., Campen, M., Herkrath, R., Kobbelt, L.: Variational tangent plane intersection for planar polygonal meshing. In: Hesselgren, L., Sharma, S., Wallner, J., Baldassini, N., Bompas, P., Raynaud, J. (eds.) Advances in Architectural Geometry 2012, pp. 319–332. Springer, Vienna (2013).

Geometrical Solution Space for Grid Structures with Double-Walled Edges

Andres Sevtsuk and Raul Kalvo

Abstract This paper introduces a method for creating double-curved grid structures made out of flat components, where fabrication is limited to only 2-dimensional cutting, making complex architectural structures accessible to a wider audience at a lower cost. The focus of the paper is to identify the limitations and to map the geometric solution-space of the method for real world construction applications. A double-walled nature of the structure enables us to significantly reduce the geometric complexity of the grid structure's nodes – instead of needing to find a combined geometric intersection for all edges meeting at a node, our solution instead requires determining a pair of adjacent planes at a time, as many times as a node's degree. But if any of these pairs of planes around a node is torsioned relative to the node's normal, then collisions might occur between different pairs of planes. This paper discusses the geometric solution-space under which such collisions are avoided, making the structural joints easy to build. As a proof of concept, we demonstrate the use of this method in a design-build pavilion that was realized at the Singapore University of Technology and Design in 2013.

1 Introduction

Grid structures are part of a larger family of thin-shell structures that have a long history of structural investigations (Mungan and Abel. 2012; Schlaich 2002, 2014). Along with funicular vaults (stone), monoliths (reinforced concrete) and membrane (cable nets, textile) shells, their effective structural properties have made grid structures attractive for constructing bridges, hangars, domes, and pavilions that require uninterrupted covered space. Grid structures can save material by using double-curved forms and concentrate more material along lines of structural thrust, thereby achieving economical, efficient and elegant structures. The modular nature of grid structures allows most of the fabrication work to be prepared

A. Sevtsuk (✉) • R. Kalvo
Singapore University of Technology and Design, 20 Dover Drive, 138682 Singapore, Singapore
e-mail: andres_sevtsuk@sutd.edu.sg; raul_kalvo@sutd.edu.sg

P. Block et al. (eds.), *Advances in Architectural Geometry 2014*,
DOI 10.1007/978-3-319-11418-7_14

in a factory, increasing the level of precision that can be achieved in such structures.

The geometric forms of membrane and funicular shell structures are dictated by the distribution of forces, where tensile structures work fully in tension and funicular shells fully in compression. Even though it is possible to cover a wide variety shapes using tensile and compression shapes (Pronk and Diminicus 2013), these structure do not provide an equivalent level of design freedom to grid structures, which can work under tension, compression as well as bending. A double-curved geometrical aspect gives grid structures more global stability and reduces material usage in comparison to structures that work predominantly in bending (e.g. wall-ceiling or column-beam structure). But in order to achieve efficient curvatures, grid structures need to contain different geometrical conditions, transitioning from a synclastic to a flat surface, for instance. Such complexities in surface curvature are typically achieved by using variable joints (Plumeyer 1995; Stephan and Collins 2007), by restricting tessellations (Pottmann et al. 2007), or by building the structure out of perpendicular beams, which is elegantly illustrated by the Metropol Parasol project in Sevilla, by J. Mayer H. architects 2011.

This paper discusses a method that allows grid structures to be constructed from a curved line network that may be regular or irregular using arbitrary n-gons while keeping all structural elements planar, allowing them to be fully fabricated on 2D cutting machines. The details of this method are discussed in (Sevtsuk and Kalvo 2014). The key benefit of the approach is that it offers great freedom in form and in the structural line-network design, while assuring that all joints and beams can be fabricated economically from two-dimensional sheet material. An additional benefit is that the structure can be entirely prefabricated and assembled in modular components on site without large space constraints and without high-precision work on site.

The geometric construction of the discussed system requires two key conditions. First, the structure needs to be composed of two parallel walls around each network edge, and second, the inner plane of each panel, facing the center of a loop, must have linear intersections with both of its immediately neighboring panels in the same loop. Both conditions are necessary to guarantee 2D fabrication. Figure 1 describes the steps involved in determining the geometry of this structure and Fig. 3 describes the naming conventions of three-dimensional elements we use below.

Figure 2 shows in plan and axonometric how a continuous grid structure is achieved through double walls around each network edge, where the planes are extruded at such angles that straight intersection lines are achieved within all interior planes meeting around a node. In the case of a single-walled solution (left), the pairs of panels on the left and right of the node cannot move any closer to the node along their own axes without starting to intersect. The single-walled solution therefore does not allow us to achieve straight intersection lines between all adjacent panels around a node, which in turn inhibits the possibility of flat-bed cutting.

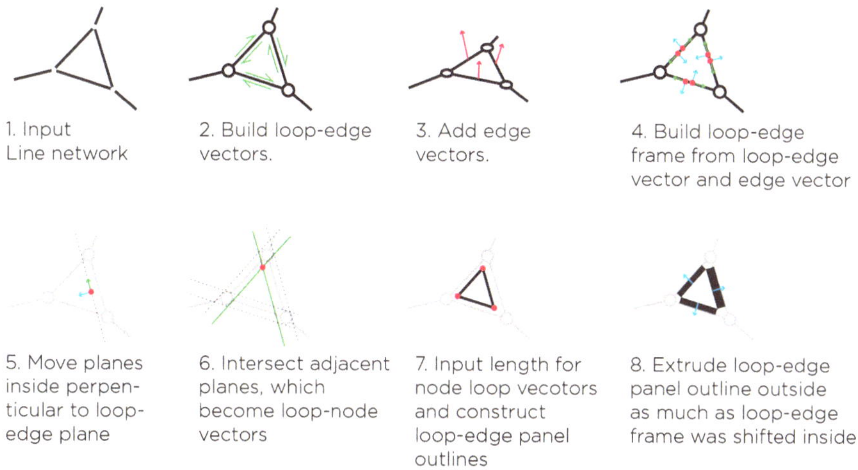

Fig. 1 Steps required for determining the geometric solution for double-walled grid structures

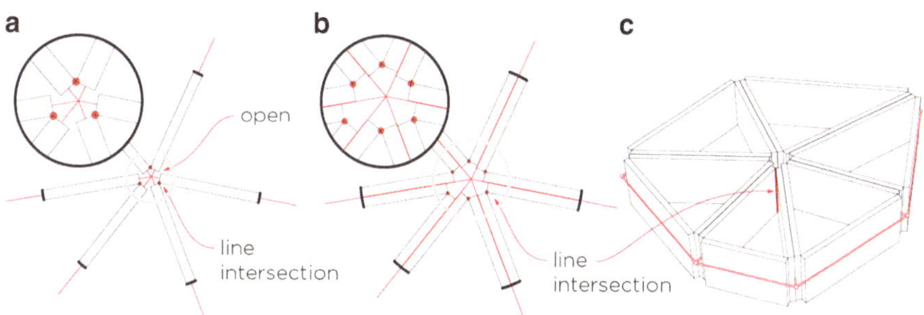

Fig. 2 *Left*: The proposed method is not feasible with single-walls on free-form line networks without creating gaps between neighboring elements or modifying the input line network. *Middle* and *Right*: A continuous grid structure is achieved through double-walls around each network edge, where the planes are extruded at such angles that straight intersection lines are achieved between all interior planes meeting around a node

The joints can be connected along a linear intersection line between two neighboring planes of grid structure's beams through a linear fastener (e.g. weld, fold, hinge etc.). These intersection lines are marked as loop-node vectors ($v_{l,n}$) in Fig. 3. The figure also introduces other notations used throughout the paper. The vertical depth of the walls becomes a structural variable that can be increased for stronger linear connections. Gaps can be formed between two parallel panels along each network edge using spacer blocks. The material thickness of the panels as well as these gaps can be useful for avoiding collisions between different panels around joints, as discussed below. The proposed solution allows a wide variety of curved line networks to be turned into a grid structures in an economical and highly flexible way.

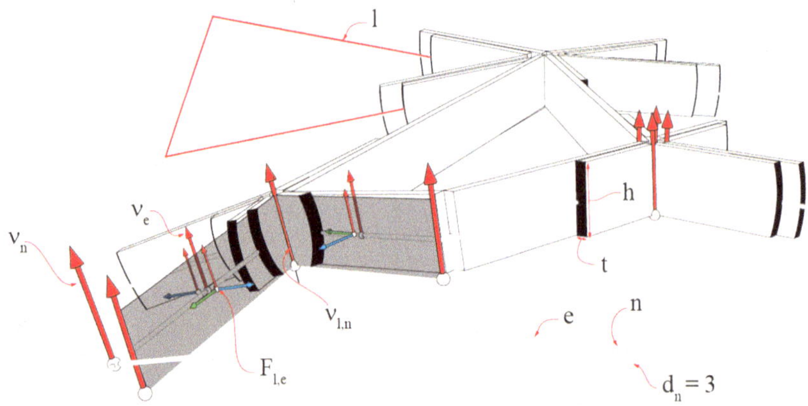

Fig. 3 Geometrical notations used: Node (n), Node degree (d_n), Node vector (υ_n), Edge (e), Edge vector (υ_e), Loop (l), Loop-node vector ($\upsilon_{l,n}$), Loop-edge frame ($F_{l,e}$). Panel height (h). Panel thickness (t)

2 Solution Space

The proposed grid structures are composed of beams with double walls along edges and a series of nodes that act as joints between different beams. The solution space of our geometric framework is the universe of geometric parameters and material dimensions under which no panels of the grid structure collide around joints. The solution space is violated when panels around node start colliding with each other, making the joints impossible to assemble. Collisions can appear for a number of reasons when a complex shape of the grid structure produces complex edge conditions around nodes (Fig. 4). It is thus important to investigating the solution space of the proposed method for two reason. First, in order to understand what geometric factors are causing collisions between panels at joints. And second, it is useful to know how geometric parameters and material dimensions can be altered so as to avoid collisions from occurring.

To find a solution space for the whole network (mesh) we start by focusing on a single node, since nodes can be solved almost independently from one another in the larger structure. Even though we show one possible sequence of determining the geometry of the grid structure loop by loop in Fig. 1, it is also possible to arrive at the same geometry edge by edge or node by node. Regardless of which way we proceed, only one geometric solution can result since only one and the same set of input lines and edge vectors can be used as the starting point. We may thus just focus on a single node.

All nodes can be described mathematically with a handful of variables, which cover all possible geometrical conditions (Fig. 5).

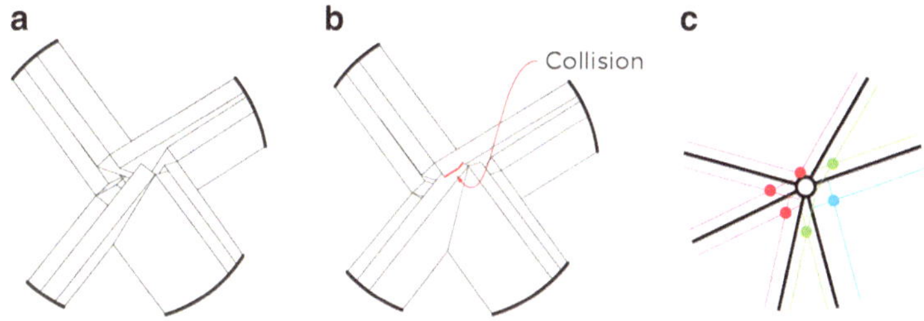

Fig. 4 (**a**). Node condition without collision. (**b**). Node condition with collision, marked with a *red line*. (**c**). When constricting the *blue* loop-node vector, both adjacent loop node vectors (*green*) are constrained not to intersect. But the blue loop-node vector is not constrained to the farther *red* loop-node vectors and therefor they can collide

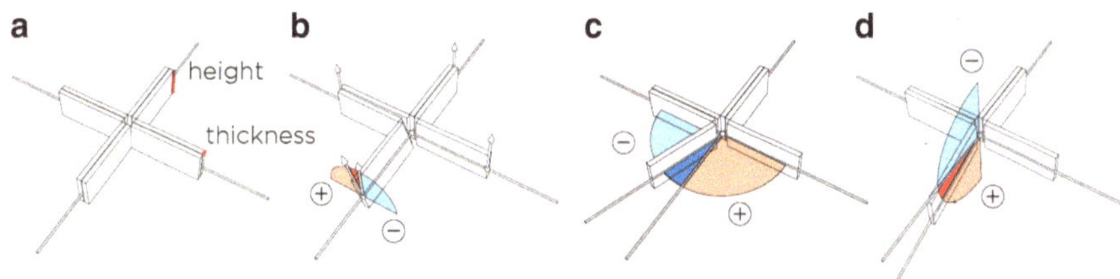

Fig. 5 (**a**). Base case of a node with degree "4", which means that four edges are connected to one node. In (**a**) none of the edges connected to the base case exhibit torsion, angle deviation or curvature. (**b**). One of the edges contains torsion. (**c**). One of the edge contains angular deviation. (**d**). One of the edges contains curvature

These variables include: a ratio between panel thickness and height (a), torsion (b), angle deviation (c), and curvature (d). Since torsion, angle deviation and curvature depend on the node's degree d_n we can collect all of them into tuples (Schneider and Eberly 2003). A node with a degree "three" has a torsion tuple t_n with components describing torsion τ_e for every edge:

$$t_n = (\mathrm{\tau}_1, \cdots \mathrm{\tau}_\mathrm{i}, \cdots \mathrm{\tau}_\mathrm{e})$$

We can apply the same idea to an angle deviation tuple a_n and its components α_e, and to the curvature tuple k_n and its components κ_e. We use φ to describe the ratio between panel thickness and height (Figs. 3 and 7). As long as the thickness-to-height ratio is constant, we can ignore the individual height and thickness values when studying the geometry around a node.

For nodes of any degree, there are three times as many variables as the node's degree plus one for the thickness-to-height ratio φ. In case of a node of degree "3" we have 10 variables, all of which can produce an effect on panel collisions. For higher degree nodes the number of variables increased linearly and there are too many variables to test individually. But for practical purposes, heuristics can be used to eliminate configurations that do not occur in realistic construction dimensions.

First, we may reduce the number of node degrees that need to be explored, since only a few degrees are commonly found in regular tessellations or mesh structures. There are three different regular tessellations – squares, regular hexagons and equilateral triangles – which respectively produce node degrees "3", "4" and "6" (Wells 1991). However, since pentagonal tessellations can also be found in architectural use, we complete the list by adding a node with degree "5".

The next heuristic simplification comes from the tuples: t_n, a_n, k_n. An exploratory study showed that if we only changed components in one tuple at a time, keeping the others constant at the base case, then only the torsion components t_n caused changes in the loop node vectors $\upsilon_{l,n}$, which can potentially cause collisions (See Figs. 3 and 6).

Angle and curvature between panels are still important in combination with torsion, but we can start by investigating collisions caused by torsion in the members. First, we need to find some realistic values for the panel thickness-to-height ratio φ, since the occurrence of collisions is related to this value (as we shall demonstrate below, Fig. 11). A base case value of $\varphi = 0.05$ corresponds in physical reality to a 12 mm thick plywood panel with a height of 240 mm. The same φ also corresponds to a 5 mm thick composite aluminum panel with a height of 100 mm or a 50 mm concrete panel with a 1,000 mm height. Figure 7 illustrates different wall thicknesses. Larger panel thickness-to-height ratios (c) tend to avoid collisions and smaller ratios (a) are more prone to collisions.

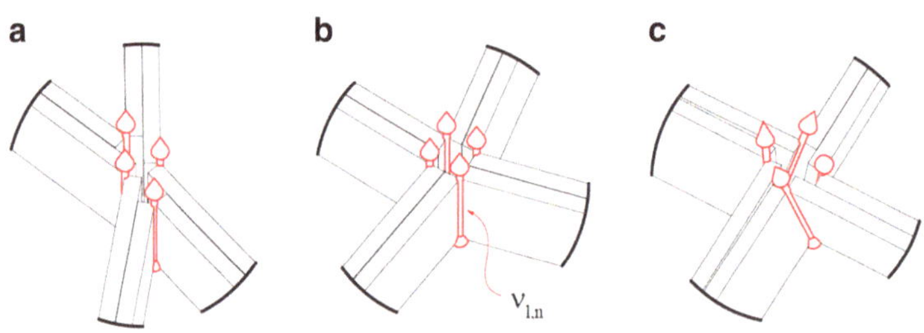

Fig. 6 (**a**). Changing values in a_n does not change $\upsilon_{l,n}$. (**b**). Changing values in k_n does not change $\upsilon_{l,n}$. (**c**). Changing values in t_n does change $\upsilon_{l,n}$

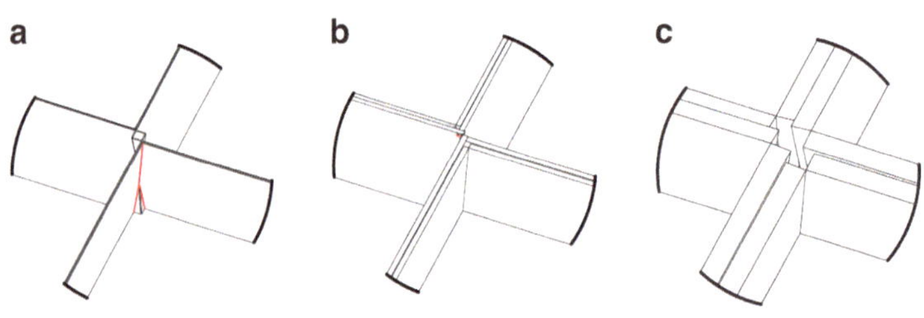

Fig. 7 Keeping $t_n = (6°, -6°, 6°, -6°)$ constant and changing $\varphi = (0.01, 0.05, 0.2)$ we see that smaller φ values produce collisions in configurations (**a**) and (**b**) (marked with *red*)

The first important, albeit intuitive, finding is that the simplest way to avoid collisions is to increase the panel thickness. This can be achieved by either using a thicker building material (e.g. 16 mm plywood instead of 12 mm) or by placing a spacer-block between two edges. Note that each pair of panels that are joined around a node share an intersection line on the inside of the loop, facing away from the node, as shown in Fig. 6. Each of the panels in a double-walled edge has a material thickness that can range from close to zero (e.g. a sheet of paper) to several centimeters (e.g. a concrete panel). Depending on the panel thickness, these intersection lines can start moving outward from the node, which reduces the hazard of collisions between different panels around the node. The outward offset is further extended if gaps are formed between the parallel panels. Manipulating panel thickness and the thickness of the spacer blocks between parallel panels can allow many curvatures and patterns of grid structures to be achieved collision-free (Fig. 8).

Having chosen a base value for φ and a node degree $d_n = 4$, we are ready to explore the effects of torsion on panel collisions. To do so, we automated the generation of t_n on a computer and tested a total 6,561 different combinations of torsion on panel collisions, using

$$t_n = (\tau_1, \tau_2, \tau_3, \tau_4)\,, \text{ where } \tau_i \in \left(-8^\circ, -6^\circ, -4^\circ, -2^\circ,\ 0^\circ,\ 2^\circ,\ 4^\circ, 6^\circ,\ 8^\circ\right)$$

We recorded each test in a table, where the first column showed if collision occurred and the next four columns showed the combinations of torsion in the four panels. Each time a collision was observed, we recorded the maximum magnitude in torsion among all four edges as well as the directionality of the torsion in each

Fig. 8 A complex shape and pattern of a grid structure is achieved with no collisions between panels around nodes by taking advantage of panel thickness and spacer gaps between parallel panels

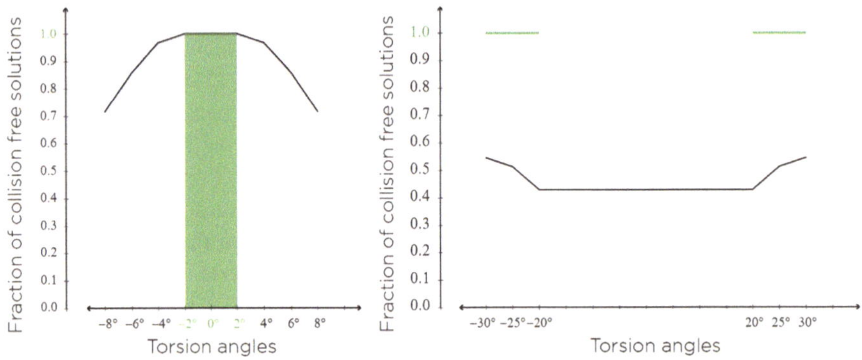

Fig. 9 *Left*: Collision simulation results for torsion. The horizontal axis denotes the maximum magnitude of torsion values among all four edges (positive and negative) and the vertical axis shows the percentage of cases where no collisions where found under such conditions. *Right*: *Lower graph* shows the same test for larger torsion magnitudes. The *upper green lines* illustrate no-collision cases with large torsion angles where all edges are torsioned in the same direction

edge. For example, a torsion combination $(-8,\ 2, 6, -2) \xrightarrow{yields} -8\ or\ (-6,\ 2, 2, 8) \xrightarrow{yields} 8\big)$. For each maximum torsion indicator, we determined the percentage of tests that resulted in a collision-free node out of the total 6,561 tests. These results are shown in Fig. 9 (Left). Plotting these results revealed a few important observations.

First, if $\tau_{\max}$ is small enough (typically $-2 < \tau_{\max} < 2$), then no collisions occur. This was expected – small torsion values do not create collisions between panels.

Second, for any given value of $\tau_{\max}$, there is no combination of $|\tau_i| < |\tau_{\max}|$ where a collision occurs. That is, collisions are always predicted by the highest absolute torsion value among all edges around a node and we do not need to keep track of the remaining lower values to detect a collision. This is important since we can now reduce the search space significantly, testing only the $t_n = (\cdots \pm |\tau_{\max,i}| \cdots)$ values. In order to make sure there is no collision for a value $|\tau_{\max}| = x$, we need to test 2^{d_n} combinations for one φ. For instance, for a node with degree "5", we need to test $2^5 = 32$ combinations. That said, it is possible to additionally use necklace combinatorics (Pólya 1937), which can reduce required combinations even further. In case of node degree "5", only 8 unique combinations are required.[1]

Third, if we limit combinations so that all components in t_n are either positive or negative, then no collisions are found. Figure 10 shows different spiral configurations achieved through torsion around a node. Collisions are avoided regardless of particular τ values, as long as they are all either positive or negative.

[1]Necklace combinatorics examines combinations in a continuous loop. For instance, combinations [a,a,b], [a,b,a], [b,a,a] are considered the same, since viewed in a continuous loop, the set does not have a start or end point.

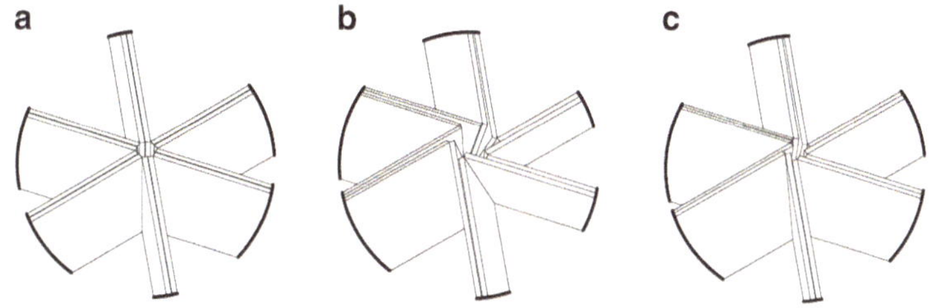

Fig. 10 (**a**) No spiral. (**b**) Constant spiral. (**c**) Variable spiral

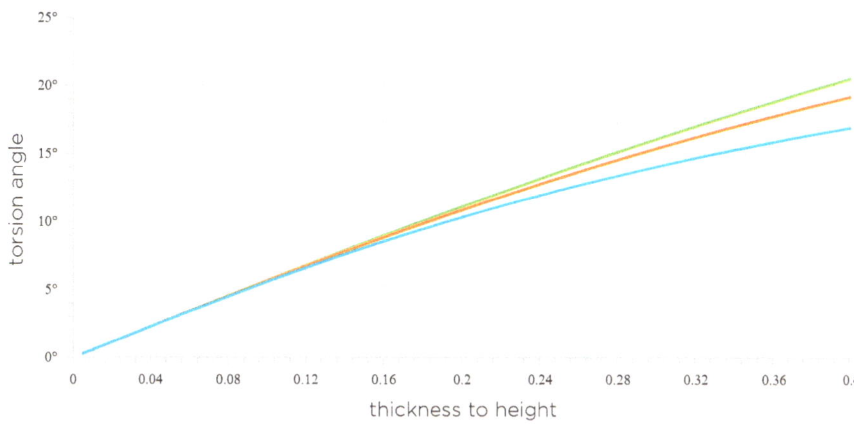

Fig. 11 Maximum collision free angle on y-axis and thickness to height ratio on x-axis. Node degree 4 (*green*), 5 (*orange*), 6 (*blue*)

This seems visually logical. To confirm this, we also conducted a series of tests with $t_n = (\tau_1, \tau_2, \tau_3, \tau_4)$, where $\tau_i \in (-30°, -25°, -20°, 20°, 25°, 30°)$. Taken by the maximum absolute torsion value alone, such large τ angles led to a lot of intersections (Fig. 9 right side, black curve), though some positive results were also found. However, when including only combinations where all τ values have the same sign, no collisions could be found at all.

We learned from the previous tests that there are some values of τ_{max} which determine whether a node with t_n faces collisions or not. We also know that this value depends on the material width-to-height ratio φ. Next, we explored the relationship between the maximum collision-free τ for different node degrees and φ ratios. In case of node degree 4, we determined the maximum τ for φ values from 0.005 to 0.4, with a test step size of 0.005. We explored a large range of thickness-to-height ratios just to see the relationship. In real world situation it is more realistic to stop around $\varphi = 0.2$, since beyond this dimension, thickness-to-height ratios appear unrealistic for real-world construction – panels whose thicknesses forms a quarter of their height or more. The results are plotted in Fig. 11, where different color graphs represent nodes of degree 4 (green), 5 (orange) and 6 (blue).

These tests reveal that if all torsion angles are kept under a certain absolute value, then no collisions are found. The trajectory of this relationship is important because it tells us that if we keep all torsion values less than the corresponding maximum absolute torsion value on the y-axis, then we can guarantee that no collisions occur, even under positive-negative torsion combinations. We denote the torsion values along this curve with $\tau_{f,\varphi}$, whose magnitudes are always positive. The value of $\tau_{f,\varphi}$ changes with φ. In other words, the collision-free maximum absolute torsion value depends on a material's thickness-to-height ratio. For example, $\tau_{f,0.05} = 2.862°$, $\tau_{f,0.01} = 0.573°$, $\tau_{f,0.1} = 5.682°$ *and* $\tau_{f,0.2} = 11.108°$. Interestingly for small thickness-to-height values we can calculate $\tau_{f,\varphi}$ for node degree 4 and use it for node degree 6. But this is not holding true for larger φ's, as shown by the spreading of the curves beyond $\varphi > 0.1$ Fig. 11.

Another important constant for different material thickness-to-height ratios is $\tau_{0,\varphi}$ (see the dash line curves in Fig. 12). If values for a particular material thickness-to-height ratio are higher than this curve, then torsion angles have to be either all positive or all negative in order to avoid collisions. There is also an operational range $\tau_{r,\varphi}$ which represents the overall limit to collision-free nodes. The proposed grid structure solution cannot handle edge torsion of 90° for instance. But in most common applications, the necessary edge torsion combinations remain well below 40° or 50°. At 50°, two adjacent edges can have a 100° difference in torsion.

We have already demonstrated that if absolute torsion angle is less than $\tau_{f,\varphi}$ then no collisions occur and any positive or negative torsion angle can be used. We also showed that if the torsion angle is greater than $\tau_{0,\varphi}$ then there only positive or negative torsion angles can be used. Now we explore what happens between $\tau_{f,\varphi}$ and $\tau_{0,\varphi}$. Figure 13 shows three graphs – green, orange and blue denoting node degrees 4, 5 and 6 respectively. We have picked one arbitrary thickness-to-height value – 0.2 in this case –, simulated positive torsion angles along the y-axis and found the

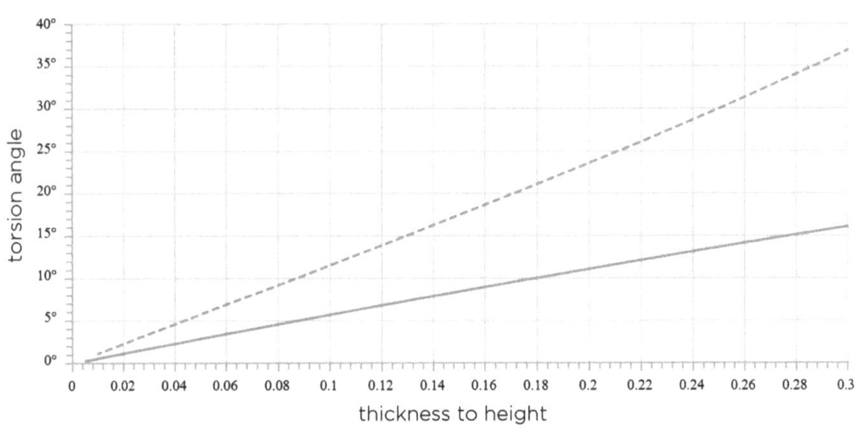

Fig. 12 Maximum collision free angle on y-axis and thickness to height ratio on x-axis. *Dashed line* denotes $\tau_{0,\varphi}$ and the *solid line* denotes $\tau_{f,\varphi}$

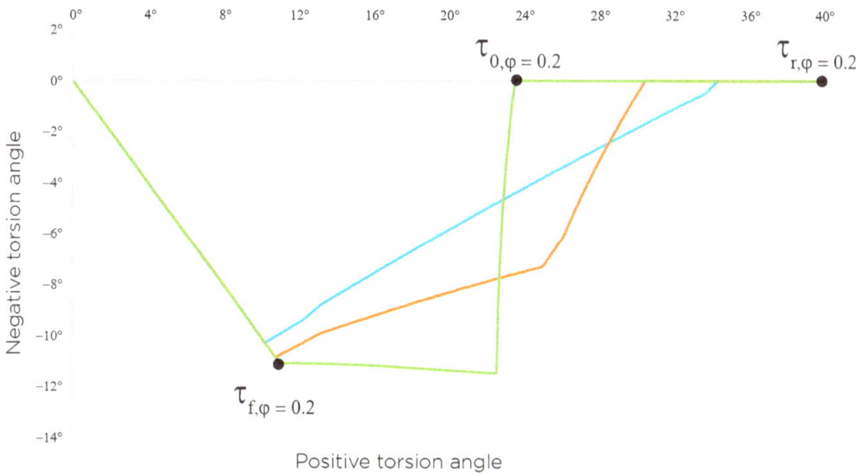

Fig. 13 Horizontal axis represents maximum positive torsion angle. Vertical axis shows the maximum negative torsion angle that corresponds to the given maximum positive torsion angle. The *green*, *orange* and *blue lines* denote node degrees 4, 5, 6 respectively. Every line can be described with three characteristic values $\tau_{f,\varphi}$, $\tau_{0,\varphi}$, $\tau_{r,\varphi}$, which divide values into a collision-free zone $\tau < \tau_{f,\varphi}$, a transmission zone $\tau_{f,\varphi} < \tau < \tau_{0,\varphi}$ and positive/negative zone $\tau > \tau_{0,\varphi}$. Note that this graph is symmetrical so that if we keep negative values on the horizontal axis then vertical axis will show maximum positive values

maximum negative corresponding torsion angle on the x-axis. For every positive τ up to $\tau_{f,\varphi}$, there is a negative τ value with the exact same absolute value (see 45° overlapping part of the graph on the left). After $\tau_{f,\varphi}$ we see that different nodes, depending on their degree, have different behavior. The maximum negative torsion value for node degree 4 (green line) keeps slowly increasing and jumps very quickly to 0 just before $\tau_{0,\varphi}$. The maximum negative torsion value for node degree 6 (blue line) starts slowly decreasing after $\tau_{f,\varphi}$. If we have a maximum positive torsion value of 20°, for instance, then for node degree 4 the maximum negative torsion angle is −11.25°, for node degree 5 it is −8.34°, and for node degree 6 it is −5.89°. If we have a node degree 4 and our maximum negative value is −11, then there is no collisions in this node.

Note that we have left out collision tests for nodes with degree 3 – e.g. hexagonal tessellations where three edges meet at each node – since degree 3 nodes do not create any collisions anyhow.[2]

We saw above that there is a certain solution space in between $\tau_{f,\varphi}$ and $\tau_{0,\varphi}$ where collisions may or may not occur, depending on the directionality of torsion,

[2]In order to see this, let us look at a node in a hexagonal tessellation as surrounded by three sectors or three loops. Since each sector shares an edge with the remaining two neighboring sectors via a double wall, no collisions are possible with the neighboring walls. At least four sectors are needed for panels to start colliding.

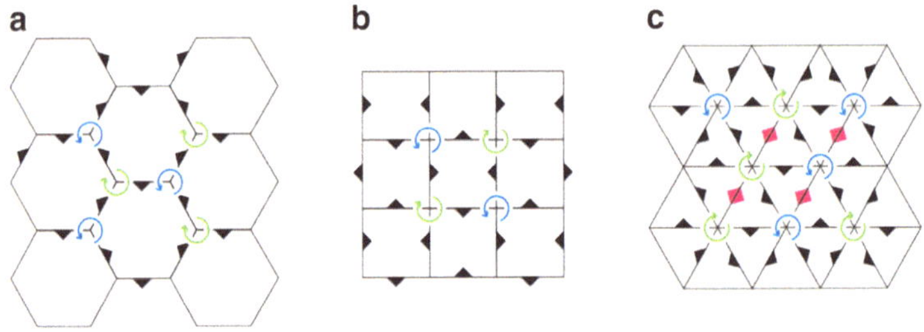

Fig. 14 Regular hexagonal (**a**), quadrangular (**b**) and triangular (**c**) tessellations and the potential use of spiraling torsion at nodes to avoid collisions. *Black arrows* indicate the torsion direction in which an edge is leaning from a top view. *Green* and *blue rotational arrows* show the direction of spirals. *Pink double arrows* indicate edges, which have conflicting leaning directions

in different edges around a node and choice of node degree. We shall now look closer at the relationship between these torsion directionalities in order to identify patterns that create collision-free solutions better than others. As mentioned above, spiral configurations, where all edges are torsioned in the same direction, allow large degrees of torsion with no collisions. It is useful to know this spiral effect, but it has practical applicability as a fix to collisions in real-world structures. It is not possible to use spiraling nodes, for instance, in continuous tessellations with node degrees 5 and 6, as shown in Fig. 14. Generally, the spiraling solutions is limited to local fixes if the tessellation has an uneven number of edges around each loop or face.

The most collision-prone patterns we found involve torsion combinations where adjacent panels fluctuate between positive and negative torsion degrees, as illustrated in Fig. 15. The figure shows how collisions are less likely when positive and negative torsion values are arranged in successive edge groups and more likely when positive and negatives alternate from one edge to the next. The worst-case pattern for collisions occurs when edge torsions alternate with extreme values on neighboring edges: $[\cdots \tau_{max}, -\tau_{max}, \tau_{max,} -\tau_{max} \cdots]$.

2.1 *Collisions Caused by Angles and Curvature*

The exploration of solution space for the double-walled grid structures showed that edge torsion was the greatest potential cause for panel collisions (see diagram in Fig. 6). Without torsion, edge angularity and curvature around a node do not produce any collisions alone. But edge angularity and curvature could still deteriorate or ameliorate the collision space if combined with torsion.

To test the effects of edge angles on nodes whose edges already have torsion, we created a dataset with a node-degree 4, which has the following parameters: $\tau_{max} = (2.8°, 2.5°, 2.0°, 1.5°, 1.0°)$, $\varphi = 0.05$ and altered the rotational angles of edges at one-degree intervals (Fig. 16). We set a constant edge rotation angle α, which increases from $1°, 2°$ up to $20°$ on the vertical axis and tested rotations in each

Fig. 15 All possible combinations using necklace combinatorics for node degree 5. *Blue* indicates negative τ and *orange* positive τ values for individal edges. *Red* diagrams denote cases where panel collisions occur

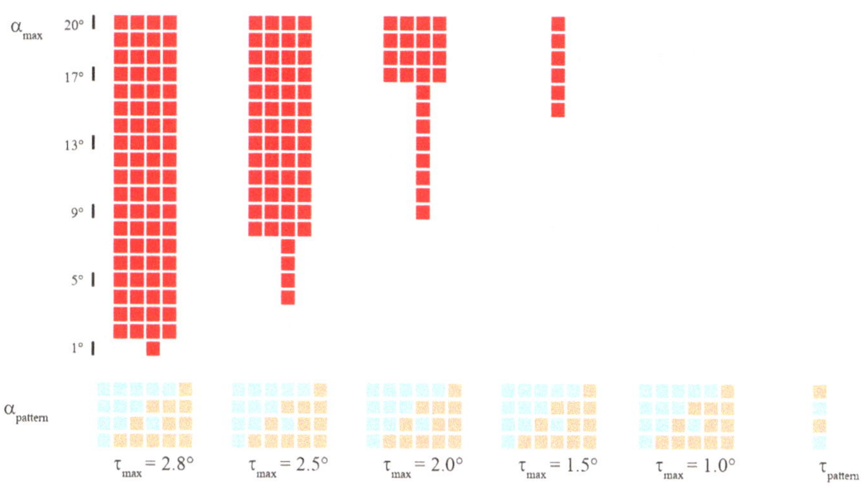

Fig. 16 *Bottom* part of the diagram shows patterns used for angles. *Blue boxes* indicate negative and *orange* positive angle directions. *Far right* column indicates the torsion directionality pattern. Every column simulates one torsion value for six different combinations of panel angle degrees, which are repeated for each degree from 1° to 20°. *Red squares* denote combinations with collisions, *gray* with no collisions. Thickness-to-height value was 0.05 and base case for $\tau_{f,0.05} = 2.5°$

of the four edges around the node in a positive or negative direction according to the given rotation angle α. In the first column $\tau_{max} = 2.8°$ there are six distinct sub-columns, where the blue and orange colors denote all possible rotation directions with the given angle α: blue squares indicate edges that were rotated in the negative direction using α and orange squares indicate edges that were rotated in the positive direction using α. The six sub-columns under each τ_{max} value exhaust all possible rotational combinations for four edges. For all of these tests, we used the worst torsion pattern, where panels have alternating strongly negative and strongly positive torsion angles $[\tau_{max}, -\tau_{max}, \tau_{max,}\ -\tau_{max}]$ (as shown in the last blue and orange column marked $\tau_{pattern}$ in the far right). The red and gray boxes show results, indicating whether collisions occurred (red) or not (gray).

These test results show that it does matter, which directional pattern the edge angles follow (see the blue and orange patterns under each column in Fig. 16). As with torsion, most collisions occur if neighboring edges alternate between negative and positive angles. If all four edges have the same rotation angle in the same direction, then the whole node rotates without affecting the collision space (left most and right most cells in each column are gray, indicating no collisions). If the four edges are rotated in different directions, then collision in $\tau_{f,\varphi}$ are indeed affected. Depending on the angular direction pattern, collisions occur more frequently for edges with alternating angle directions than grouped angle directions.

But the findings also allow us to detect the solution space where angular rotations create no collisions at all despite the presence of torsion. The second column in Fig. 16 illustrates that a four-edge node with very thin panels (small thickness-to-height ratio) $\tau_{f,0.05} = 2.5°$ can contain edge rotation angles up to $-8°$ and $8°$ and stay collision free as long as these rotation angles are not alternating between each neighboring edge, but rather grouped sequentially by the direction of the rotation. For instance, both angular sequences (−8°,−8°, 8°,8°) and (−8°,8°, 8°,8°) are collision free even under the worst alternating torsion pattern.

Finally, we also performed similar studies on the effects of curvature (Fig. 5), but found somewhat surprisingly that curvature has no effect on collisions produced by edge torsion. This is logical at a second glance, since curvature does not affect loop-node vectors $v_{l,n}$ and therefor also does not create any additional collisions (see Fig. 6).

Conclusions

The paper has investigated a new method for realizing grid structures, which relies on double walls along each network edge. A double-walled nature enables us to significantly reduce the geometric complexity of the grid structure's nodes – instead of needing to find a combined geometric intersection for all edges meeting at a node, our solution instead requires determining a pair of adjacent planes at a time, as many times as a node's degree.

(continued)

The key benefit of this method is that it makes grid structures possible on a wide variety of curved surfaces and on a large selection of line-network patterns with relatively simple geometric operations. The entire structure can be made out of planar components that are fabricated only using two-dimensional cutters, such as laser-cutters, blade-cutters, 3-axis routers, etc., making the fabrication significantly cheaper than traditional grid structures.

While the method is fairly robust and flexible with thicker panels (or thinner panels with spacer blocks between them), collisions can occur between edges around a node when panels are thin or the geometries of the nodes complex. Such collisions can prevent the realization of the structure due to joints that do not fit. It was therefore important to investigate the solution space that exists for such structures without any collisions under different geometrical and material parameters.

First, we found that the easiest solution to avoiding panel collisions around nodes involves adding thickness to the panels (or placing spacer blocks between parallel panels). This simple move can make relatively complex structural geometries collision free (Figs. 17 and 18). But this fix is not always applicable – for either engineering or aesthetic reasons, small thickness-to-height ratios may be important or desirable for a structure.

Second, we found that panel collisions around nodes are almost entirely caused by torsion in the edges. The angles and curvature of intersecting edges alone had no effect on collisions, they mattered only in conjunction with torsional panels. The paper thus mainly investigated the effects of torsion on collisions. We found the following.

Fig. 17 Completed SUTD gridshell project by the SUTD City Form Lab and ARUP engineering

(continued)

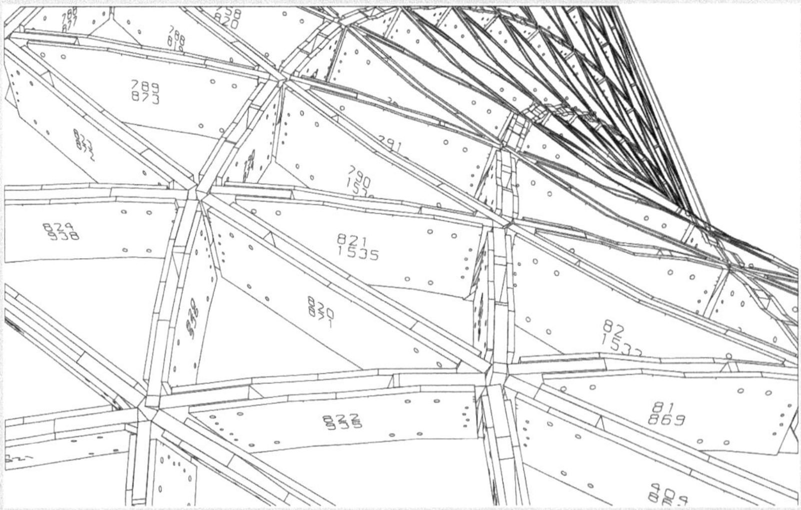

Fig. 18 Close up view from 3D model

The maximum magnitude of torsion observed among all edges converging on a node has a significant effect on collisions. The absolute value of the most torsioned edge $|\tau|_{max} = \max(|\tau_1|, \cdots |\tau_i|, \cdots |\tau_e|)$ in t_n can alone predict whether or not collisions occur. If the highest torsion angle does not create collision, then none of the lower torsion angles in edges around the same node $|\tau_{i \in t_n}| < |\tau|_{\max}$ produce collision.

A maximal collision-free torsion magnitude $|\tau|_{\mathrm{m}}$ for a node of degree $d_n = i$ applies very similarly to any other node degree $d_n \in [4, \mathbb{I}]$. This means that if we find a collision-free torsion limit for a node with four edges ($d_n = 4$), then we can use the same limit for $d_n = 6$ and so on.

There are three distinct zones in the collision-free solution space for torsion values under all material thickness-to-height ratios. First, smaller values under $\tau_{f,\varphi}$, which can contain any combination and direction of torsioned panels until the limiting torsion angle $\tau_{f,\varphi}$. Second, large torsion values above $\tau_{0,\varphi}$ where panels around a node must all be torsioned as a spiral in the same direction (either negative or positive direction). And third, collision-free solutions can also be found with intermediary torsion angles between $\tau_{f,\varphi}$ and $\tau_{0,\varphi}$ where the pattern of torsion directionality becomes decisive for collisions. For instance, an alternating patter of maximum torsion – minimum torsion – maximum torsion between adjacent edges around a node is most collision prone, but if the edges with the same torsion direction are grouped next to each other, collisions can be avoided. Particular numeric limits for allowable

(continued)

torsions angles for each material thickness-to-height ratio were determined as part of the tests. Torsion angles between $\tau_{f,\varphi}$ and $\tau_{0,\varphi}$ are closely related to node degrees.

These findings make it possible to detect potential collisions at a grid-structure's nodes computationally, using the torsion, angle, curvature and material thickness-to-height ratios. Furthermore, the findings can also help solve these collisions computationally, by adjusting the edge geometries around each node such that all nodes fall in the solution space. Whereas we have so far produced a RhinoPython library for generating the above grid-structures, it will remain a research and development task for the future to include automated collision detections and collision fixes in the library.

The presented method of grid structure generation was used to build an experimental pavilion at the Singapore University of Technology and Design (Fig. 17). At the time of construction, the intricacies of the geometric solution spaces were not known to us. As a result, we relied heavily on increasing the gap size between panels in order to avoid collisions at nodes. The new findings can enable analogous structures to be created out of very thin panels with no gaps between them.

References

Mungan, I., Abel, J.: Toward lightness in concrete: some 20th century shells and bridges. J. IASS. **53**, 75–82 (2012)

Plumeyer, K.: Joint connection-system for planar or three-dimensional trusses. US Patent 5,399,043 (1995)

Pólya, G.: Kombinatorische Anzahlbestimmungen für Gruppen, Graphen und chemische Verbindungen. Acta. Mathematica. **68**(1), 145–254 (1937)

Pottmann, H., Liu, Y., Wallner, J., Bobenko, A., Wang, W.: Geometry of multi-layer freeform structures for architecture. ACM Trans. Graph. **27**(3), 65 (2007)

Pronk, A., Diminicus, M.: 84 ways to manipulate a membrane. J. IASS. **612**, 257–270 (2013)

Schlaich, J.: On some recent lightweight structures. J. IASS. **43**(139), 69–79 (2002)

Schlaich, J.: On Architects and Engineers, Shell Structures for Architecture – form Finding and Optimization, pp. VII–XI. Routledge (2014)

Schneider, P., Eberly, D.: Geometric Tools for Computer Graphics, pp. 14–15. Morgan Kaufmann Publishers, Amsterdam (2003)

Sevtsuk A., Kalvo, R.: A Freeform Surface Fabrication Method with 2D Cutting. In: Proceedings of symposium on simulation for architecture and urban design, SimAud 2014, pp. 109–116. Tampa (2014)

Stephan, S., Collins, I.: Multiple node junction structure. US Patent App. 11/487,113 (2007)

Wells, D.: The Penguin Dictionary of Curious and Interesting Geometry. Penguin, London (1991). pp. 121, 213, and 226–227

Designing Symmetric Derivatives of the Miura-ori

Pooya Sareh and Simon D. Guest

Abstract The Miura fold pattern, or the Miura-ori, is a flat-foldable origami tessellation with a wide range of engineering applications. In particular, the Miura-ori has been applied to the folding of deployable structures for various architectural applications, e.g. folding roofs and shelters. In recent years, researchers have proposed variations on the Miura-ori which change both geometry and functionality of the pattern.

We introduce a framework for the symmetric generalisation of the Miura-ori. We study the Miura crease pattern as a 'pmg' wallpaper pattern. We reduce the symmetry of the Miura-ori to obtain new patterns while preserving the flat-foldability condition at each node. We conclude that we are able to use the Miura-ori to systematically design a variety of novel patterns, through appropriate design variations on the original pattern.

1 Introduction

The Miura-ori (2006) is a developable, flat-foldable and rigid-foldableorigami tessellation which has been applied to the folding of engineering structures for various architectural applications such as folding roofs and shelters. In recent years, researchers have proposed variations on the Miura-ori which change both geometry and functionality of the pattern (Nojima 2002; Tachi 2009; Sareh and Guest 2012). As a 2D tessellation, it can be studied from a symmetry perspective.

Mathematicians have shown (see, e.g., Schwarzenberger (1974)) that there are exactly 17 distinct *plane symmetry groups*, also known as *wallpaper groups*. Each group has a unique *unit cell*, and a certain number of *symmetry elements*, placed in certain positions relative to the unit cell. The symmetry elements include centres of *rotation*, axes of *reflection*, and axes of *glide reflection*. Schattschneider (1978) presents the international notation for the 17 plane symmetry groups.

P. Sareh (✉) • S.D. Guest
Department of Engineering, University of Cambridge, Trumpington Street, CB2 1PZ Cambridge, UK
e-mail: pooya.sareh@gmail.com; sdg@eng.cam.ac.uk

P. Block et al. (eds.), *Advances in Architectural Geometry 2014*,
DOI 10.1007/978-3-319-11418-7_15

This paper develops a geometric framework for the symmetric generalisation of the Miura-ori. There are two general assumptions we use throughout this study. Firstly, the 'fold direction' or 'mountain-valley assignment' of the fold lines does not affect the symmetry of a pattern; secondly, two crease patterns are considered to be 'similar' if a uniform scaling or rigid motion of one of them can make it coincide with the other one.

In the next section, the symmetry of the Miura-ori is investigated.

2 Symmetry of the Miura Fold Pattern

The symmetry of an infinitely repeated Miura crease pattern is *pmg* in the international notation. For a typical Miura-ori crease pattern shown in Fig. 1a, the 'starting parallelogram' of the pattern, P, from which the entire pattern can be generated by symmetry operations, is shaded in green. It has an acute angle α, and two side lengths b and l, where b is along the horizontal lines of the pattern, i.e. along the y-direction.

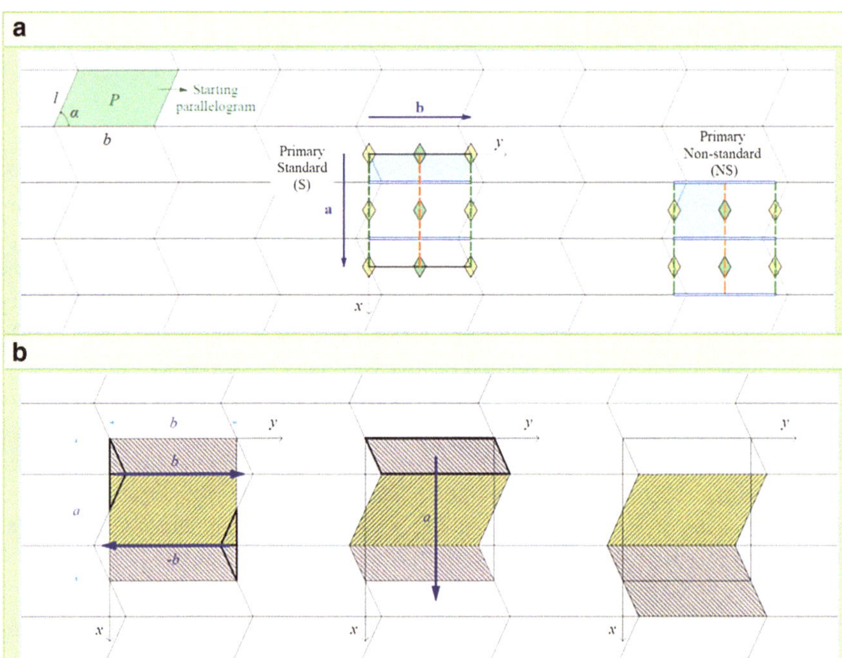

Fig. 1 (**a**) Two choices for the smallest *pmg* unit cell for the Miura-ori. Double and *dashed lines* represent reflection and glide reflection axes, respectively. 2-fold axes are shown by rhombuses. The *blue shaded* area shows the fundamental region of the pattern. The starting parallelogram of the pattern is shaded in *green*; (**b**) recomposition of the Miura-ori unit cell to parallelograms

Two choices for the smallest *pmg* unit cell for the Miura-ori are shown in Fig. 1a. Only the unit cell on the left matches the standard *pmg* unit cell used in the International Tables for Crystallography (Hahn 2005). The blue shaded area shows the 'fundamental region' of the pattern. Different colours for a symmetry element represent different classes of that element in the pattern.

We primarily use the unit cell shown on the left along with its axes throughout this work. We call this the *primary standard* choice for the smallest unit cell of the Miura-ori.

From Fig. 1b we can see that the smallest unit cell of the Miura-ori contains two parallelograms. The left hand section of this figure shows how to move the triangular fragments (bordered by bold lines) formed by the crease lines and the borders of the unit cell to obtain the figure in the middle; it is then sufficient to translate the bordered parallelogram on top of the middle figure in the x-direction to form two complete parallelograms. We use this recomposition process to find out the number of quadrilaterals within the unit cell of a crease pattern throughout this study.

The recomposition process gives us the number of quadrilaterals 'in each direction' in the unit cell of a pattern. As Fig. 1b shows, the smallest unit cell for the Miura-ori contains two parallelograms in the x-direction and one parallelogram in the y-direction. We call this crease pattern a $pmg_{2,1}$ pattern. We classify the crease patterns obtained by applying variations on the Miura fold pattern using the following definition:

Definition A tessellating convex quadrilateral mesh designed by displacing the nodes of the Miura fold pattern is called $G_{i,j}$, where G is the name of its maximal plane symmetry group, and i and j are the number of quadrilaterals in the x- and y-directions, respectively, within the unit cell of the pattern. The y-direction is the direction of the parallel fold lines in the Miura fold pattern before applying variations.

We use this definition throughout this study. In the next section, we present example patterns generated using this framework.

3 Example Patterns

3.1 Isomorphic Variations

Applying the proposed variations to the Miura-ori, we have developed an extensive family of isomorphic derivatives for this pattern. The derivations of some of these variations have already been presented in Sareh and Guest (2013). In the sake of brevity, we do not explain the design details here, but we only present an example fold pattern designed using the framework.

Figure 2 shows an example for a flat-foldable $pmg_{6,2}$ variation of the Miura fold pattern alongside the assigned mountains and valleys. The pattern consists of three

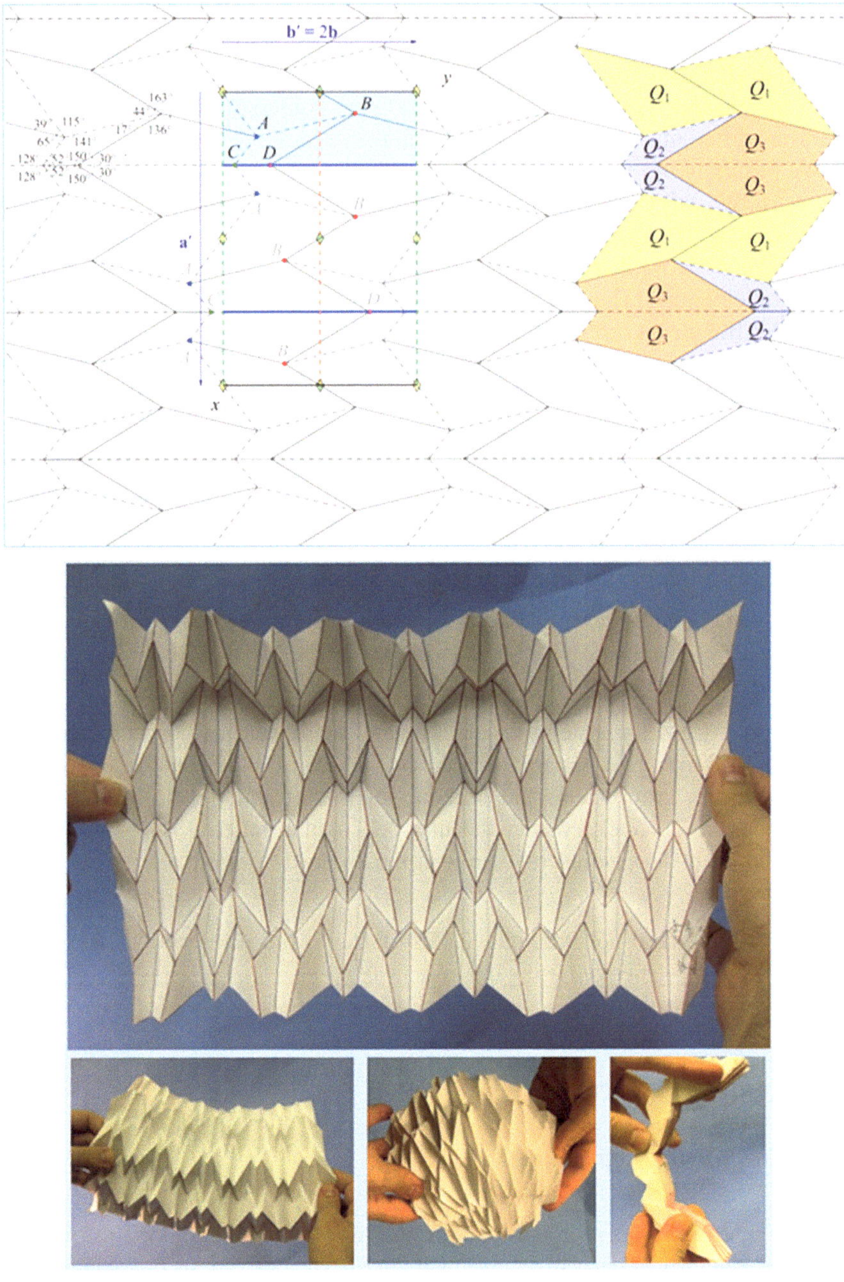

Fig. 2 *Top*: an example for a flat-foldable $pmg_{6,2}$ variation of the Miura-ori. It consists of three different starting convex quadrilaterals, shown as Q_1, Q_2 and Q_3. *Solid lines* show mountain fold lines, while *dashed lines* represent valley fold lines; *bottom*: a cardboard model of the fold pattern depicted in the previous figure; the picture on the *bottom right* shows the model in its flat-folded condition. The other pictures show the model in a partially folded state, deformed or non-deformed

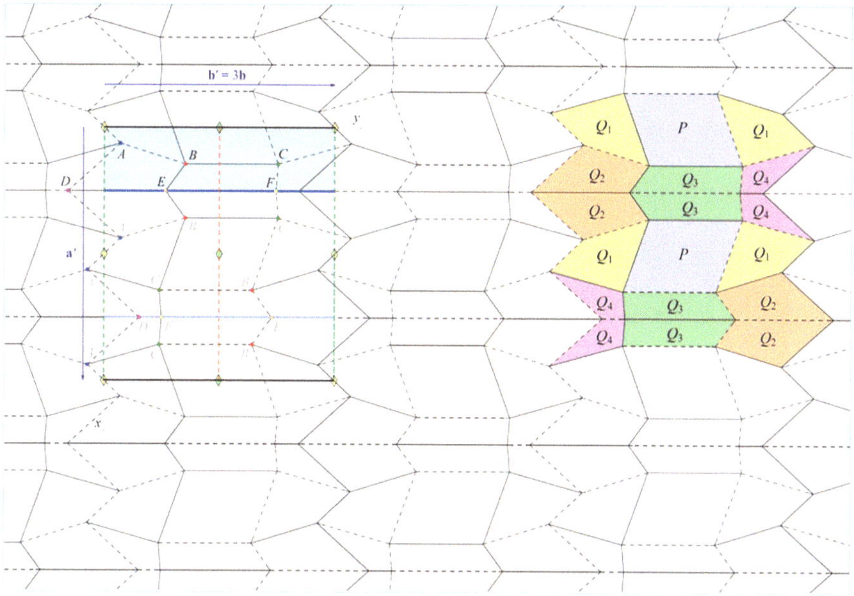

Fig. 3 An example for a flat-foldable $pmg_{6,3}$ variation of the Miura-ori. It consists of 4 different starting convex quadrilaterals, shown as Q_1, Q_2, Q_3 and Q_4, as well as a single parallelogram. *Solid lines* show mountain fold lines, while *dashed lines* represent valley fold lines

different starting convex quadrilaterals, shown as Q_1, Q_2 and Q_3. The lower part of the figure shows a cardboard model of the fold pattern; the picture on the bottom right shows the model in its flat-folded condition. The other pictures show the model in a partially folded state, deformed or non-deformed. As we see in this figure, the model is globally curved in the y-direction.

Figure 3 depicts a $pmg_{6,3}$ variation of the Miura-ori which consists of four distinct starting quadrilaterals as well as a single starting parallelogram. The folding process of the pattern shown in Fig. 3 is represented in Fig. 4a using a cardboard model. Figure 4b shows the profile of the folded pattern which reveals that the tessellation is globally planar. Figure 4c depicts the pattern in a partially folded state with a light source behind it.

Generalising the Miura-ori isomorphically, we have shown that in a $pmg_{m,n}$ variation of the Miura-ori, where $m = 2(2p-1)$, the following statements are valid (p and q are natural numbers):

- If n is even, i.e. $n = 2q$, there are $q(2p-1)$ distinct convex quadrilaterals in the fold pattern. In the case $m = 2$, i.e. $p = 1$, all the convex quadrilaterals must be trapezoids.
- If n is odd, i.e. $n = 2q - 1$, there are $q(2p-1) - p$ distinct convex quadrilaterals in the fold pattern in addition to a single parallelogram. In the case $m = 2$, i.e. $p = 1$, all the convex quadrilaterals must be trapezoids. In the case $m = 2$, i.e. $p = 1$, and $n = 1$, i.e. $q = 1$, the pattern is the Miura-ori which consists of a single parallelogram.

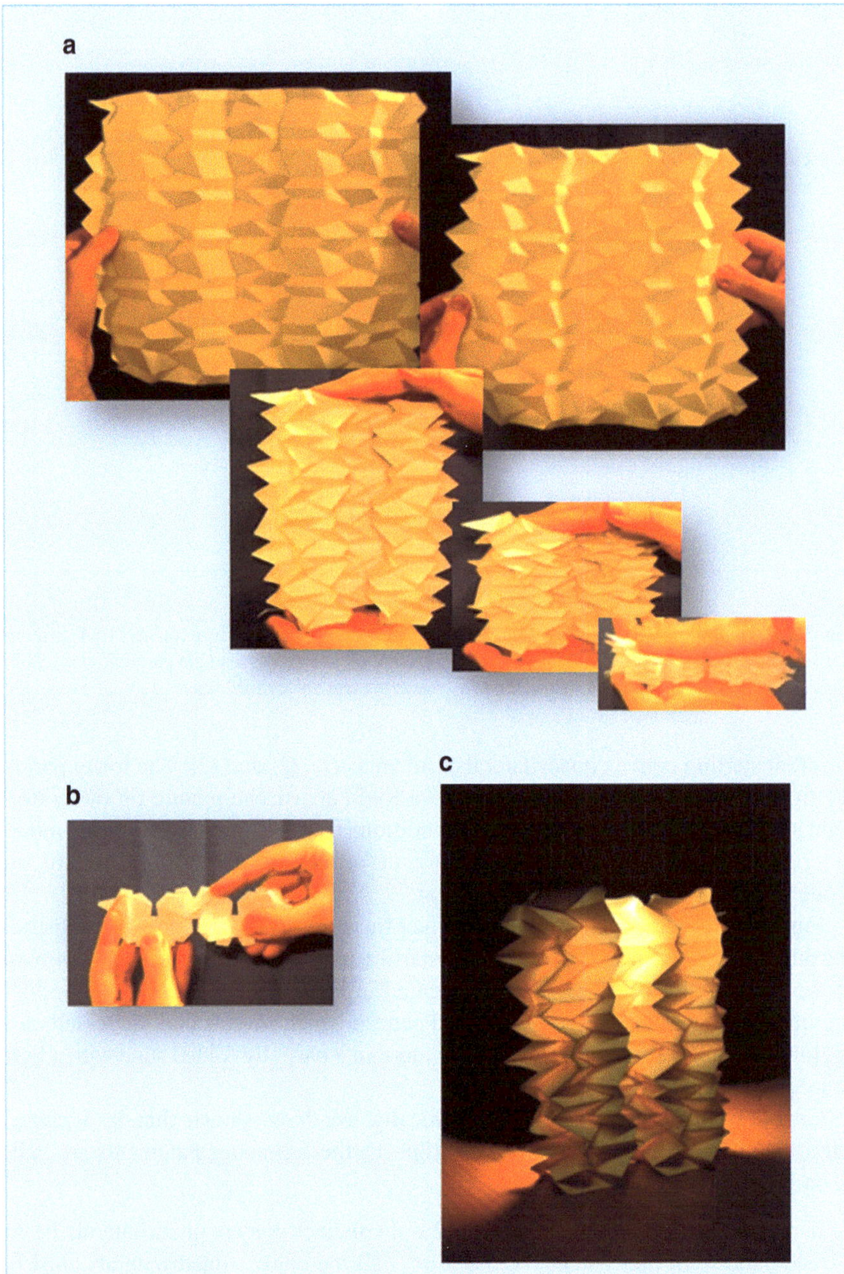

Fig. 4 (**a**) The folding process of a cardboard model of the fold pattern depicted in the previous figure;(**b**) the profile of the model in the fully-folded state; (**c**) the model in a partially folded state with a light source behind it

3.2 Non-isomorphic Variations

Applying the proposed variations to the Miura-ori, we have developed an extensive 'non-isomorphic family' for this fold pattern (see Sareh (2014) for details). Again we start with the Miura-ori, but reduce the symmetry by migrating from *pmg* to its subgroups, which may also include the enlargement of its unit cell. For brevity, we only present an example pattern for the non-isomorphic variations of the Miura fold pattern. Figure 5 shows an example for a flat-foldable $pgg_{6,2}$ variation of the Miura fold pattern alongside the assigned mountains and valleys. The pattern consists of three different starting trapezoids, shown as T_1, T_2 and T_3.

The $pgg_{6,2}$ variation of the Miura-ori is a globally planar pattern, without having any (globally) straight longitudinal polyline. In other words, we have managed to replace the straight longitudinal lines in the Miura fold pattern by a pair of alternating polylines, while preserving the global planarity of the pattern. Figure 6 shows a polypropylene model of the fold pattern depicted in Fig. 5.

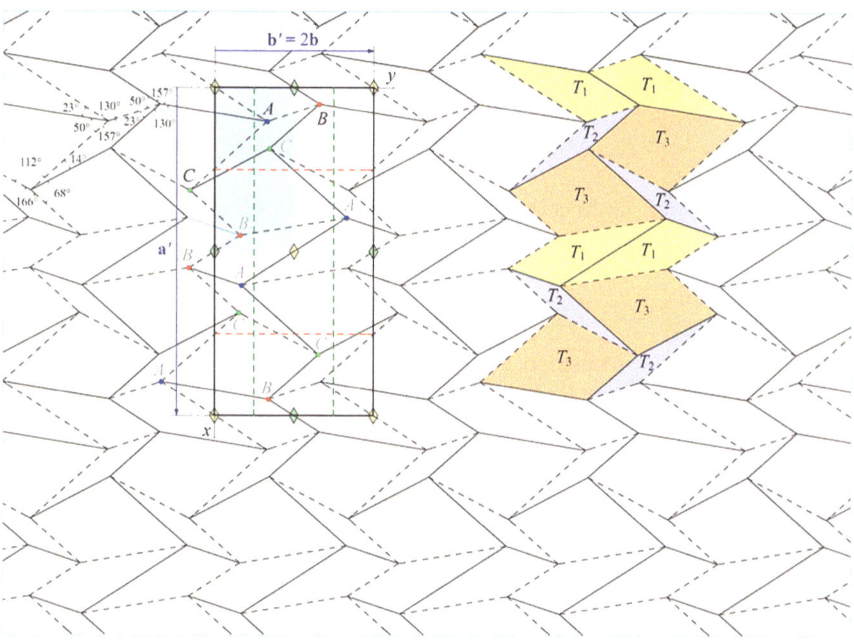

Fig. 5 An example for a flat-foldable $pgg_{6,2}$ variation of the Miura-ori. It consists of three different starting trapezoids, shown as T_1, T_2 and T_3. *Solid lines* show mountain fold lines, while *dashed lines* represent valley fold lines

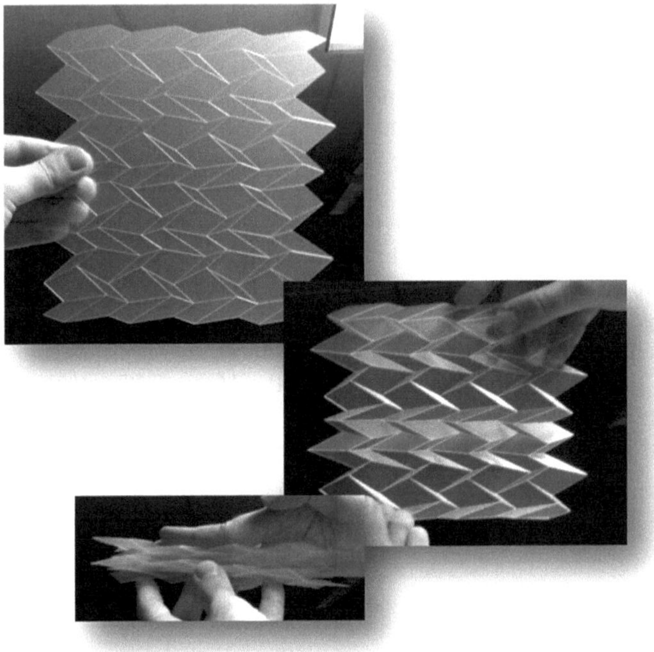

Fig. 6 A polypropylene model of the fold pattern depicted in the previous figure

Conclusions

Starting with the Miura-ori as a flat-foldable *pmg* wallpaper pattern, we developed a framework for the systematic generation of symmetric derivatives of this pattern. We looked at the Miura-ori as a globally planar origami tessellation based on a single starting parallelogram. Using wallpaper groups, we showed that through appropriate design variations on the original pattern, it can be used to systematically design flat-foldable tessellations with a variety of quadrilateral facets, which are either globally planar, or globally curved.

References

Hahn, T.: International Tables for Crystallography, Volume A: Space-Group Symmetry. Springer, Dordrecht (2005)

Miura, K.: The science of Miura-ori: a review. In: Pasadena, C.A. (ed.) Fourth International Conference on Origami in Science, Mathematics, and Education (4OSME), Pasadena, California (2006)

Nojima, T.: Modelling of folding patterns in flat membranes and cylinders by origami. JSME Int. J. C **45**(1), 364–370 (2002)

Sareh, P., Guest, S. D.: Tessellating variations on the Miura fold pattern. In: ASS-APCS Symposium, Seoul (2012)

Sareh, P., Guest, S. D.: Minimal isomorphic symmetric variations on the Miura fold pattern. In: The First International Conference on Transformable Architecture (Transformable 2013), Seville (2013)

Sareh, P.: Symmetric descendants of the Miura-ori. Ph.D. Dissertation, Engineering Department, University of Cambridge, UK (2014)

Schattschneider, D.: The plane symmetry groups: their recognition and notation. Am. Math. Mon. **85**(6), 439–450 (1978)

Schwarzenberger, R.L.E.: The 17 plane symmetry groups. Math. Gaz. **58**, 123–131 (1974)

Tachi, T.: Generalization of rigid-foldable quadrilateral-mesh origami. In: Proceedings of the International Association for Shell and Spatial Structures (IASS), Valencia (2009)

Algorithmic Optimization of the Cross-Section Distribution Across a Steel Framework Structure

Lucas Lombard, Jérôme Lalande, and François Consigny

Abstract The main goal of the present paper is to introduce an algorithmic method that can be used in order to automatically optimize large spanning truss structures, such as stadium roofs and other large steel frameworks. Implementing such a method allows design to be freed from its dependence on the engineer's work accuracy, not to mention the time saved. Therefore, it has proven to be very useful when it comes to comparing the influence of many parameters within a short lapse of time, allowing several people to work on the same project and compare results.

1 Introduction

The design method illustrated in this article primarily applies to large spanning truss structures, featuring thousands of steel bars. When working on such a project, choosing the right cross-section for each bar can prove itself to be a very time-consuming activity, which could sometimes lead to under-optimized cross-section distributions due to being pressed for time.

During Preliminary Project Design phases, one wants to be able to compare as accurately as possible as many different options as required, in order to find the most suitable geometry/structural scheme. This can be carried out simply by having a qualitative look at the behavior of each structure, and trust the engineer's experience to pick the right one. Sometimes however, a quantitative comparison may be required. When needed, comparing the resulting weight of all the structures after optimization is probably what gives the best glimpse of the performance gap between two options. In this case, it is essential to use a method that is able to size the models quickly (especially when there is a lot of models to compare) with the exact same precision so that the comparison is meaningful. As far as Preliminary Project Design phases are concerned, we look for efficiency rather

L. Lombard (✉) • F. Consigny
Egis Concept, 4 rue Dolorès Ibarruri, TSA 80006, F93188 Montreuil Cedex, France
e-mail: l.lombard@elioth.fr

J. Lalande
ENPC, 4 rue Dolorès Ibarruri, TSA 80006, F93188 Montreuil Cedex, France

P. Block et al. (eds.), *Advances in Architectural Geometry 2014*,
DOI 10.1007/978-3-319-11418-7_16

than comprehensiveness, and therefore the algorithm shown here opts for increased calculation speed and model simplicity rather than the most tricky design code rules.

If the problematic had to be summed up in one sentence, it would be the following one: How can we choose the best combination of steel cross-sections among the ones available in the catalogue, in order to minimize the weight of a structure when the geometry has been set?

In a perfect world, one could just set the perfect cross-section for each bar (i.e. the lightest one able to withstand the current efforts in the bar). However, that would lead to a collection of tenths, perhaps hundreds of different sections, which is not even realistic. On the other hand, picking manually a few sections from the catalogue appears to be a really laborious task.

We will go through a method that uses dynamic programming to solve the problem, implemented through a VBA code. That way, it can perform iterative optimization until the cross-section distribution has converged.

2 Dynamic Programming and Iterative Design Process

2.1 Dynamic Programming

The method itself is based on several assumptions that restrain its range of use:

- Design optimization has to care only about stresses and instabilities, not about displacement criteria (indeed it is quite hard to know which bars need to be reinforced when a deformation criteria is not satisfied)
- The available cross-sections of the catalogue must be strictly sorted by resistance (i.e. a cross-section classified as more resistant than another one must be more resistant according to all criteria, including buckling concerns)

The first step is to get the effort distribution in the model, as a result out of a finite-elements calculation program. Then we get a combination of efforts (Axial force, shear force y and z, bending moment y and z, torsion moment if any). From these efforts, an optimal cross-section can be defined, as the lightest section satisfying to all the resistance criteria, according to the considered effort combination. The result of this process can be presented as a graph, featuring the distribution of the different bars in the model, classified by ideal cross-section. At this point, all the sections available in the catalogue can be chosen. Figure 1 below represents such a graph, matching each of these ideal cross-sections (represented by its area A) with the cumulated length (L) of all the bars supporting such a cross-section.

These cross-sections (the ones that are at least ideal for one bar) form the base pool of choices available, from which the final selection will be made. Let K be the number of elements in this pool.

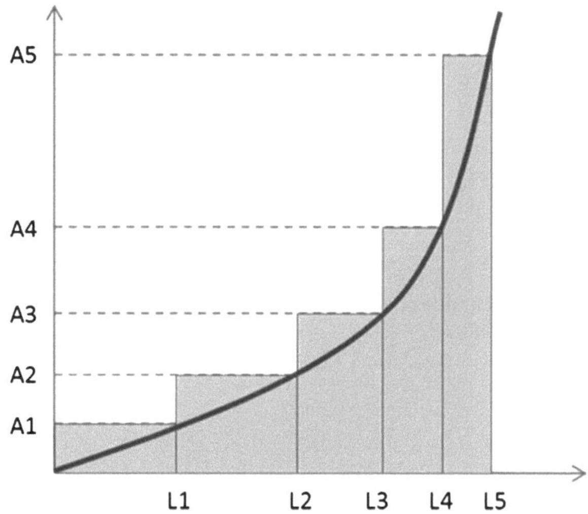

Fig. 1 Distribution graph of the optimal cross sections

Table 1 Number of operations necessary to test every solution

n\K	10	20	50	100	150
3	120	1,140	19,600	161,700	551,300
5	252	15,504	2.1E+06	7.5E+07	5.9E+08
10	1	184,756	1.0E+10	1.7E+13	1.2E+15
15	–	15,504	2.3E+12	2.5E+17	1.6E+20
20	–	1	4.7E+13	5.4E+20	3.6E+24

From this point, we have to choose the m different cross sections that will eventually compose the structure, among this pool of K possible choices. The number of theoretical different possibilities being $\binom{K}{m}$, Table 1 below offers a quick glimpse of how big this number can get, when K and n rise slightly.

As a suitable algorithm seems mandatory in order to explore the solution tree, dynamic programming proves itself to be very helpful for this kind of problem. Dynamic programming algorithms are well suited for problems involving a sequence of decisions to take, in order to minimize the cost of a function (in our case the total weight of the structure). The impact of each decision on the global cost depends on previous decisions, which makes the problem not easy to solve. The decision process can be seen as a succession of m steps to go from the initial stage to the final one, each step j matching the decision d_j. The aim of the method is to determine the best sequence of decisions $(d_1, \ldots, d_m)$ in order to minimize the global cost:

$$C = \sum_{j=1}^{m} c_j \tag{1}$$

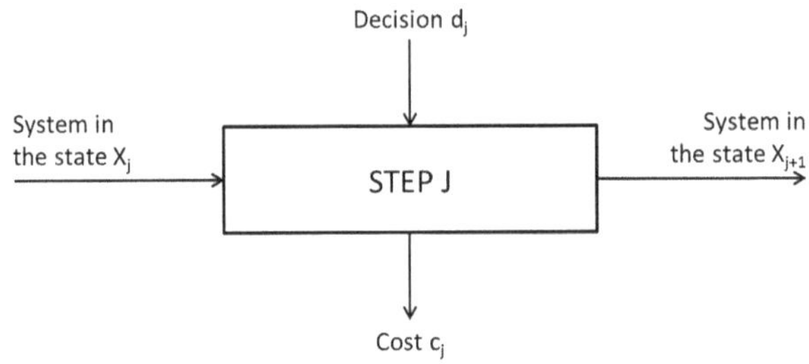

Fig. 2 Dynamic programming typical step

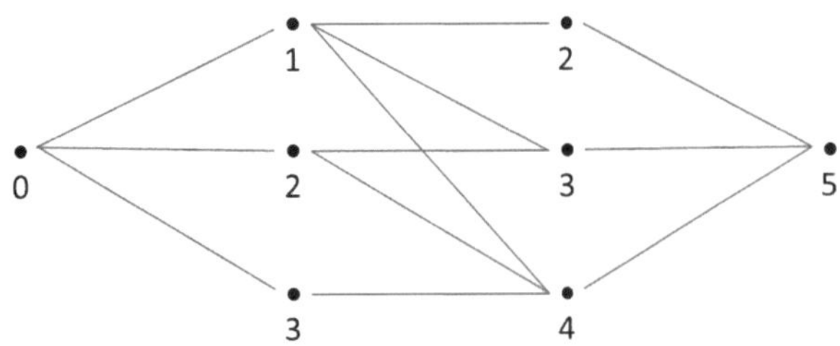

Fig. 3 Illustration of all the possible solutions

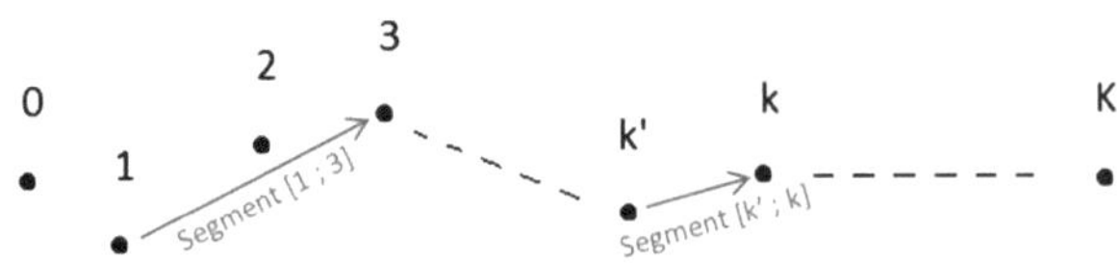

Fig. 4 Analogy with an oriented graph

This process can be illustrated as follows (Fig. 2):

This method relies on the "principle of optimality", formulated by Richard Bellman in 1957. This principle states the following:

If the decision sequence $(d_1, \ldots, d_m)$ *is optimal for a global problem involving* m *steps, then if we suppose that state* x_j *is reached after the* j*th step, the decision sequence* $(d_{j+1}, \ldots, d_m)$ *is optimal for the subproblem involving* $(m - j)$ *steps and starting at state* x_j*.*

In other words, the optimal sequence of decisions is composed of subsequences that are optimal for the corresponding sub problems. Thus a complex problem can be broken up into smaller subproblems that can be solved once and for all, their solutions being stored and re-used when needed in order to solve the larger one.

The solution space is composed of all the possible paths linking two given points (the starting point being the null cross-section and the final point being the maximum required cross section) in m steps ($m - 1$ intermediate cross-sections being chosen). The optimal solution is the shorter path (i.e. the path for which the global weight of the structure is minimal).

The solution space is illustrated in Fig. 3, in the case which the maximal required cross section is 5 and the number of steps is 3.

The list of the K sections contained in the catalogue can be sorted by their area and represented with K points on an oriented graph (Fig. 4):

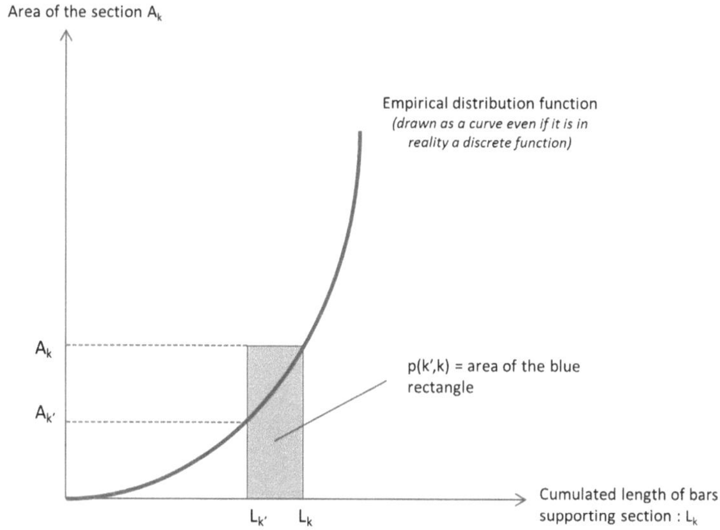

Fig. 5 Graphical visualization of the "length" $p(k', k)$

The "length" (or "cost") of the segment $[k'; k]$ is defined as the weight of all the bars for which k is a satisfying cross section but k′ is not. It can be written as:

$$p(k', k) = (L_k - L_{k'}) \cdot A_k \cdot \rho_s \tag{2}$$

This "length" can be visualized on the representation of the function $A = f(L)$, which gives the relation between the area of the cross section and the cumulated length of bars supporting this section (Fig. 5).

The total weight of the structure is then equal to the sum of the areas of the blue rectangles, illustrated in Fig. 6 for six different cross-sections:

When implementing this method through an algorithm, it actually turns into filling in two matrixes. The first one is a K^2 sized matrix, with all the $p(k'k)$ for $k' < k$.

The second one, referred to as matrix P, is a $K * m$ matrix in which each element $P(k, j)$ is equal to the "length" of the shortest path linking points 0 and k in exactly j steps (i.e. with $j-1$ intermediate points). The concrete equivalent of this definition is that $P(k, j)$ is the minimal weight of the structure with j sections chosen when the maximum required section is section k. This matrix can be filled in step by step using the dynamic programming equation (3) starting with $P(k, 0)$, then $P(k, 1)$ and so on until $P(k, m)$ ... the information of interest at the end being $P(K, m)$.

The dynamic programming equation can be written as follow:

$$P(k, j) = min(P(k', j-1) + p(k', k)) \qquad \forall k' < k \tag{3}$$

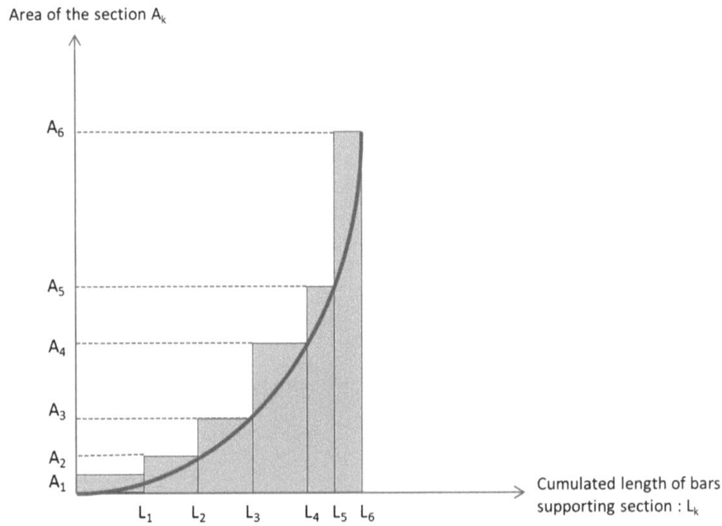

Fig. 6 Graphical visualization of the weight of the structure

Table 2 Number of operations required by the algorithm

n\K	10	20	50	100	150
3	150	600	3,750	15,000	33,750
5	250	1,000	6,250	25,000	56,250
10	500	2,000	12,500	50,000	112,500
15	–	3,000	18,750	75,000	168,750
20	–	4,000	25,000	100,000	225,000

The number of computations required to fill in such a table is $K^2 * m/2$, which is much more reasonable than $\binom{K}{m}$ as shown in Table 2.

The dynamic programming method is summed up in Fig. 7.

These results have been compared with two control methods:

- The first one chooses cross-sections so that they are equally spaced in terms of area: if A_{max} is the area of the maximal required cross section and m the number of sections we have to pick, the first section will have an area as close as possible to A_{max}/m, the second to $2 \cdot A_{max}/m$ and so on.
- The second one selects cross-sections so that the cumulated length for each section is the same: if L is the cumulated length of bars in the whole model, each section will be selected so that its cumulated length is equals to L/m.

The comparison was carried out using a simple square structure (20×20 m) made of square hollow sections, uniformly loaded so that the maximal required cross section will be the 14th in the catalogue (corresponding to a SHS 220). Six different sections have to be chosen (Fig. 8).

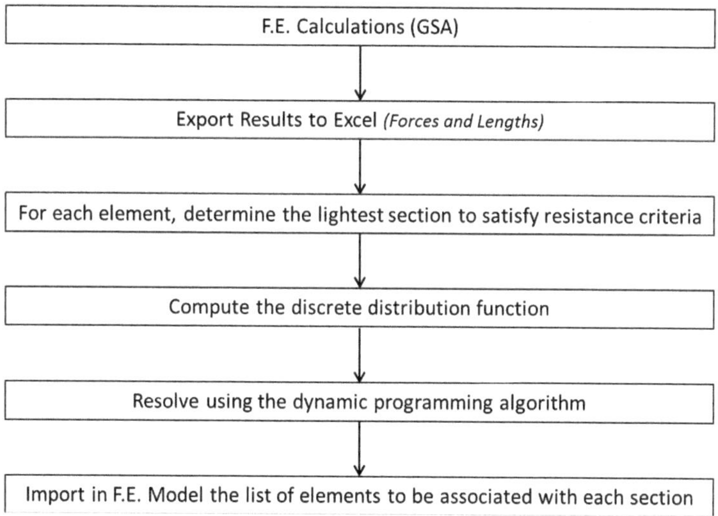

Fig. 7 Summary of the dynamic programming algorithm

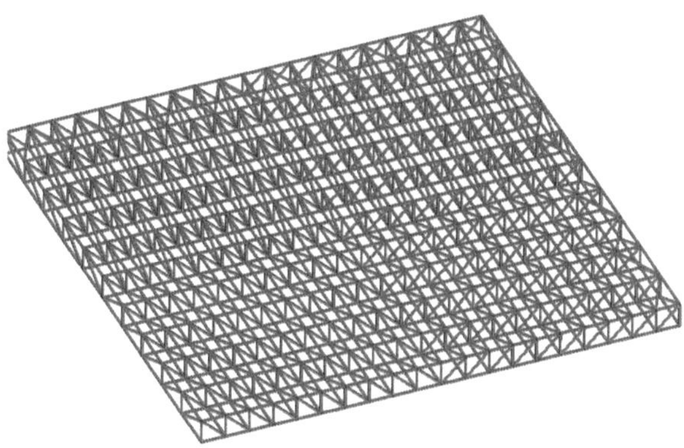

Fig. 8 Truss structure used for the comparison

Our algorithm lead to a final truss weight of 17.09 tons, while the first and second control methods respectively lead to 18.06 tons and 19.09 tons. This means a gain of 6 % or 12 % on the global weight, which can be really significant when dealing with large spanning truss structures weighing several thousand tons.

2.2 *Iterative Process*

The dynamic programming algorithm, as explained before, can find out the optimal distribution of cross-sections over the structure, for a given effort distribution.

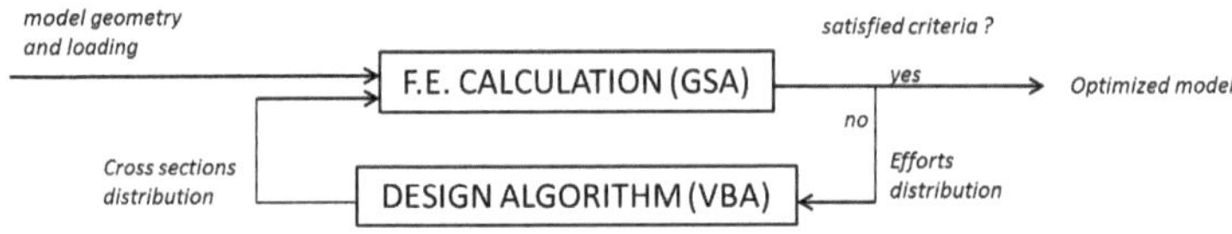

Fig. 9 Iterative process diagram

However, this newly determined cross-section distribution will make this effort distribution obsolete for two main reasons:

- Part of the loading is due to the weight of the structure itself, so changing cross-sections will affect these loads and therefore make efforts in the bars change. It is particularly true in large spanning truss structures for which self weight represents an important part of the loading.
- If the structure is hyperstatic, which is often the case for large truss structures, then the distribution of efforts depends on internal stiffness, and then on the cross-sections distribution.

This implies that we have to implement the cross-section choice into an iterative process, sketched in Fig. 9, that loops through both FEM calculation and cross-section design as many times as necessary until the distribution converges.

As far as the convergence criteria is concerned, it depends on the amount of time available for computation. The number of different solutions being finished, there will be a moment when two following steps are strictly identical, and then the distribution will then be frozen forever. However, this moment can take some time to happen. In order to reduce the computing time, it's possible to set some arbitrary criteria, such as stopping iterations when the difference between two successive steps gets upper-bounded by a certain value.

One can evaluate this difference by several means:

- Number (or %) of bars of which the cross-section has been changed during last iteration.
- Difference on the effort distribution between the two last iterations.
- Overall weight difference between the two last iterations.

On the example presented below, the last one was chosen as it avoids the risk of stopping the calculation after modifying the cross-section list, and therefore disrupted the distribution, even if the consequences on the global weight are small.

3 Testing the Model Sensibility and Application to the FFR Stadium Roof Design

The first model the method was applied to is a simple square-shaped steel framework, supported on its four sides by pin supports. The framework is 20 m in width, 1.5 m thick, with S235 steel hollow sections. Loading is composed of self weight

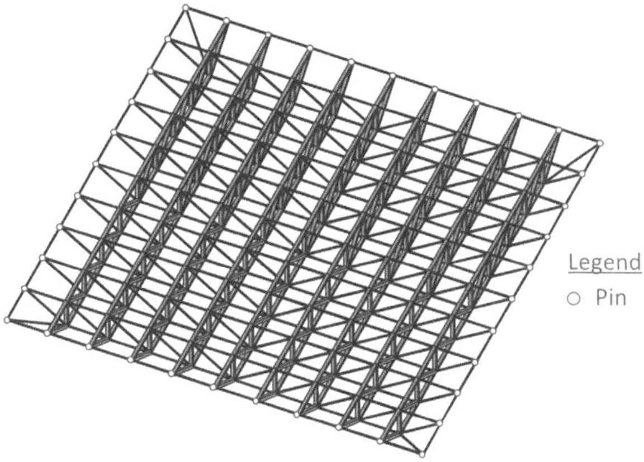

Fig. 10 Testing model illustration

and an additional live load, chosen so that the entire spectrum of cross-section can be used (Fig. 10).

3.1 *Initial Conditions*

The main goal of this test was to quantify the impact of the initial distribution on the final result. Intuitively, one can figure out that if the structure is particularly hyperstatic or if self weight is a major part of the loads, the initial cross-section distribution (set up by the engineer), will impact the convergence speed, and perhaps even the limit. Therefore, several tests were carried out on this modal, implementing different initial configurations and looking at the results.

- Testing different initial uniform distributions (same cross-section all over the model) for a model with mostly live load (gravity $\times$ 1+10 kN/m^2 live load)
- Testing different initial uniform distributions (same cross-section all over the model) for a model with mostly self weight load (gravity $\times$ 13+5 kN/m^2 live load)
- Testing random initial distributions for both models.

Figures 11 and 12 show the results for uniform initial distribution. When the self weight is secondary, each initial configuration leads to the exact same system with a global weight of 15.04 tons. But when the self weight is dominant, some differences appear and the gap between two tests can reach 4 %. Figure 12 also shows that the lightest initial systems will lead to the lightest final systems and reciprocally.

Concerning random initial distribution, a statistical distribution of the results is displayed on the graph below, for both models (Fig. 13).

We have here an expected limit weight of 16.98 tons, with a standard deviation of 0.21 tons. Based on these results, the worst found minimum (17.45 tons) is at least

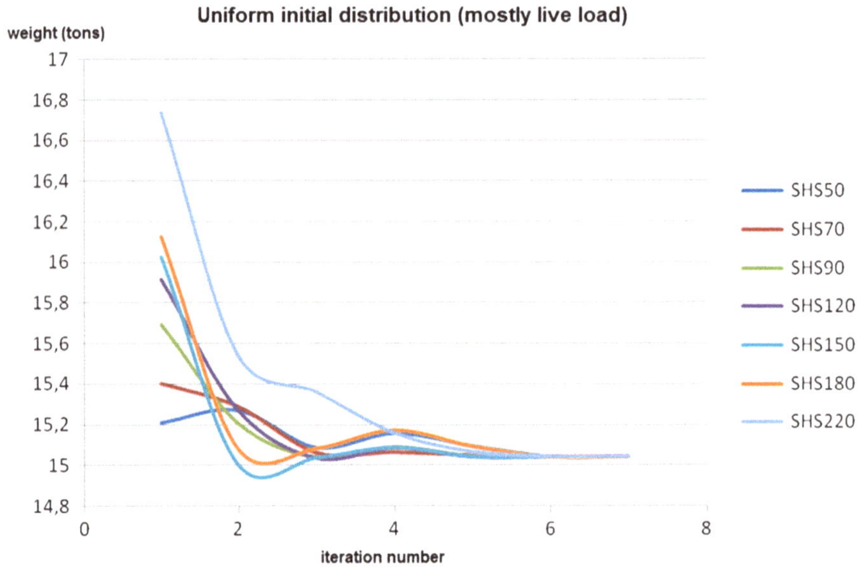

Fig. 11 Convergence starting from uniform distributions (mostly live load)

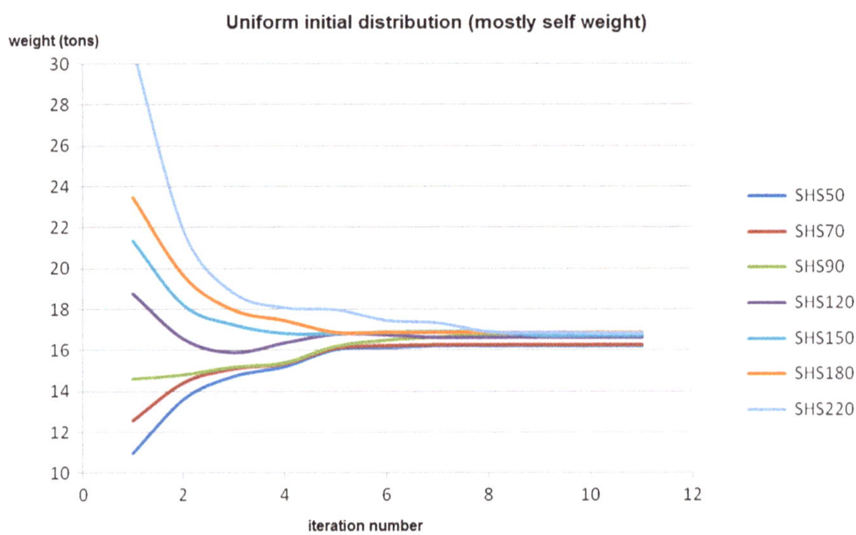

Fig. 12 Convergence starting from uniform distributions (mostly self weight)

0.95 tons far from the real minimum, since some models converged to a weight of 16.65 tons. That represents a deviation from the minimum of 5.7 % of the latter. It reveals the importance of the initial distribution in the process.

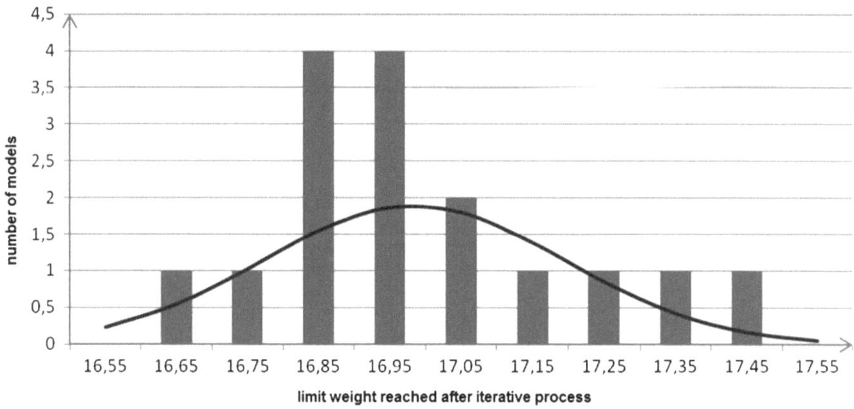

Fig. 13 Statistical distribution of the results for random initial distribution

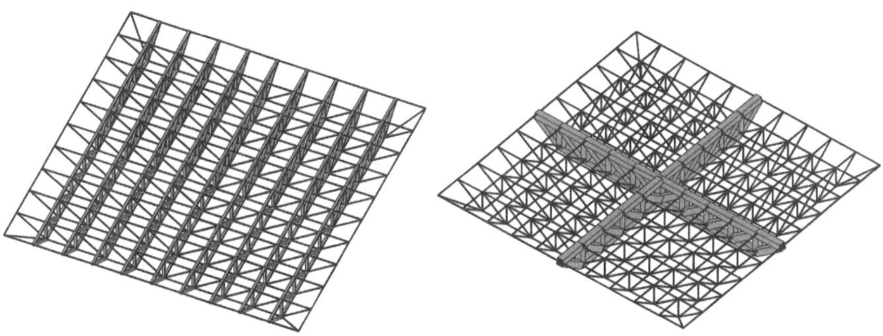

Fig. 14 Initial configuration influencing the static behavior of the truss

3.2 Initial Static Behavior

Another test was to study the influence of the initial static behavior on the final system. We chose three configurations: a bidirectional truss, an unidirectional truss and a truss supported by two crossed beams (Fig. 14).

The algorithm shows that the initial static behavior will influence the final weight of the structure, as illustrated on the figure below. Therefore, it is important for the engineer to choose an initial configuration that is as close as possible to the final system he imagines (Fig. 15).

3.3 FFR Stadium Application

This method was then applied to the optimization of geometry parameters on a real project: the roof of the future French Rugby Union stadium.

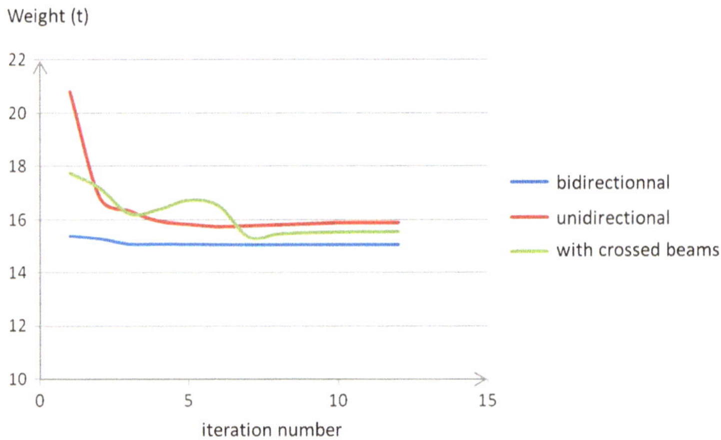

Fig. 15 Influence of the initial static behavior

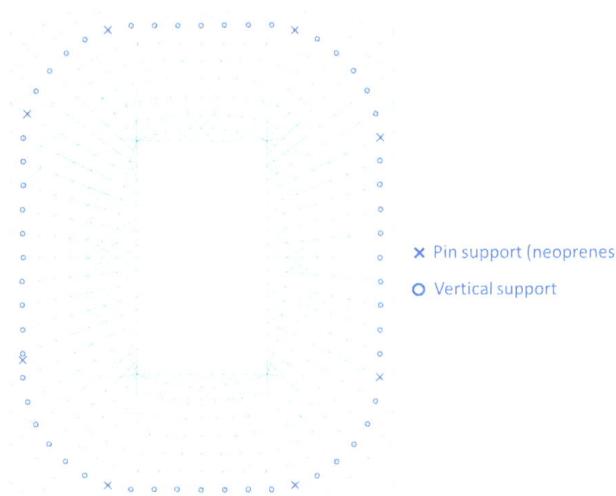

Fig. 16 Top view of the roof and supports

The roof itself consists of a square-shaped steel truss framework, following a system of parallels and meridians, as illustrates the top view on Fig. 16.

This framework works together with a tension system made of a tension ring, connected to the inside and outside edges of the framework via struts and ties (Fig. 17).

The algorithm was first use to compare different truss typologies, illustrated in Fig. 18. Since the sizing is done automatically, it is perfectly rational and so completely independent from the work of the engineer. However, we have to keep in mind that the algorithm is only optimizing thc global weight of the structure. The structures that came out of this method were then analyzed to check other criteria

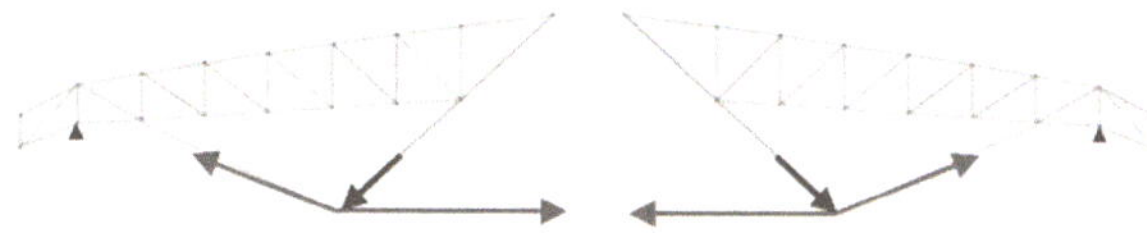

Fig. 17 Section view of the roof and tension system

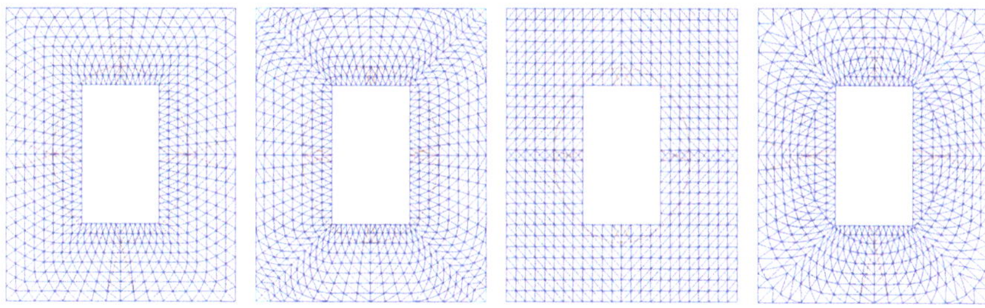

Fig. 18 Truss typologies studied

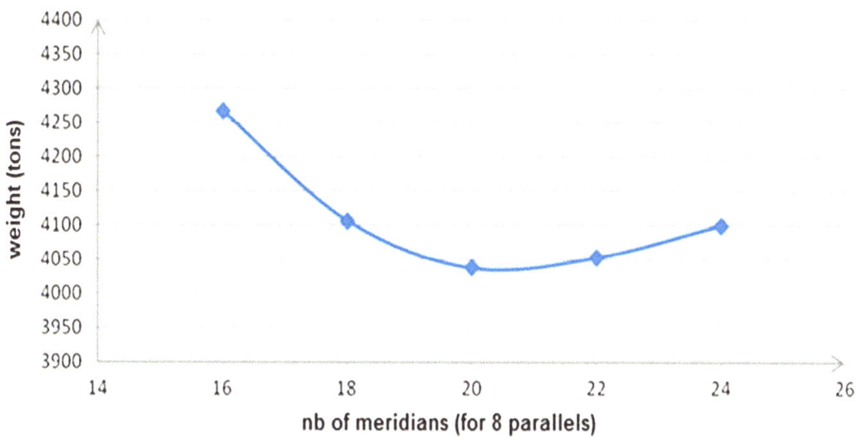

Fig. 19 Finding the number of meridians minimizing global weight

such as deformations, buckling modes and buckling load factors, dynamic behavior, construction simplicity... We finally adopted the configuration on the right.

Once the typology was chosen, the algorithm was then used to optimize geometric parameters, such as the number of meridians, the number of parallels and heights of the inner/outer rims. For each value of a parameter, an iterative process is carried out, leading to a value minimizing the structure's weight.

For example, Fig. 19 presents the optimization of the number of meridians (number of parallels being set).

Once again, the algorithm is based on one single parameter: global weight of the structure and this can lead to unrealistic results. We used an iterative process to determine the optimized number of parallels and meridians. We first analyzed what would be the optimized number of meridians for a given number of parallel (equal to 8): the minimal weight was obtained for 20 meridians. We then analyzed what would be the optimized number of parallels for a given number of meridians (equal to 20): the minimal weight was obtained for 11 meridians. We then analyzed again what would be the optimized number of meridians for a given number of parallels (equal to 11): the minimal weight was obtained for 24 meridians. As we drew the structure with 11 parallels and 24 meridians, we realized that it was unrealistic due to the number of assemblies involved and we changed our selection criteria to include this number in the process.

4 Results Discussion and Further Applications

This method is able to give relevant results in a limited lapse of time, although its precision could be improved. The dependency on the initial distribution can get quite high in the case of a structure with few extra loads (up to a 6 % deviation in the example presented above). However, in practice, it is very easy to find a reasonable initial condition, based on the way the structure should work and preliminary assessments. Moreover, it seems that starting with a weaker structure leads closer to the optimum than starting with oversized cross-sections.

Rather, the tool has been mainly used for comparison purposes, as illustrated in the example above, concerning the FFR stadium roof design. In this kind of problem, the initial distribution can be set up so that it is equivalent in each model. That way the dependency on the initial conditions is no longer relevant since relative optimal weights matter.

The main limit faced by this method is the impossibility of taking a deformation criteria into account during the cross section design step. Since each bar is designed separately, and deformation is a matter of global stiffness, it has to be checked afterwards and solved by the engineer, updating the cross-section distribution manually.

Conclusion

This design tool was created in order to allow fast preliminary design step comparisons. It proves itself to be very useful for this purpose, in spite of the limits inherent to the method itself. It can also provide a good starting point for more advanced design.

The accuracy of the design can in theory get pushed to a higher point, as the algorithm supports every design calculation worksheet. It is mainly a matter of calculation time available. However, the more criteria used, the smaller the cross section pool gets, as it needs to be strictly sorted for all criteria at the same time.

(continued)

Furthermore, the algorithmic design tool can be associated with any form finding program, such as genetic solvers, in order to perform more complex geometric optimizations. This can be done by implementing the code into any parametric 3D modeler that can loop with a finite elements program.

Planar Panelization with Extreme Repetition

Mathieu Huard, Michael Eigensatz, and Philippe Bompas

Abstract No satisfactory general solution exists today to enable high repetition of elements in architectural freeform structures. In this paper we investigate why and propose several solution approaches to design highly repetitive structures with the simplest (and usually most affordable) geometric base elements: planar polygons. We explore the potential and limitations of these approaches and show that there is a large class of structures achievable by repeating even only a single polygon. We discuss how the proposed techniques reach beyond the topics of repetition and panelization and could be part of a more diversified investigation of "freeform" in architectural geometry.

1 Introduction

Repetition is a key factor in reducing complexity and cost in the design, engineering, construction, fabrication, and logistics of an architectural project. Trying to achieve a high repetition of elements in a freeform structure, however, is a challenging endeavour. Often, fundamental issues inherent to the problem formulation even prevent a practically useful repetition: for example, if we try to reduce the number of unique panels in the given planar panelization of a freeform surface, even a mild clustering of the panels can quickly lead to significant gaps between the panels or can unpleasantly distort the aesthetical qualities of the panelization, leading to undesirable zig-zag wiggling along the panel seams (Fig. 1).

M. Huard (✉)
Evolute GmbH/Schwindgasse 4/10, A 1040 Wien, Austria
e-mail: huard@evolute.at

M. Eigensatz
Evolute GmbH, Schwindgasse 4/10, A 1040 Wien, Austria

TU Wien, Wien, Austria
e-mail: eigensatz@evolute.at

P. Bompas
RFR Group/Paris 4 rue d'Enghien 75010, Paris, France

P. Block et al. (eds.), *Advances in Architectural Geometry 2014*,
DOI 10.1007/978-3-319-11418-7_17

Fig. 1 Planar panelizations with repeating a single equilateral triangle

For this paper, we tried to approach architectural freeform structures with a high repetition from a different angle. In particular, we set our focus on the following goals:

- Extreme repetition, i.e., the use of only a handful of different elements.
- Planar polygonal elements, possessing the most basic and usually most affordable geometry.
- Exact planarity and exact repetition of the elements without tolerances, visible gaps, or aesthetical distortions.
- Flexible and interactive design possibilities, enabling both: starting from scratch (bottom-up, pre-rationalization) and from a smooth design surface (top-down, post-rationalization), or even a combination of the two.

This work started with the investigation of repetition in planar panelizations but we quickly realized that there is a bigger picture: as mentioned, the fundamental issues of repetition optimization often lie in the formulation of the problem itself. Taking a step back we realized that a major restriction comes from the requirement of smoothness. Due to the increasing popularity of NURBS modelling tools, and probably aided by the mathematical beauty of connections between discrete differential and classical differential geometry, research in architectural geometry has put a strong focus on the approximation of smooth freeform surfaces. In practice, however, architects, builders, and designers have always been intrigued by non-smooth structures as well. Even in nature, one finds appealing and inspiring non-smooth "architectures", interestingly often involving repetition. In the subsequent sections we will show that dropping the smoothness constraint is a key ingredient of our approach to designing freeform structures with extreme repetition. We believe that this concept of gaining significant degrees of freedom by considering non-smooth structures can be seen in a bigger context and deserves further investigation in architectural geometry. While structures that are not smooth in the classical sense might be more difficult to understand and control mathematically, they may offer the potential to optimize for further challenging functional or environmental objectives.

We call our results panelizations because they all describe true discrete surfaces where all polygons perfectly match up to form a single sheet, without holes or intersections. However, our results are not limited to panelizations in the strict sense. They could represent facades, space frames, mosaics, landscapes, etc. We therefore kept many of our illustrations abstract, trying to highlight the general design potential.

2 Previous Work

2.1 Practice

Working with repeating elements for practical or aesthetical reasons has a long tradition in architecture. Figure 2 lists a small subset of related work. The principle of surface discretization can be found in the earliest constructions such as the great pyramid of the Giza or the Mayan Pyramid. More than being only a technological answer to the act of fabricating and erecting huge constructions from smaller repetitive elements, these discretizations have a design potential to create new tectonics serving the building aesthetics. In the end of the fifteenth century, Ferrara Biagio Rosseti used facetted stones to define the appearance of the Palazzo dei Diamanti. More recently, crystalline or diamond like structures have been used as poetic reference. The Harpa concert hall of the Henning Larsen architects in Reykjavik and the Oxford Street in London facade by Future Systems architects are good examples of both facetization and heavy repetition. In the early 1950s, the research by Konrad Wachsman and Buckminster Fuller on large span structures led to what is now called space frames. These structures were conceived as fully prefabricated systems with a few number of component types, all identical up to their scale. They also take advantage of the structural potential of highly triangulated geometries. A more recent project, the Beijing Water Cube by PTW architects is based on a Wearie-Phelan voxelization of space, which is related to our Voxel Meshes, though no repetition was used to our knowledge. A third family of projects should be mentioned, starting with the seminal project built in Montreal, Habitat 67 by Moshe Shafdie for the world expo. Constituted of prefabricated blocks the size of a room, interlocked between one another, the habitation building is like a gigantic Lego game. Similar projects were built by representants of the Japanese avant-garde, the Metabolist group, which designed megastructures based on prefabricated elements in the early 1970s, i.e., the Capsule hotel in Tokyo of Kisho Kurokawa. Finally, the french architect Jean Renaudie had a similar approach than Habitat 67 for his habitation project in the center of the new town of Ivry built from 1971 to 1980. The building is spread horizontally and vertically in order to maximize the sun exposure and to provide inhabitants with large terraces and private gardens.

While based on different geometric techniques than presented in this paper, all of these works are based on related principles such as repetivity, prefabrication, and non-smoothness, and constitute a great source of inspiration for our work

Fig. 2 Inspiration for our work from nature and architecture. Top left to bottom right: Giant's Causeway, Ireland – Honeycomb – Beijing National Aquatics Center, PTW Architects, CSCEC, CCDI, and Arup, Bejing, China – Pyramids, Egypt – Lego lion, Nathan Sawaya – Palazzo dei Diamanti, Ferrara Biagio Rossetti, Italy – Harpa Concert Hall, Henning Larsen Architects, Iceland – Spaceship Earth (Epcot), Walt Disney Imagineering, U.S.A. – Oxford Street facade, Future Systems, England – Habitat 67, Moshe Safdie, Canada – Nakagin Capsule Tower, Kisho Kurokawa, Japan – Habitation, Jean Renaudie, France

2.2 Research

Methods for optimizing the rationalization of architectural designs have received significant attention in the recent past. A method for minimizing general production costs (Eigensatz et al. 2010) was achieved using a global optimization framework, minimizing the number of needed molds with respect to quality constraints. This approach gives good approximations for any type of initial design but focuses on panelizations with a mixture of planar, single curved, and double curved panels. The repetition is not enforced in the panel boundaries but only in the shapes of the molds generating the panels.

A few works have been more specifically focusing on reducing the number of planar panels taking into account their boundaries. This is usually done by the means of clustering methods. By operating small vertex displacements to an initial panelization, the aim is then to fit all panels within a finite number of classes (clusters), then obtaining as few unique panels as there are classes. In Fu et al. (2010), the authors do so in the case of planar quad panels, leading to visually satisfying results, but high number of clusters when one wants a faithful approximation of the initial surface. Another method working with triangle panels has been proposed in Singh and Schaefer (2010), resulting in possibly faithful approximations but often at the price of gaps or aesthetical deficiencies. In all cases, those clustering operations result in a number of unique panels $k \gg 1$. Despite constraining the results to be smooth and close to the initial surface, they will necessarily lose those properties when k becomes sufficiently small.

Designing shapes with a limited set of repetitive elements is a problem sometimes encountered when studying origami. Some specific regular tessellations of the plane lead to aesthetical folding patterns, like the waterbomb tessellation. It has been noted (Tachi 2010) that the basic equilateral tessellation of the plane can be used to design highly corrugated shapes, producing "crumpled paper"-like results.

To our knowledge, there has only been one attempt at researching architectural freeform design using only equilateral triangles. The architect Alain Lobel (http://www.equilatere.net) has studied and classified hundreds of shapes built exclusively with those triangles, which he named *Lobel Frames*. By introducing singular points and playing with edge angles, Lobel managed to create smooth-looking, visually interesting surfaces. We will discuss Lobel Frames and their relation to our work in Sect. 3.3.

Our work on equilateral triangles also has some relations to Isenburg et al. (2001), where the authors investigate what geometry information is encoded purely in the connectivity of an approximately equilateral triangle mesh.

3 Equilateral Triangle Meshes

As it represents the most elementary scenario, we start our investigation with freeform structures achievable using only a single polygon: the equilateral triangle. We will see that by repeating even just this one simplest possible polygon we can achieve a large variation of freeform panelizations. We divide the results into three different classes: first, we investigate what class of structures we can obtain when we still require smoothness and what the limits are (Sect. 3.2). Then, we extend this class by allowing a distinct set of non-smooth points in the structure, leading to semi-smooth results (Sect. 3.3). Finally we drop the smoothness entirely, reaching a more free but also less tamed class of structures (Sect. 3.4).

Before we dive into these three classes, we briefly explain the optimization framework underlying all of them in the next section. Readers more interested in applications may skip to Sect. 3.2 upon first reading.

3.1 Optimizing Equilateral Triangle Meshes

We present in this section an optimization based framework for approximating and designing shapes with equilateral triangles. Starting from an initial mesh, the aim is to apply local displacements to each initial vertex so as to converge towards a mesh matching the desired panelization criteria. This is done by minimizing the target functional

$$f = \omega_1 f_{edges} + \omega_2 f_{prox} + \omega_3 f_{fair}, \tag{1}$$

where ω_i are the weights of the different criteria. The edge length constraint is enforced by the term

$$f_{edges} = \sum_i (\text{length}(e_i) - l)^2,$$

where e_i are the edges of the optimized mesh, and l is the target common length all the edges of the result panelization should fulfill. In the context of approximation (or design using a control shape), the proximity to the reference surface is guided by

$$f_{prox} = \sum_k \text{dist}(\mathbf{x}_k, T_k)^2,$$

where $\mathbf{x}_k$ are the vertices of the optimized mesh, and T_k are the tangent planes at the points $\mathbf{y}_k$ of the reference surface which are closest to $\mathbf{x}_k$. Finally, when targeting smooth surfaces, the fairness of our result is guided by the term

$$f_{fair} = \sum_k ||\mathbf{x}_i + \mathbf{x}_j - 2\mathbf{x}_k||^2,$$

for all the pairs of opposite neighbouring vertices (i, j) of the vertex $\mathbf{x}_k$.

Each of the three classes presented in the following three sections require the minimization of the functional f, although involving different strategies regarding the weights ω_i. We minimize f using an iterative optimization algorithm based on a Gauss-Newton method, and describe in each case the initialization step, which is well-known to strongly influence the accuracy and convergence speed of such algorithms.

Gauss-Newton, being a least-squares solver, does not guarantee reaching the exact same edge lengths for all triangles. By iteratively reducing the importance of all weights except for the target edge length, however, the results can basically be arbitrarily exact down to floating-point precision. For all our results, the edge lengths are identical within practical tolerances (<0.001 mm), making the results exact equilateral structures for the architectural application.

3.2 Smooth Structures: Approximating Developable Surfaces

What can we achieve with repeating a single equilateral triangle if we require the result to be smooth? Note that when we talk about smoothness of polygonal structures we always mean smoothness in a discrete sense. For an equilateral triangle mesh to look smooth, all vertices of the panelization should have six neighbours. More or less neighbours will lead to high angles around the vertex and a "kinked" impression as shown in Fig. 3. We can easily obtain a planar shape with equilateral triangles and all six-neighbours vertices. Think of this shape as a discretization of a planar sheet of paper: we could start to design curved shapes by "bending" the initially planar shape (Fig. 3). This provides us with an immediate intuition of the set of classes achievable with equilateral triangles, namely the developable surfaces.

Our experiments show that if the triangles are small enough with respect to the curvature, we are indeed able to smoothly approximate any developable surface with an equilateral triangle mesh. This is a highly useful class of surfaces in architecture, as we know from their impressive use by architects like Frank Gehry. We are not aware that this general connection between equilateral triangle meshes and developable surfaces has been investigated so far in architecture.

To describe the connection on a more theoretical level: if all triangle vertices have six neighbours (vertices have valence 6), the mesh is said to be a regular mesh. Since all the angles equal $\frac{\pi}{3}$, the sum of incident angles to each vertex equals 2π, which means the mesh represents a *discrete developable* surface (its discrete Gaussian

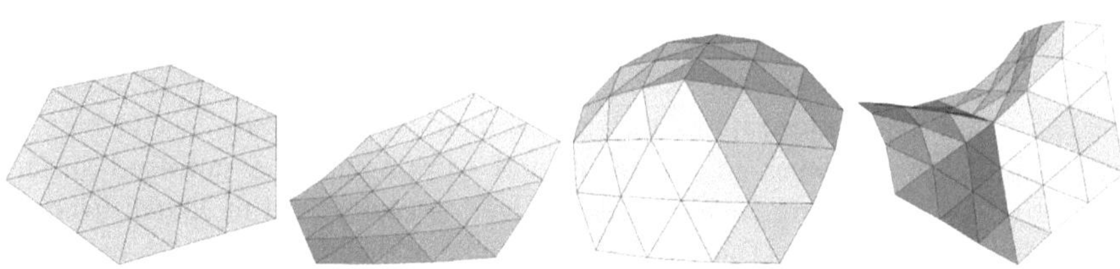

Fig. 3 Patches with equilateral triangles. Left to right: Planar tiling – A regular patch with all valence 6 vertices can "bend" to approximate smooth developable shapes – Introducing a singularity of valence <6 will automatically lead to a non-smooth impression – The same is true for singularities of valence >6

curvature is vanishing everywhere). In other words, *any* regular equilateral triangle panelization is unfoldable on the plane without stretching or cutting the panels.

Unfortunately, there is no exact equivalence between discrete and continuous developable surfaces. Not all developable surfaces can exactly be approximated by a regular equilateral triangle mesh. One main problem is that the curvature of the developable surface might get too high for the resolution of the triangle mesh to approximate it well (i.e. the triangles are too big). A second issue lies in the singular points of developable surfaces like the apex of a cone, which can be seen as an extreme case of the curvature becoming infinitely high. A third problem is posed by closed manifolds with zero or more than one boundary. When wrapping around a cone, for example, the triangles will not generally match up, leaving a seam line of unmatched triangles. Only three circular cones can be perfectly discretized by introducing a singularity at the apex, corresponding respectively to a vertex of valence 3, 4 or 5. Introducing singularities to obtain those particular cones and more general shapes is further addressed in Sect. 3.3.

If the triangles are small enough, however, and the surfaces are not closed, our tests indicate that smooth developable surfaces can be nicely approximated. Figure 4 demonstrates this for three basic developable patches: a cylinder, a cone and a more general single curved surface. It was surprising to us that, as nicely demonstrated by the cylinder example, it does not only work when the mesh edges nicely follow the rulings of the developable surface. Basically any orientation of the mesh can achieve a smooth approximation of the developable surface.

Figure 5 shows an architectural example employing equilateral triangles to smoothly approximate large scale developable surface patches. Each triangle mesh patch could be unfolded in its entirety to a planar net without distortion, which could have further practical benefits besides the fact that all panels have the same shape.

The examples in this section were obtained using our optimization framework discussed in Sect. 3.1. Starting from an initial triangular mesh, we minimize the function (1) using only the weights ω_1 and ω_2. Since we approximate smooth surfaces, it is not necessary to use the fairness term. The weights are chosen such that $\omega_1 \gg \omega_2$. This ensures that the equilateral criterion is enforced as a hard constraint, then using the closeness term to approximate as well as possible the reference surface.

In order to converge to a good solution, we initialize the optimized mesh as follows. Since we approximate developable surfaces (see Fig. 6a for an example), we can define parameterizations that are isometric to the plane. It can be seen as unfolding the surfaces in the plane and working on the unfolded surface (Fig. 6b). Distances along such a parameterization correspond to actual distances embedded on the surface. Therefore, by picking equidistant points with respect to the parameterization, we can link these points so as to form an equilateral triangle planar mesh. This mesh can easily be computed using a Delaunay triangulation. Although the vertices of the obtained initial mesh are equidistant along the surface (e.g. along the unfolded surface, see Fig. 6c), they aren't exactly in the 3D space (Fig. 6d), but it still provides an accurate initialization from which the optimization process can potentially converge to a nice result.

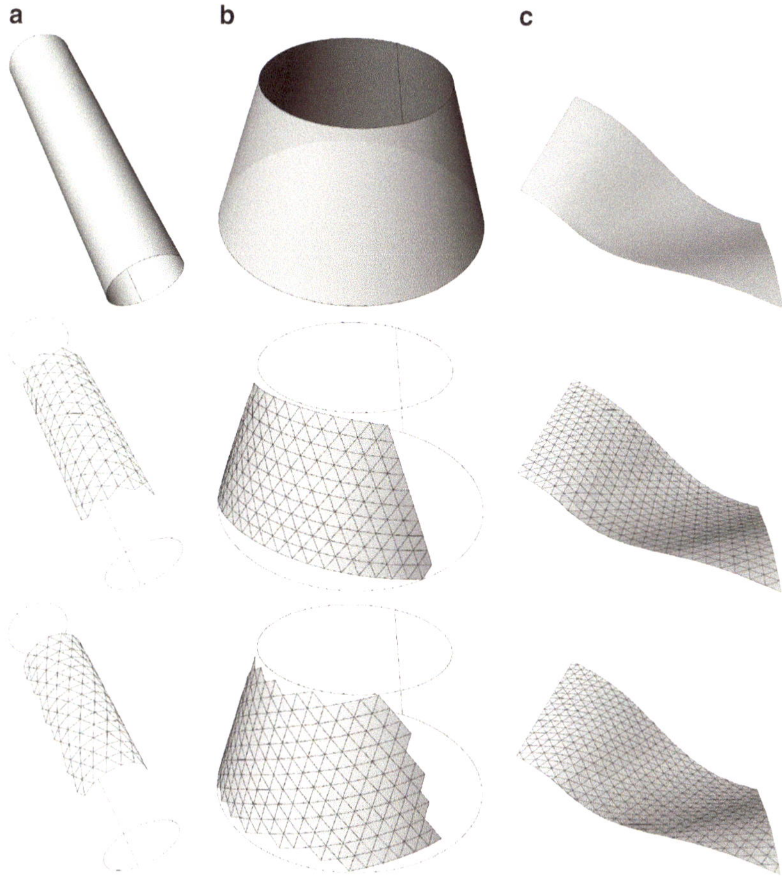

Fig. 4 Patches of smooth developable surfaces approximated with equilateral panelizations: (**a**) Circular cylinder. (**b**) Circular cone. (**c**) General developable surface

Fig. 5 Architectural design with large scale developable patches that are realized entirely by repeating a single, planar, equilateral triangle

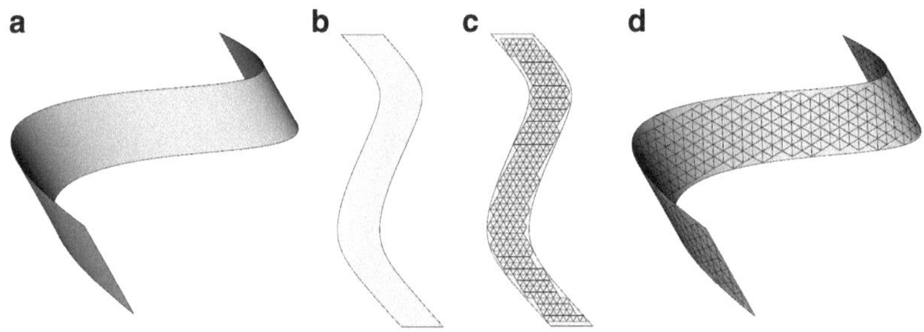

Fig. 6 Initialization for the optimization framework to approximate smooth developable surfaces with equilateral triangle meshes

Interestingly, it seems to be possible to some extent to also generate regular equilateral triangle meshes that appear to be slightly double curved. The discrete gaussian curvature will still be zero and the mesh will still be perfectly unfoldable.Whether this effect is simply due to accumulated tolerances for the edge lengths or to some inherent property of even perfectly exact equilateral triangle meshes remains to be further investigated.

3.3 Introducing Singularities: Lobel Frames and Connectivity Shapes

As indicated in the previous section, as soon as we introduce a vertex that has more or less than six neighbours in an equilateral triangle mesh, we obtain a kinked result (as shown in Fig. 3). We call such irregular vertices *singularities*. While the resulting structure might not look entirely smooth anymore, it might still be appealing. The architect Alain Lobel, as well as the computer scientists Isenburg et al. (2001) investigate exactly this setting. Lobel approached the problem in a very constructive manner, trying to classify numerous families of equilateral triangle meshes with singularities according to their geometric properties. He called these structures *Lobel Frames*. Isenburg et al. touched this topic coming from an entirely different direction (and with different applications in mind): they realized that a lot of the geometry of an equilateral mesh is given, once we know where the singularities are. They called their results *Connectivity Shapes*. Introducing singularities induces strong connectivity constraints and significantly restricts the space of obtainable surfaces as compared to the regular meshes analysed in the previous section. Building on these insights and using our optimization framework, we show that Lobel-Frame like structures can be very easily obtained through optimization, avoiding intricate construction rules.

Figurc 7 illustratcs our approach: We can start with a flat embedding of a non-equilateral triangle mesh, mostly regular but possessing a few singularities

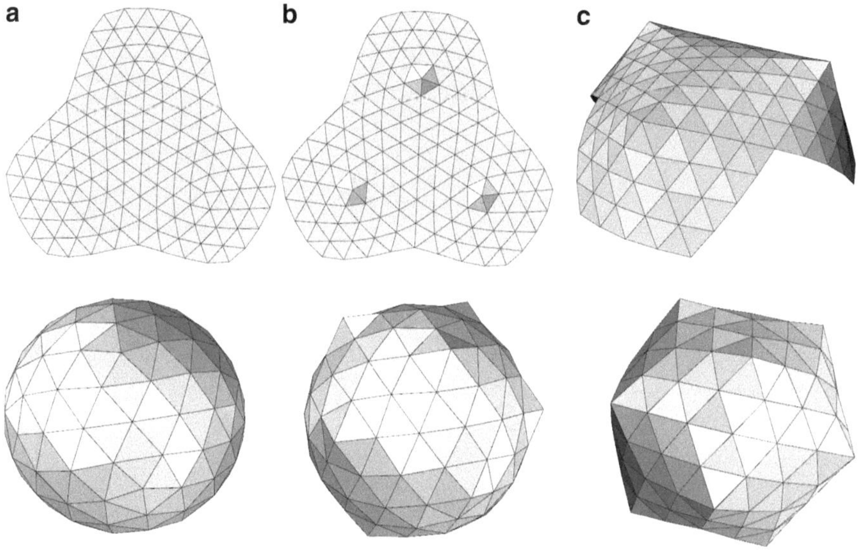

Fig. 7 Initialization and optimization of equilateral triangle meshes with singularities: (**a**) Initializing a non-equilateral mesh in the plane or on the sphere. (**b**) Manually moving the singularities to indicate how the geometry around the singular points should evolve. (**c**) Automatically deduced equilateral triangle structures from the inputs shown in (**b**)

(Fig. 7a). For a closed mesh we can start with a sphere. The work of Isenburg et al. states that most of the geometry of an equilateral triangle mesh is defined through its connectivity, leaving very few degrees of freedoms left. For example, the singularities can evolve towards either side of the tangent plane. In order to fix these remaining degrees of freedom, we manually move the singularities onto the sides we would like them to be in the final result, indicating the direction the optimization should take (Fig. 7b). If we additionally state that the result should be as smooth as possible, the mesh will snap to a unique Lobel-Frame like structure by the click of a button (Fig. 7c). In contrast to the developable shapes from the last section, no initial design surface is required for this approach. Thus it can be seen as a bottom-up, pre-rationalization design approach, as opposed to the top-down, post-rationalization approach used for the developable structures. Figures 1 and 8 show further structures obtained using our approach.

Some specific comments about the optimization: for this setting, we only use the weights ω_1 and ω_3. Even after this initialization step, the optimization won't always reach a global fairness minimum if we set $\omega_1 \gg \omega_3$ from the start. Therefore we adopt a different strategy. The first optimization iteration is done with $\omega_1 = \omega_3$, which implies that the edge lengths will slightly differ from their target value l. From there, ω_3 is decreased during the subsequent iteration steps until the edge length constraint is fulfilled within desired tolerance. We then reach a global minimum, resulting in a semi-smooth equilateral structure.

Fig. 8 An architectural structure generated by prescribing a few distinct singularities into an otherwise regular equilateral triangle mesh. All panels are the same

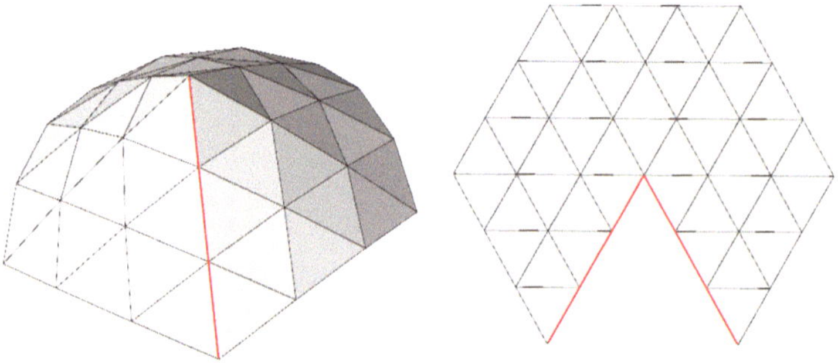

Fig. 9 Even in the presence of singularities, equilateral triangle meshes can be unfolded distortion free onto the plane by introducing one or more cuts (shown in *red*)

In practice, it is interesting to note that such panelizations, although not discrete developable in their entirety, can easily be unfolded onto the plane with a limited amount of cuts along edges adjacent to the singularities (see Fig. 9 for a basic example).

3.4 Dropping Smoothness

Although not globally smooth, the Lobel Frame like structures of the last section were still smooth except at the singular points. During our investigation we realized that the smoothness requirement heavily limits the shapes we can obtain with equilateral triangles. We therefore set out to explore the obtainable shapes when dropping the classical smoothness constraint. The idea is to loosely approximate general surfaces, far beyond usual tolerances. The reference surfaces are only used as control shapes to guide the results towards the desired features. Figure 10 shows experiments with different resolutions on a wavy surface. Since the reference surface is only followed very roughly, this approach is not a post-rationalization in the strict sense, but rather a hybrid approach, where the design surface serves as a control geometry enabling a user-guided design.

Fig. 10 Study of a non-smooth equilateral triangle mesh following a wave at different resolutions. Dropping the smoothness requirement leads to increased design freedom but makes it more challenging to control the aesthetics of the result

Fig. 11 Without the smoothness requirement it becomes possible for regular equilateral triangle meshes to approximate double curved shapes, which automatically leads to a crumpled impression

Similarly as when trying to cover a sphere with a piece of paper, as soon as we try to approximate a double curved surface with equilateral triangles, the result will start to crumple. This effect is demonstrated in Fig. 11.

Since we want to use any type of surfaces as guides, it is in general difficult to provide an accurate initialization similar to the one we used in Sect. 3.2. Instead, we propose a pre-optimization step: starting from any equilateral triangular mesh, we compute a few optimization iterations with the parameters $\omega_1 \ll \omega_2, \omega_3 = 0$. This results in a mesh approximating closely the reference surface, while still loosely controlling the edge distortion from the target length l. We then move on to the actual optimization process and set the parameters such that $\omega_1 \gg \omega_2 = \omega_3$. Even though smoothness isn't our goal, it is necessary to give it a non-zero weight to prevent the results from self-intersecting.

The main challenge is to control the aesthetics. Without smoothness, it is natural to look for a substitute criterion, ensuring the results still have an aesthetical value. In the case of corrugated surfaces, one could look at the existence of regular patterns along the surface, or symmetries. These properties are more global than standard smoothness properties and are thus difficult to enforce through optimization algorithms. This makes the results hard to control and while some results are very appealing, others look off. To the human eye, a surface featuring almost but not quite perfect regularity has in general even less aesthetical value than a result lacking any regularity. In that sense, this subjective effect recalls the famous "uncanny valley" encountered in robotics and computer graphics.

Our efforts to bring back some structure and control into the design of non-smooth equilateral triangle meshes have lead us to the concept of Voxel Meshes, which we will introduce in the following section.

4 Voxel Meshes

In the last section we have seen that dropping the highly restrictive smoothness requirement provides us with significantly improved design freedom, at the cost of less control. The results can quickly become very chaotic and random, lacking symmetry or structure that would help humans to read the design. In this second part, we introduce a more general approach, that enables the design of equilateral triangle meshes but also many other classes of repetitive polygonal panelizations. It stems from the will to give up smoothness in exchange for design freedom but at the same time it enforces some structure, symmetry, and pattern by construction.

The central idea of Voxel Meshes is depicted in Fig. 13: we start with some regular polyhedra (voxels) that can be repeated to fill the entire three dimensional space. Since we only use a few different voxels, only a few different polygonal shapes will fill (or better scatter) the entire space in a very structured way. We can then cut through this space and "carve out" a connected (2-manifold) sheet of polygons that defines a panelization with a repetition of only those initial few polygonal shapes used to fill the space. We explain this concept in more detail in the next section. Readers more interested in the application may skip to Sect. 4.2, where we demonstrate the generality of this approach by various results.

4.1 Voxel Mesh Generation

The theory behind Voxel Meshes can basically be divided into three steps: (1) Voxelization of the 3D space. (2) Extraction of a manifold sheet (the Voxel Mesh) by cutting a dual graph. (3) Providing of a means to manually update and refine the Voxel Mesh. In this section, we demonstrate these three steps at hand of a 2D example illustrated in Fig. 12. The theory naturally extends to 3D, where it is much harder to visualize.

1. *Voxellization of the space*. We define a basic block of voxels that can fill the entire space. In 3D, these voxels are polyhedra. In 2D, the voxels are polygons. Figure 12a shows a 2D voxelization with only 2 different voxels: an equilateral triangle and a square (so called Snub-Square tiling). The fewer elements we use to fill the entire space, the higher the repetition of elements we will have in our final panelization. Note that in our 2D example, the repeating elements are line segments of the same length, filling up the space. As we will see later in this

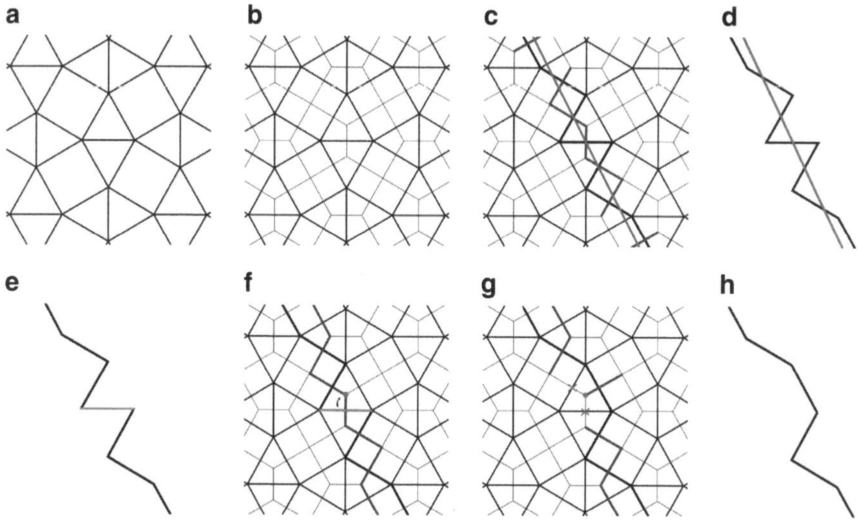

Fig. 12 2D illustration of the basic principles of Voxel Mesh generation. (**a**)–(**d**) Show the generation of a Voxel Mesh sheet: (**a**) Voxelization of the entire space using a block of two voxels (a triangle and a square) as the generator. Thus the space is filled by line segments of equal length. (**b**) The dual graph (*blue*) of the voxelization. (**c**) The dual graph is cut with a reference surface (a curve in 2D, here a simple line in *red*). The edges of the dual graph that are cut by the line are highlighted in *thicker blue*. Each of those edges has a corresponding voxel line segment, highlighted in *thicker black*. The collection of all those highlighted line segments build the final Voxel Mesh sheet (a polyline in 2D) out of highly repetitive elements, shown in (**d**), following the design surface. The bottom row (**e**)–(**h**) shows a local update operation: (**e**) The user selects one of the elements (line segments in 2D) to update, highlighted in *green*. The Voxel Mesh sheet can now be locally updated in one of the two possible directions defined by the corresponding dual graph edge. The dual vertex indicating the update direction is shown as a *green point* in (**f**). All dual edges incident to this vertex flip their state: active edges become inactive and vice versa, shown in (**g**). The new set of active edges (highlighted in *thick blue*) define the updated Voxel Mesh sheet (**h**)

section, the choice of this base voxel block (we call it the *Generator*) heavily influences the aesthetics of the resulting Voxel Mesh.

2. *Extracting a manifold sheet.* We can now cut the voxelized space with any given reference surface. In our 2D example, a reference surface is a curve. We will use a simple line. To extract a sheet closely following this reference surface, we construct a graph dual to the voxelized space. That is, every voxel centre represents a vertex of the dual graph, and two dual vertices are connected by a dual edge if the corresponding voxels are direct neighbours. The dual graph is shown in blue in Fig. 12b. Figure 12c shows the reference line in red. Whenever the line intersects a blue edge, we add the dual black line to the resulting sheet. The final Voxel Mesh is thus a sheet defined through the edges of a dual graph cut. The final sheet approximating the reference line is shown in Fig. 12d. Since our initial voxelization of the space only consisted out of line segments with the same length, the final Voxel Mesh sheet is a curve consisting of line segments

of the same length, i.e. a structure repeating a single element. Also note that the sheet is a non-smooth, continuous curve, without gaps or intersections. In 3D, this sheet will be a surface instead of a line, consisting of repeating planar polygons instead of line segments.

3. *Manual local updates.* The steps (1) and (2) can be used to quickly and easily design a planar, highly repetitive panelization. Similar to the approach in Sect. 3.4, these two steps enable a user-guided design that resembles post-rationalization. But with Voxel Meshes, since we have such a clear and confined definition of our possible design space, we can also let the user directly edit the design by locally shifting the Voxel Mesh sheet backward or forward. Figure 12 illustrates the underlying operations in 2D: The user can select an element in the sheet where it needs to be updated, indicated by a green line segment in Fig. 12e. The dual edge corresponding to this line segment now instructs us on how to update the sheet: we take the starting or ending vertex of the dual edge (depending in which direction the user wishes to modify the sheet). At that vertex, we flip the state of all incident edges: if they are in the sheet (thick blue) we move them out of the sheet (thin blue) and vice versa (Fig. 12f, g). This new set of thick blue dual edges define the updated sheet shown in Fig. 12h. This technique can even be used to design entirely in a pre-rationalized manner, starting with a completely flat sheet and refining the Voxel Mesh sheet step by step.

4.2 Designing with Voxel Meshes

Generating the Voxel Mesh sheet is purely combinatorial and there is no numerical optimization involved. Therefore, the resulting polygonal mesh is a perfect repetition of identical and perfectly planar polygons, without any tolerances. As detailed in the previous section, results can be generated using a top-down, post-rationalization approach by sculpting the result with a guiding smooth design surface, as well as using a bottom-up, pre-rationalization approach, by iteratively building a Voxel Mesh sheet from scratch with repeating local modifications. Or a combination of the two.

The number of different polygons in the final structure depends on the chosen voxel base block (the Generator). For the results in this paper we used four different Generators, shown in Fig. 14. We can again just work with one equilateral triangle (Fig. 14a), or add further polygons like a second triangle (Fig. 14b), a square (Fig. 14c), or a larger selection of different polygons (Fig. 14d).

Since the underlying structure is always a highly regular voxelization of the space, the resulting panelization will automatically depict some structure and regularity, while still offering great design freedom. Various design parameters are available:

- The Generator: Choosing different voxel base blocks will entirely change the aesthetics and structure of the result, offering limitless exploration possibilities.

- The Reference Surface: The reference surface enables guided sculpting of the Voxel Mesh sheet. Not only the shape of the surface but also its position and orientation within the space will heavily influence the result. See e.g. Fig. 15.
- The Scale: Varying the scale of the reference surface or the generator will change the nature of the structure. See also Fig. 15.
- Local Modifications: The Voxel Mesh can be locally refined as described in Sect. 4.1. This provides a fine-grained control on the design.

Figures 13–17 demonstrate the versatility of Voxel Meshes.

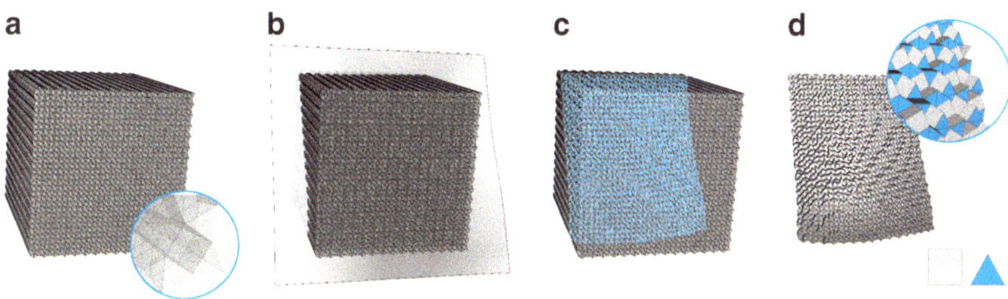

Fig. 13 The basic workflow of Voxel Mesh design guided by a reference surface: (**a**) The entire space is filled by a few different voxels. For this example, the SnubSquare generator is used (see Fig. 14c). (**b**) The voxelized space is cut by a smooth surface. This is the same reference surface that is used to generate the results of Fig. 17. (**c**) The reference surface cuts out a Voxel Mesh sheet using the process described in Sect. 4.1. (**d**) The resulting panelization consisting only of two different polygons (an equilateral triangle and a square)

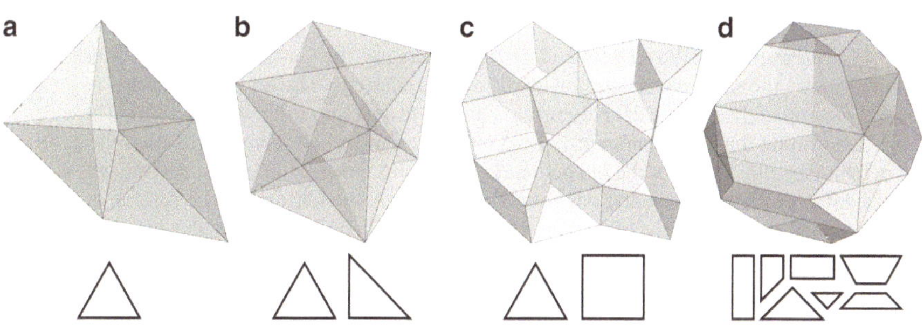

Fig. 14 Different voxel base blocks (generators) that can fill the entire space. Four generators (**a**) OctaTetra, (**b**) CutUpCube, (**c**) SnubSquare, and (**d**) TruncatedOctahedra were used to produce the results in this paper. They produce panelzations consisting of only 1, 2, or 7 different planar polygons. Many other generators exist

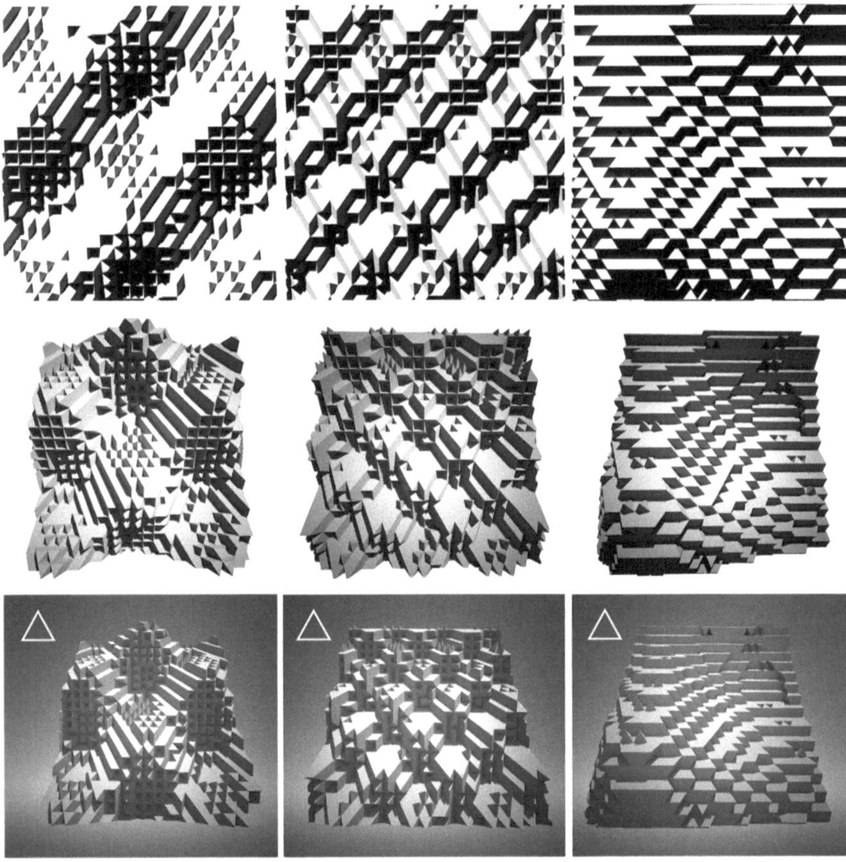

Fig. 15 Three different Voxel Mesh sheets demonstrating the versatility of Voxel Meshes: each column shows the same panelization with different renderings. The results can appear as anything between completely abstract 2D patterns, to tectonic textures, to fully 3D structures. These results also demonstrate the design influence of the reference surface: all three results used the generator OctaTetra (see Fig. 14a). For the *left* and *middle columns*, the same reference surface was used but with a different scale and orientation. For the *right column*, the same reference surface as in Fig. 13 was used, and the shown structure is actually exactly the same structure as shown in Fig. 17 right. All three panelizations repeat only a single equilateral triangle

Fig. 16 Landscape design study, using the generator CutUpCube. For the landscape design in Fig. 1, the generator OctaTetra was used (see Fig. 14 for a list of generators)

Conclusion and Future Work

We have proposed Equilateral Mesh Optimization and Voxel Meshes as two complementary approaches to design planar freeform panelizations with extreme repetition. Both approaches make use of the insight that relaxing or even dropping the requirement of a smooth structure can significantly increase the degrees of freedom, enabling flexible design with only a handful of different elements. The resulting polygonal structures exhibit exact repetition and exact planarity and are discrete manifold surfaces without gaps or intersections.

This work focuses on the geometric concepts. For any practical application, the engineering and manufacturing aspects of the structures would have to be considered as well and there are certainly interesting related questions that can be investigated in these fields.

The class of surfaces that can be smoothly approximated with equilateral triangles is yet to be described exactly. Another important question concerns the additional freedom resulting from the relaxation of the equilateral constraint within construction tolerances. The integration of further optimization constraints like coplanarity of vertices can be investigated. The relation between developable surfaces and regular equilateral triangle meshes extends to any regular tiling of the plane, so other regular structures smoothly approximating developable surfaces are probably possible.

The Voxel Meshes offer a large potential for future work: since they define a highly structured design space, they could well serve as a basis to optimize

(continued)

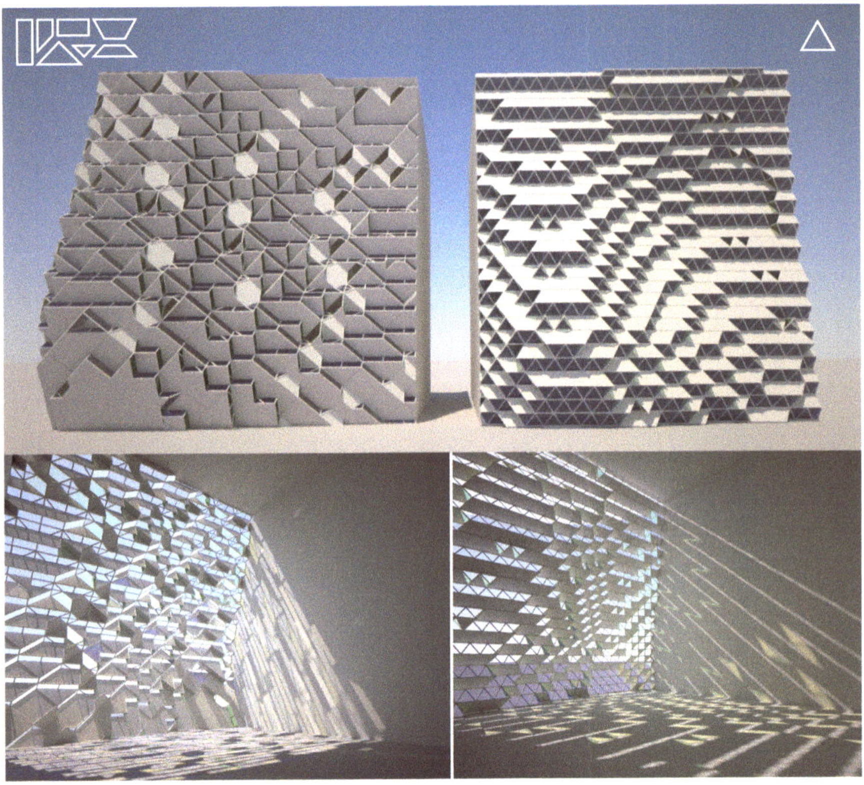

Fig. 17 Two facade studies designed using the same design surface (surface shown in Fig. 13) but two different generators: TruncatedOctahedra (*left*) and OctaTetra (*right*) (see Fig. 14 for a list of generators). The *bottom row* shows the corresponding inside views. These results indicate the potential of Voxel Meshes for environmental optimization: Due to the high regularity by construction, we can easily select all panels facing a given direction to control how light enters the building

for environmental or functional objectives via combinatorial optimization. This potential is already shown in some of the results: for example, the strong underlying structure makes it easy to select all panels facing a given direction, and making them transparent or opaque, optimizing for lighting conditions. Approaching this in a more systematic and automated manner offers a strong potential for lighting and shading optimization. Additionally, instead of just using a single sheet, two or more sheets could be extracted to form a volumetric space frame structure with identical members.

Acknowledgements This work has received funding under the European Union's Seventh Framework Programme FP7/2007-2013/ under REA grant agreement PIAP-GA-2011-286426 – GEMS (http://www.geometrie.tuwien.ac.at/ig/gems/index.php).

References

Eigensatz, M., Kilian, M., Schiftner, A., Mitra, N.J., Pottmann, H., Pauly, M.: Paneling architectural freeform surfaces. ACM Trans. Graph. **29**(4), 45 (2010). Proc. SIGGRAPH

Frey, W.H.: Boundary triangulations approximating developable surfaces that interpolate a closed space curve. Int. J. Found. Comput. Sci. **13**, 285–302 (2002)

Fu, C.-W., Lai, C.-F., He, Y., Cohen-Or, D.: K-set tilable surfaces. ACM Trans. Graph. **29**(4), 44 (2010). Proc. SIGGRAPH

Isenburg, M., Gumhold, S., Gotsman, C.: Connectivity shapes. In: Proceedings of the Conference on Visualization '01 (VIS '01), San Diego, pp. 135–142. IEEE Computer Society (2001)

Pottmann, H., Asperl, A., Hofer, M., Kilian, A.: Architectural Geometry. Bentley Institute Press, Exton (2007)

Schiftner, A., Raynaud, J., Baldassini, N., Bo, P., Pottmann, H.: Architectural freeform structures from single curved panels. In: Advances in Architectural Geometry, pp. 45–48. TU Wien, Vienna (2008)

Singh, M., Schaefer, S.: Triangle surfaces with discrete equivalence classes. ACM Trans. Graph. **29**(4), 46 (2010) Proc. SIGGRAPH

Tachi, T.: Freeform variations of origami. J. Geom. Graph. **14**(2), 203–215 (2010)

Interlocking Folded Plate: Integrated Mechanical Attachment for Structural Wood Panels

Christopher Robeller, Andrea Stitic, Paul Mayencourt, and Yves Weinand

Abstract Automatic joinery has become a common technique for the jointing of beams in timber framing and roofing. It has revived traditional, integrated joints such as mortise and tenon connections. Similarly, but only recently, the automatic fabrication of traditional cabinetmaking joints has been introduced for the assembly of timber panel shell structures. First prototypes have used such integrated joints for the alignment and assembly of components, while additional adhesive bonding was used for the load-bearing connection. However, glued joints cannot be assembled on site, which results in several design constraints.

In this paper, we propose the use of dovetail joints without adhesive bonding, on the case study of a timber folded plate structure. Through their single-degree-of-freedom (1DOF) geometry, these joints block the relative movement of two parts in all but one direction. This presents the opportunity for an interlocking connection of plates, as well as a challenge for the assembly of folded plate shells, where multiple non-parallel edges per plate must be jointed simultaneously.

1 Introduction

Architectural designs have often been inspired by folded shapes such as origami, however the folding principle can rarely be applied to building structures directly. Alternatively, folded plates can be cast as concrete thin-shells (Trautz and Herkrath 2009), which requires complex formwork and an elaborate vertical support structure. Constructions with discrete elements have been realized with fiber-reinforced plastics (Pizzi 2003) or tubular steel.

A prefabricated folded plate built from cross-laminated timber panels has been proposed by Buri (2010). It combines the elegant and efficient shape of folded plate shells with the advantages of structural timber panels, such as CO_2 storage and a low weight-to-strength ratio. However, a major challenge in the design of a

C. Robeller (✉) • A. Stitic • P. Mayencourt • Y. Weinand
EPFL Timber Construction Laboratory IBOIS, Station 18, CH 1015 Lausanne, Switzerland
e-mail: christopher.robeller@epfl.ch; andrea.stitic@epfl.ch; paul.mayencourt@epfl.ch; yves.weinand@epfl.ch

P. Block et al. (eds.), *Advances in Architectural Geometry 2014*,
DOI 10.1007/978-3-319-11418-7_18

timber folded plate is presented by the joints: Since timber panels cannot be folded, a large amount of edgewise joints has to provide two main functions. One of these functions is the load-bearing behaviour, where *Connector features* of the joints must provide a sufficient stiffness and rigidity. The second main function of the joints is the assembly of the parts, where *Locator features* of the joints are essential for a precise and fast positioning and alignment of the parts.

Hahn (2009) examined the structural behaviour of a first timber folded plate shell built from plywood and assembled with screwed miter joints, concluding that the load-bearing performance could be improved significantly with more resistant connections.

Inspiration for such improvements may be found in *integral mechanical attachment* techniques, the oldest known technique for the jointing of parts, where the geometry of the parts themselves blocks their relative movements (Messler 2006). Such integrated joints have recently been re-discovered by the timber construction industry. Beginning in 1985, mortise-and tenon joints have been repatriated in timberframe and roof constructions (Hundegger 2014). Only very recently, integrated joints have also been proposed for the edgewise jointing of timber panels. La Magna et al. (2013) and Krieg et al. (2014) have applied finger joints to plywood panels and Robeller et al. (2014b) have demonstrated an application of dovetail joints for cross-laminated timber panels (CLT). In these prototype structures, the integrated joints have played an important role for the assembly of the components. They have also participated in the load-bearing connection of the parts, but additional adhesive bonding was needed. With few exceptions (HESS 2014), such glued joints cannot be assembled on site, because they require a curing period with a specific temperature and humidity (Purbond 2011). Therefore, their application is limited to off-site assembly of larger components, which complicates both transport and handling while still requiring additional connectors for the final assembly.

In this paper, we propose the use of dovetail joints without additional adhesive bonding, on the case study of a timber folded plate shell (Fig. 1).

Fig. 1 Timber folded plate built from 21 mm LVL panels, assembled with single-degree-of-freedom dovetail joints without adhesive bonding. Components interlock with one another

Through their single-degree-of-freedom (1DOF) geometry, these joints block the relative movement of two parts in all but one direction. This presents the opportunity for an interlocking connection of plates, as well as a challenge for the assembly of folded plate shells, where multiple non-parallel edges per plate must be jointed simultaneously.

1.1 Dovetail Joint Geometry and Mechanical Performance

Using polygon mesh processing, we describe an edgewise joint based on its edge E. From the mesh connectivity, we obtain the edge vertices p and q and the adjacent faces F_0 and F_1 with their face normals n_0, n_1. We use the polygon mesh to represent the mid-layer of timber panels with a thickness t and offset F_1 and F_2 at $\pm\frac{t}{2}$ to obtain the lines L (Fig. 2a). From a division of E, we obtain the points X_j for a set of reference frames $\{u_1, u_2, u_3\}$, where $u_1 \parallel \vec{pq}$ and $u_2 \parallel n_0$ (Fig. 2b). A finger joint geometry is obtained from an intersection of planes located at X_j, normal to u_1, with the four lines L.

Without additional connectors, finger joints are *planar joints* with three degrees of freedom (3DOF). They can resist shear forces parallel to the edge and in-plane compressive forces. However depending on the plate geometry, thickness and most of all rotational stiffness of the connection detail, bending moments are also transferred between the plates. Also, due to the rotation of the plate edge caused by bending, in-plane traction forces perpendicular to the edge line appear and their magnitude increases under asymmetrical loads. Such forces, which occur as a result of out-of-plane loading, cannot be supported only by shear and in-plane compression resistant joints (Fig. 3).

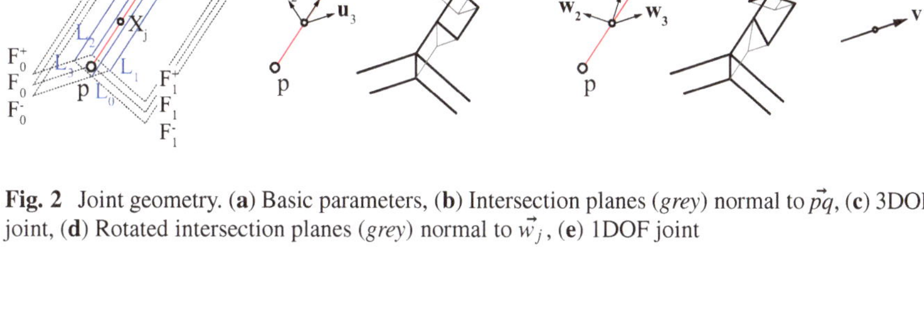

Fig. 2 Joint geometry. (**a**) Basic parameters, (**b**) Intersection planes (*grey*) normal to $\vec{pq}$, (**c**) 3DOF joint, (**d**) Rotated intersection planes (*grey*) normal to $\vec{w}_j$, (**e**) 1DOF joint

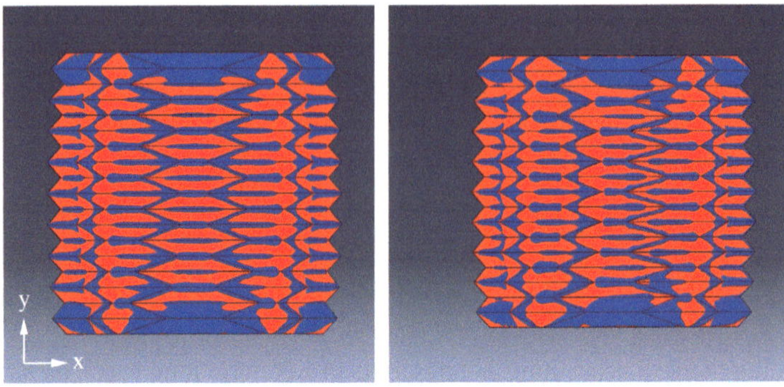

Fig. 3 FEM analysis (top view) of a 3 × 3 m, 21 mm Kerto-Q folded plate thin shell assuming fully stiff joints. Distribution of traction (*red*) and compression (*blue*) stresses in the y direction. *Left side*: gravity load case. *Right side*: asymmetric snow load

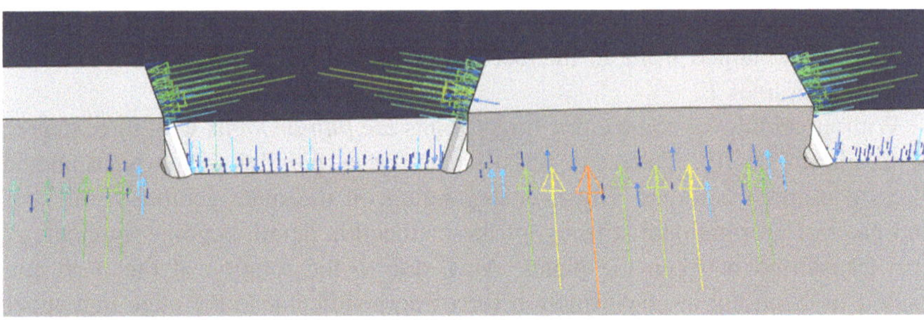

Fig. 4 FEM simulation of bending on a dovetail joint connecting two Kerto-Q 21 mm LVL panels. The bending moment applied is transformed into compression, normal and shear forces parallel to the inclined contact faces

On a dovetail joint (Fig. 2d, e), the intersection planes on the points X_j are normal to a rotated vector w_1. It is obtained from a rotation of the reference frame $\{u_1, u_2, u_3\}$ about u_3 at an alternating angle $\pm\theta_3$. The resulting rotated side faces reduce the dovetail joints degrees of freedom to one translation $\vec{w_3}$ (1DOF). Sebera and Simek (2010) have suggested $\theta_3 = 15°$ for spruce plywood panels. Such prismatic joints can only be assembled or disassembled along one assembly direction $\vec{v} = \vec{w_3}$. In addition to the finger joints resistance to shear and compressive forces, dovetail joints can, without adhesive bonding, also resist bending moments and traction forces which are not parallel to $\vec{v}$. Due to the inclination of the side faces of the joint, resistance to these forces can be improved significantly. In that way the inclined faces take over the role that the glue would have in a finger joint (Fig. 4).

1.2 Fabrication Constraints

One of the main reasons for the resurgence of finger and dovetail joints is the possibility of automatic fabrication. However, the mechanical performance of the joints depends on fabrication precision. At the same time, fast machine feed rates are important for a time-efficient production. We have fabricated such joints with a robot router and a gantry router, achieving higher precision with the gantry machine, which is more stiff and provides a higher repeat accuracy.

The variability of the machine-fabricated joints is enabled by the 5-axis capability of modern routers: Although traditional edgewise joints in cabinetmaking were used for orthogonal assemblies, both the finger and dovetail joint can also be applied for non-orthogonal fold angles, which was essential for the reference projects mentioned before. However, there are certain fabrication-related constraints for machine-fabricated dovetail joints. In order to integrate the joint fabrication directly with the panel formatting, we use a side-cutting technique (Koch 1964), which is limited to a tool inclination β_{max}. We obtain this limit from the specific geometry of the tool, tool-holder and spindle used for the joint fabrication (Fig. 5).

The parts can be assembled in two ways, as shown in Fig. 5, which allows to address a larger range of interior fold angles φ.

$$\varphi = \arccos \frac{(\vec{n_0} \times \vec{pq}) \cdot (\vec{n_1} \times \vec{pq})}{\|(\vec{n_0} \times \vec{pq})\| \cdot \|(\vec{n_1} \times \vec{pq})\|}$$

From this we obtain the fabrication-constrained most acute fold $\varphi_{min} = 90° - \beta_{max}$ and most obtuse fold $\varphi_{max} = 90° + \beta_{max}$. With standard routing tools, this technique allows for the jointing of acute folds up to $\varphi = 50°$, which is ideal for folded plate structures. Very obtuse fold angles $\varphi \geq 140°$, which might be required for small-tiled geodesic dome structures, cannot be fabricated with this method.

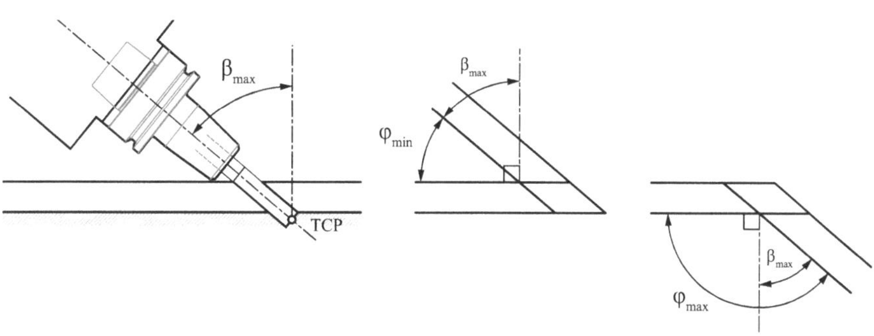

Fig. 5 Fabrication Constraints. Side-cutting technique used for the automated fabrication of 1DOF edgewise joints with common 5-axis CNC routers. The maximum tool inclination β_{max} results from the tool and the tool holder geometry. From this we obtain the range of possible fold angles $\pm\varphi$ between panels

1.3 Simultaneous Assembly of Multiple Edges

The assembly of doubly-corrugated folded plates requires the *simultaneous* jointing of multiple edges per component (Fig. 6), which has implications on both the shell and the joint geometry.

For multiple 1DOF-jointed edges E, simultaneous assembly is only possible if the individual assembly directions $\vec{v}$ are parallel. With a normal dovetail joint geometry, where $\vec{v} = \vec{n_0} \times \vec{pq}$ (Fig. 2d, e), this is not the case: A simultaneous assembly is only possible for parallel edges, which allows only for rectangular assemblies, such as drawers or a cabinets.

In order to simultaneously join non-parallel edges, we must rotate the assembly direction v of the joints to make them parallel. This possibility is known from Japanese cabinetmaking (Tetsuya 2007), where certain joints, like the Nejiri Arigata Joint are assembled along the external bisector of the fold.

We extend the Japanese technique to a cardan rotation of the frames $\{u_1, u_2, u_3\}$ about u_1 at the angle θ_1 and about u_2 at the angle θ_2 (Fig. 7a). Each of these rotations is constrained to a maximum value. $\theta_{1,max}$ is set by the fold interior angle $180° - \varphi_i$. The range is large for acute fold angles and small for obtuse fold angles. $\theta_{2,max}$ is fabrication-constrained through the maximum tool inclination β_{max} (Fig. 5). The specific limit relates to φ and θ_3. For our setup, $\theta_{2,max}$ was approximately $\pm 20°$. From these constraints, we obtain a pyramid-shaped window S for every edge E (Fig. 7b). This window illustrates all possible assembly directions for E.

For the simultaneous jointing of three edges E_1, E_2, E_3, we overlay the three rotation windows S_1, S_2, S_3. If there is an intersection $S_1 \bigcap S_2 \bigcap S_3$ the edges can be jointed simultaneously. The assembly direction must be chosen within the intersection. As a result of these limited rotations, the angle between neighbouring, simultaneously jointed edges cannot be very acute. Folded plate patterns like the Herringbone, the Diamond, or the Hexagon pattern, which we chose for our prototypes, (Buri 2010) (Fig. 6) work well for our joining technique. Another

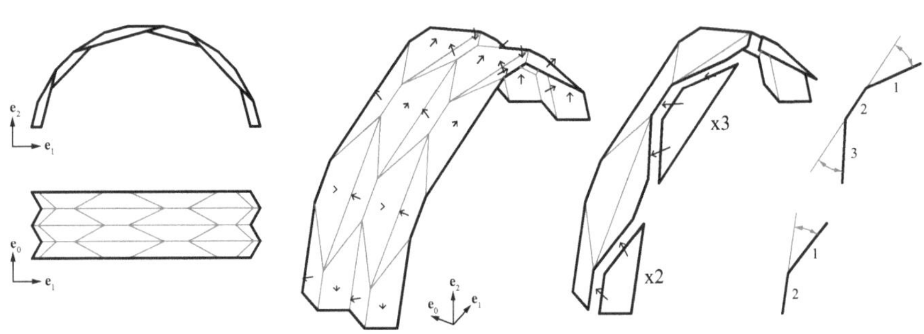

Fig. 6 The assembly of a folded plate from discrete elements (*left side*) requires the *simultaneous* assembly of non-parallel edges. (*right side*) We rotate the insertion direction of our 1DOF joints, to make the insertion vectors of simultaneously jointed edges parallel. We chose a hexagon reverse fold pattern which requires only moderate rotations

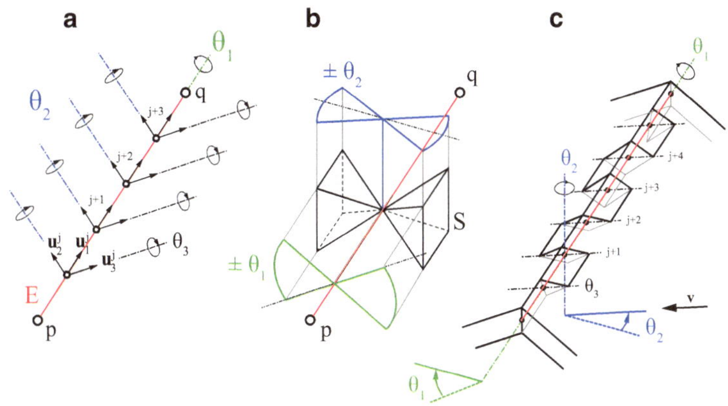

Fig. 7 (**a**) Rotation of the frames $\{u_1, u_2, u_3\}$ about the three Euler angles θ_1 (*green*), θ_2 (*blue*), and θ_3. (**b**) Rotation window S, illustrating all possible assembly directions. (**c**) Dovetail joint with assembly direction rotated about $\theta_1 = 20°$ and $\theta_2 = 20°$

essential feature provided by these reverse-folds are the acute fold angles, which easily satisfies the fabrication-constrained range of $\varphi_{min} = 50°$ to $\varphi_{max} = 140°$.

2 Interlocking Arch Prototype

In an assembly of multiple components (Fig. 8), a step-by-step sequence must be planned for the assembly of the parts. The completed structure can only be disassembled piecewise in the reverse order of assembly. In this way, the elements interlock with one another like a burr puzzle (Wyatt 1928).

Each joint consists of two parts, which must be parallel during assembly. We therefore chose a folded plate geometry with relatively short edges. The manual assembly of long edges may be more difficult but can be simplified with a modified joint geometry. It is important to know the approximate direction of insertion for each part, as this is not easily visible through the joint geometry. Deformations of the arch during the assembly should be minimised. We have assembled this first prototype lying on the side. However larger assembly may require temporary punctual supports. Although the in-plane dimensional stability of the Kerto-Q panels is very high, panels may be slightly warped and some force may be necessary during assembly. While we have simply used a rubber hammer, more advanced techniques could be applied.

To understand the mechanical behaviour of the built prototype, we have applied a vertical load at mid-span of the arch and measured the vertical deflection at the same point. The total load of 821N was applied in two identical load cycles consisting of four loading/unloading sub-cycles. First, a vertical load of 117N was applied in seven steps, after which the load of the last four steps was removed. The loading

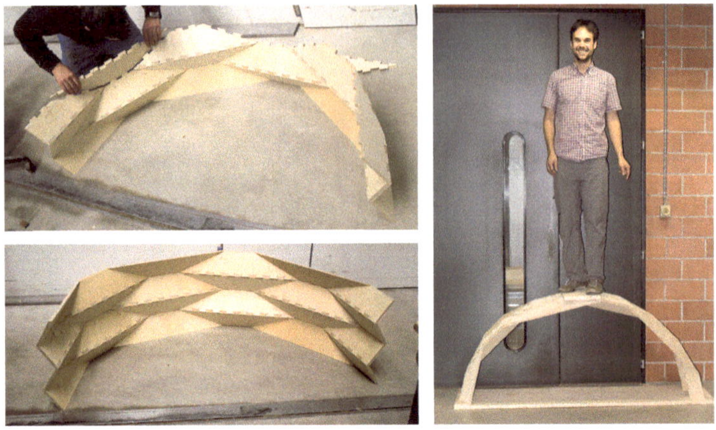

Fig. 8 Folded-plate arch prototype built from 12 mm birch plywood (9-layer, I-I-I-I-I). Assembled without adhesive bonding or metal fasteners. Span 1.65 m, self-weight 9.8 kg

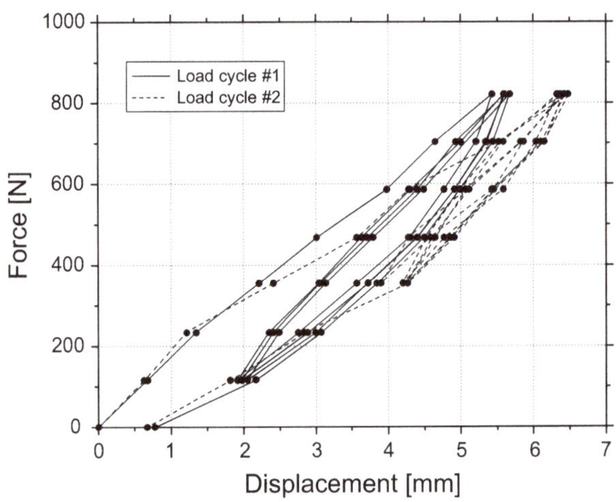

Fig. 9 Series of 3-point flexural tests on the small scale interlocking arch prototype built from Metsawood 12 mm birch plywood panels

and unloading of the last four steps was repeated three more times, after which the complete load was removed and the residual deflection was measured (Fig. 9).

Under a vertical load equal to the arch's dead weight of 9.8 kg (98N), the deflection measured at mid-span was 2 mm. From this we obtain a span-to-deflection-ratio of $L/750$ and the arch's structural efficiency which reaches 8.6 when loaded with 821N (ratio of the maximal load over the dead weight of the arch).

3 Interlocking Shell Prototype

3.1 Automatic Geometry Processing

Using the RhinoPython application programming interface, we have developed a computational tool which lets us instantly generate both the geometry of the individual components and the machine G-Code required for fabrication. The tool processes arbitrary polygon meshes, and generates 1DOF joints for all non-naked edges where the fold angle φ is larger than φ_{min} and smaller than φ_{max} shown in Fig. 5 (non-smooth meshes). It also requires an input of edge identifier tuples identifying those edges which must be jointed simultaneously, as well as the thickness of the LVL panels. Exploiting this geometrical freedom, we have tested our computational tool on the design of a folded plate shell prototype with an alternating convex-concave transversal curvature. The shell spans over 3 m at a thickness of 21 mm, using Kerto-Q structural grade LVL panels (7-layer, I-III-I) (Fig. 10).

Comparing this doubly-curved folded plate with a straight extrusion (as tested by Buri 2010), it can be concluded that the slight double-curvature proves to be very beneficial when it comes to global deflections, for example those caused by wind loads. Deflections for the doubly-curved shell geometry in the vertical direction are up to 39 % smaller and up to 13 % smaller in the lateral direction than the ones for the straight extrusion one.

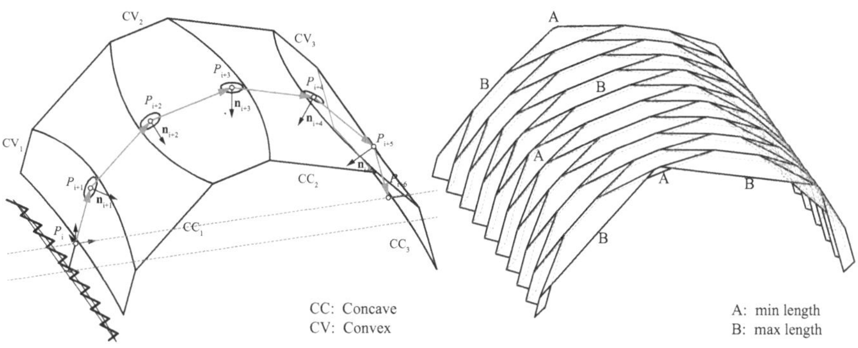

Fig. 10 Doubly-curved folded-plate: The radius ($R = 17m$) of the transversal curvature is determined by the folded plates maximum amplitude h (Buri 2010), which is inversely proportional to the number of segments m of the cross-section polyline (*grey*). We obtain this polyline from a circular arc divided into segments of equal length. The interior angle $\gamma = ((m - 2) * 180) - k$ of this polyline is proportional to all fold angles φ. The geometry of our prototype was fabrication-constrained to a maximum component length $B \leq 2.5m$

3.2 Assembly

Figure 11 shows a part of the connectivity graph of the doubly-curved folded plate shell prototype. It illustrates the hierarchy, direction and order in which the 107 components with their 239 edgewise 1DOF joints must be assembled.

Figure 12 shows the components from Fig. 11 in 3D, demonstrating how the component based on mesh face F_{86} is being inserted. Its three edgewise joints E_{41}, E_{68} and E_{89} must be assembled simultaneously. The three assembly vectors of the edges $\vec{v_{41}}$, $\vec{v_{68}}$ and $\vec{v_{89}}$ have been rotated to be parallel. The same applies for the adjacent edges on the left side of the faces F_{67}, F_{69}, F_{88}, F_{103} and F_{105} (see Fig. 11: all faces with multiple outgoing arrows). The roman numerals on top of Fig. 11 show the interlocking sequence. All joints marked with I. (E_{18}, E_{99}, E_{100}, E_{44}, E_{66}, E_{91}, E_{59}, E_{43}) can be assembled individually and independently in a first step. Within the rotation window of the edge, we can freely rotate $\vec{v}$ for these edges (the greater the angle between $\vec{v}$ and the main direction of traction e_1, the better). After the connection of the joints marked II. (E_{35}, E_{65}, E_{98}), all connections left of this point cannot be disassembled any more without disconnecting II. The components interlock with one another, similar to a burr puzzle (Wyatt 1928). The rest of the assembly follows the same logic.

3.3 Completed Shell Prototype and Load Test

Figure 13 shows the completed folded plate prototype, with a span of 3 m and a shell thickness of 21 mm. Boundary conditions that restrain displacements of the supports in every direction, but allow rotations, were applied on both sides. A longitudinal line load was introduced along the top of the shell and vertical displacement was measured at center point (Fig. 14).

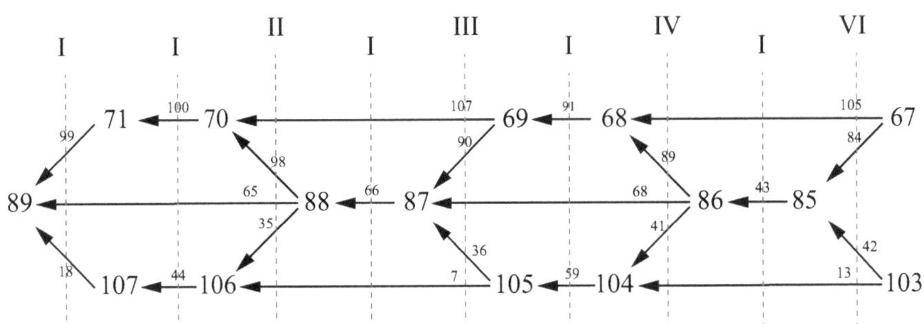

Fig. 11 Partial connectivity graph of the folded plate shell prototype. (Left-to-right side assembly) Large numbers represent mesh Faces F, small numbers represent mesh edges E. Roman numerals illustrate the interlocking hierarchy

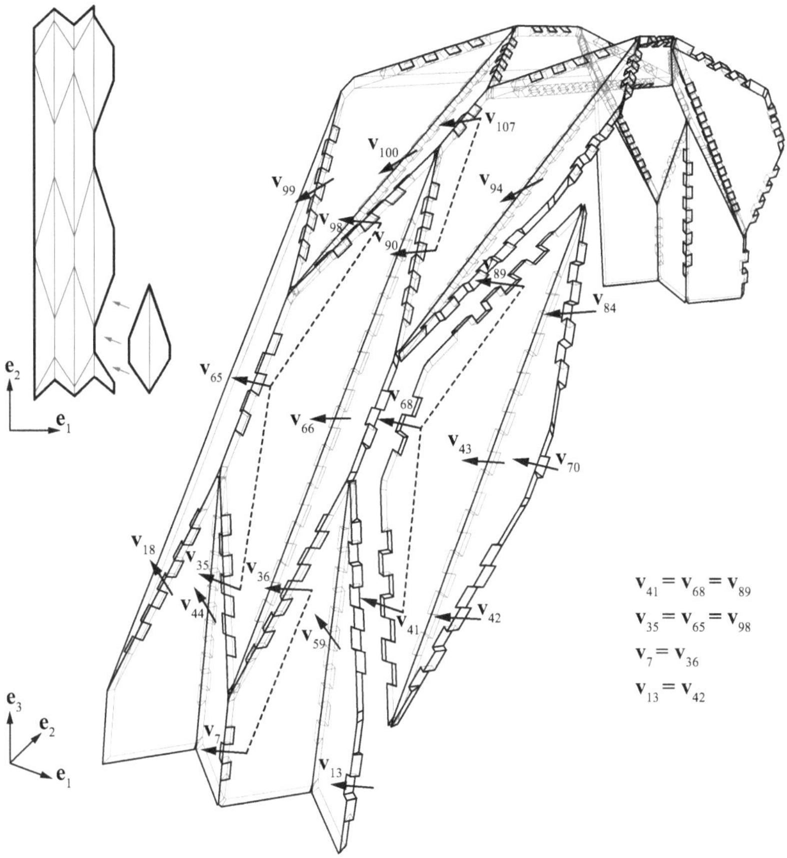

Fig. 12 Left-to-right-side assembly of the Interlocking folded plate shell prototype. Built from Kerto-Q structural grade LVL panels (7-layer, I-III-I)

The prototype structure was also modelled in FE analysis software (Abaqus) and loaded in the same way. The plates were modelled using shell elements, where the mid-surface is used to represent the 3D plate and transverse shearing strains are neglected. Connections between the plates were considered as completely rigid in order to obtain minimal displacements of the structure. By comparing the displacements of the structure with infinitely stiff joints with the ones measured on the prototype, we obtained information about the actual semi-rigidity of the joints. The results obtained from the testing of the large scale prototype showed that the load of 25 kN, that corresponds to the proportional limit of the load-displacement curve, causes a vertical displacement of 23 mm. In the FE model, the load applied in the same manner caused a vertical displacement of 2.6 mm.

Fig. 13 Folded-plate shell prototype, built from 21 mm LVL panels. With a self-weight of 192 kg, the prototype with a span of 3 m was tested with a line-load up to 45 kN

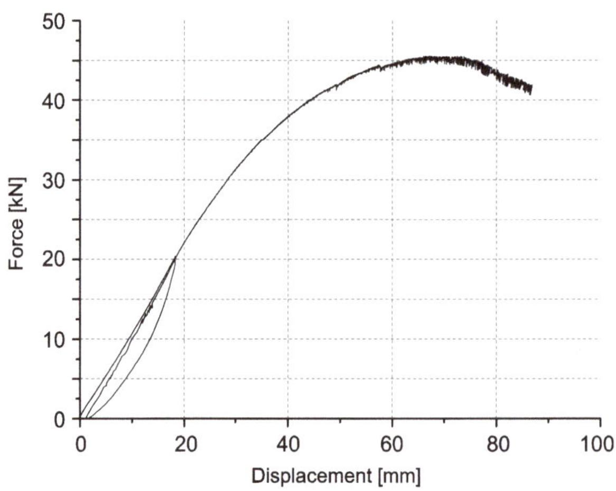

Fig. 14 Load-displacement curve of the shell prototype. A longitudinal line load was introduced along the top of the shell. Vertical displacement was measured at the center point

Conclusion

A timber folded plate shell combines the structural advantages of timber panels with the efficiency of folded plates. However, in such discrete element assemblies, a large amount of semi-rigid joints must provide sufficient support

(continued)

for the adjacent plates in order to ensure an efficient load-bearing system. This remains a challenge with much potential for improvements (Hahn 2009).

Integrated edgewise joints present an interesting addition and an alternative to state-of-the-art connectors: Compared to adhesive bonding, such joints can be assembled rapidly on site. Also, compared to costly metal plates and fasteners, which are typically required in large quantities (Neuhaus 2004), the fabrication of integrated joints is not more expensive. The replacement or reduction of metal fasteners with an integrated mono-material connection includes advantages such as improved aesthetics, ease-of recycling or a homogenous thermal conductivity of the parts, which can reduce condensation and decay. Graubner (1986) Another particular advantage is the possibility to join thin panels: The current technical approval for the Kerto-Q panels does not permit screwed joints on panels with a thickness of less than 60 mm (DIBt 2011).

Recent experimental projects have already demonstrated first applications of integrated edgewise joints for timber panels. This paper followed up on these projects, examining the particular advantages, potential and challenges of 1DOF joints for timber folded plate shells. We have demonstrated how this joint geometry helps resisting the forces which occur in such structures. In addition to the load-bearing connector features, the joints provide locator features, which allow for precise positioning and alignment of the parts through the joint geometry. This improves both accuracy and ease of assembly. Furthermore, we have presented a solution for the simultaneous assembly of multiple edges per panel, which is essential for the application of 1DOF joints in a folded plate shell structure. The per edge "rotation window" integrates the joint constraints related to assembly and fabrication. It can be processed algorithmically and give instant feedback on wether or not a set of non-parallel edges can be jointed simultaneously. This provides a tool for the exploration of a variety of alternative folded plate shell geometries.

The prototypes presented in this paper already suggest possible patterns and demonstrate the reciprocal relationship between the geometry of the plates and the joints. Two built structures allowed us to test and verify the proposed methods for fabrication and assembly while providing valuable information about the load-bearing capacity of the integrated joints.

For the application in a large-scale building structure, further research is required to determine if the integrated joints can replace additional connectors entirely or reduce their amount. A possible combination of integrated joints with additional metal fasteners has been demonstrated recently in the LaGa Exhibition Hall (ICD/ITKE 2014). Another possibility would be a combination of the 1DOF joints with integrated elastic interlocks (Robeller et al. 2014a; Simek and Sebera 2010)

Acknowledgements We would like to thank Stéphane Nicolas Roche for the discussions, as well as Gabriel Tschanz for assisting with the fabrication and assembly of the prototypes. We also thank Jouni Hakkarainen and the Metsa Group for the supply of information and materials.

References

Buri, H.: Origami – folded plate structures. Doctoral thesis, EPFL, Lausanne (2010)
DIBt: Allgemeine bauaufsichtliche Zulassung Kerto-Q Z-9.1-100, Paragraph 4.2 and Attachment No 7, Table 5. Deutsches Institut für Bautechnik (2011)
Graubner, W.: Holzverbindungen,Gegenüberstellung von Holzverbindungen Holz in Holz und mit Metallteilen. Deutsche Verlags-Anstalt Stuttgart (1986)
Hahn, B.: Analyse und Beschreibung eines räumlichen Tragwerks aus Massivholzplatten. Master thesis, EPFL, Lausanne (2009)
HESS: Hess limitless (2014). http://www.hess-timber.com/de/produkte/
Hundegger: Hundegger (2014). http://www.hundegger.de/en/machine-building/company/our-history.html
ICD/ITKE: Laga exhibition hall (2014). http://icd.uni-stuttgart.de/?p=11173
Koch, P.: Wood Machining Processes. Wood Processing. Ronald Press, New York (1964)
Krieg, O., et al.: Hygroskin: Meteorosensitive pavilion. In: Fabricate 2014 Conference, Zurich, pp. 272–279 (2014)
La Magna, R., et al.: From nature to fabrication: biomimetic design principles for the production of complex spatial structures. Int. J. Spat. Struct. **28**, 27–40 (2013)
Messler, R.W.: Integral Mechanical Attachment: A Resurgence of the Oldest Method of Joining. Butterworth Heinemann, Burlington (2006)
Neuhaus, H.: DIN EN 1995 (Eurocode 5) – design of timber structures. DIN Deutsches Institut fur Normung e. V. (2004)
Pizzi, E.: Renzo Piano. Birkhaeuser, Basel (2003)
Purbond: National Technical Approval Z-9.1-711/Single-component polyurethane adhesive for the manufacture of engineered wood products. DiBT. based on the test results of the Otto-Graf-Institute (MPA, University Stuttgart) (2011)
Robeller, C., Mayencourt, P., Weinand, Y.: Snap-fit joints: integrated mechanical attachment of structural timber panels. In: Proceedings of the 34th International Conference of the Association of Computer-aided Design in Architecture ACADIA, Los Angeles (2014a)
Robeller, C., Nabaei, S.S., Weinand, Y.: Design and fabrication or robot-manufactured joints for a curved-folded thin-shell structure made from CLT. In: Robotic Fabrication in Architecture, Art and Design 2014, pp. 67–81. Springer, Dordrecht (2014b)
Sebera, V., Simek, M.: Finite element analysis of dovetail joint made with the use of CNC technology. Acta Univ. Agric. Silvic. Mendel. Brun. **58**, 321–328 (2010)
Simek, M., Sebera, V.: Snap-fit joints: integrated mechanical attachment for structural timber panels. In: Proceedings of the International Convention of Society of Wood Science and Technology and United Nations Economic Commission for Europe, Geneva (2010)
Tetsuya, K.: Dento kanehozo kumitsugi : Sekkei to seisaku no jissai. LLP Gijutsushi Shuppankai, Seiunsha, Tokyo (2007)
Trautz, M., Herkrath, R.: The application of folded plate principles on spatial structures with regular, irregular and free-form geometries. In: Proceedings of the International Association for Shell and Spatial Structures (IASS) Symposium 2009, Valencia (2009)
Wyatt, E.: Puzzles in Wood. Bruce Publishing, Milwaukee (1928)

The Ongreening Pavilion

John Harding, Will Pearson, Harri Lewis, and Stephen Melville

Abstract This paper describes the work of Ramboll Computational Design during the design and construction of the Ongreening Pavilion timber gridshell. The structural approach involved form-finding bending-active timber laths, connected at intersections to form a doubly curved shell. The resulting form was simple to fabricate and assemble, realised using 6.5 mm thick Finnish birch plywood laths that could achieve high curvature while maintaining desired strength. Due to the random nature of the final lath topology, the resulting structure was extremely stiff in spite of its low material weight, acting similarly to a continuous monocoque. The fully demountable shell was first erected at Ecobuild 2014 in London.

1 Introduction

In late 2013, Ramboll Computational Design (RCD) were approached by Ongreening Ltd to assist with the design of a temporary pavilion to mark the launch of their new web-based platform for green building. The pavilion structure was required to be demountable, made from timber and fit within a given 10×8 m plot secured at Ecobuild 2014 in London, a public exhibition on sustainable building.

J. Harding (✉) • W. Pearson
University of Bath, Claverton Down, BA2 7AY Bath, UK
Ramboll Computational Design, Ramboll UK, 60 Newman Street, W1T 3DA London, UK
e-mail: john.harding@fastmail.fm; will.pearson@ramboll.co.uk

H. Lewis
Specialist Modelling Group, Foster and Partners, Riverside, 22 Hester Road, SW11 4AN London, UK
e-mail: hlewis@fosterandpartners.com

S. Melville
Ramboll Computational Design, Ramboll UK, 60 Newman Street, W1T 3DA London, UK
e-mail: stephen.melville@ramboll.co.uk

P. Block et al. (eds.), *Advances in Architectural Geometry 2014*,
DOI 10.1007/978-3-319-11418-7_19

During the initial design phase, Ongreening Ltd expressed an early desire for a single-layer gridshell formed from pre-stressed timber members of constant bending stiffness that form so-called elastica curves (Levien 2008). Although typical structures attempt to avoid bending behaviour, bending-active structures instead utilise bending to give rise to form in a pre-buckled state.

Traditional pre-stressed timber gridshells have a lattice with a fixed topology which is first laid out flat and then pushed into shape. The form is manipulated by allowing an in-plane rotation at each connection which is then fixed when the desired shape is found (Harris et al. 2004). Such methods have successfully been used on The Mannheim Multihalle (Burkhardt et al. 1976), The Downland Gridshell (Harris et al. 2003) and The Savill Building (Harris et al. 2008).

For the Ongreening Pavilion, due to the size restrictions of the site it became clear early on that a similar method would not be possible, and also restrict the solution space during design exploration whilst attempting to meet client requirements.

Instead, a method of bending each individual lath *sequentially* was adopted. This meant that a wider variety of forms could be explored but at the cost of requiring a more complex assembly. Due to the small size of the pavilion, this compromise was deemed to be acceptable. The final approach adopted was similar to that used on The Faraday Pavilion (Nicholas et al. 2013), the Smartgeometry 2012 gridshell (Kudless 2012) and recent developments in using bending-active elements for tension structures (Van Mele et al. 2013) whereby the grid topology is not pre-prescribed, but rather emerges as part of a simulation.

Bending active gridshells often have a rotationally symmetry cross-section, so that any torsion in the member does not effect the orientation of the connection details at member intersections. For example, the Faraday Pavilion was constructed using fibre reinforced polymer tubes. For The Ongreeening Pavilion however, flat timber laths were proposed with a particular grain direction, thus adding an additional design constraint.

The overall process of generating the geometry is shown in Fig. 1. This first involved the form-finding of planar curves that defined a freeform surface. This surface was then used to derive a set of primary laths based on geodesic lines. Secondary elements were then woven around these primaries, again by generating geodesics this time from a more random set of seed points that were refined manually. Finally, splice points along the laths were located, with the final geometry then exported for detailed design.

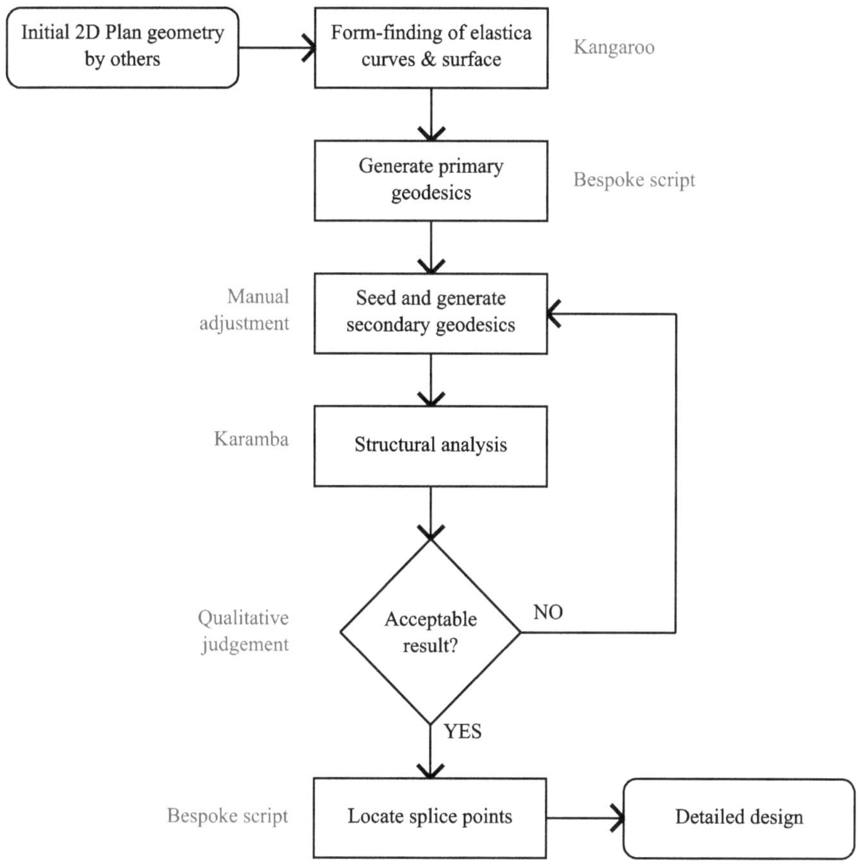

Fig. 1 Schematic diagram showing overall process in generating the final geometry

2 Initial Form-Finding

In theory, any doubly curved shape can be discretised into flat, straight laths so long as their paths follow surface geodesics (Pirazzi et al. 2006). However, during the process of construction it is advantageous for each lath to take its required shape using a minimal amount of effort. For example, by pushing the ends of a single lath with constant material properties and allowing it to deform a particular curved geometry is found for each single lath which can then be used to set out the shell. It was therefore decided to derive a set of *primary* laths that during the first-phase of assembly would take up a particular bending-active shape when constrained at their ends.

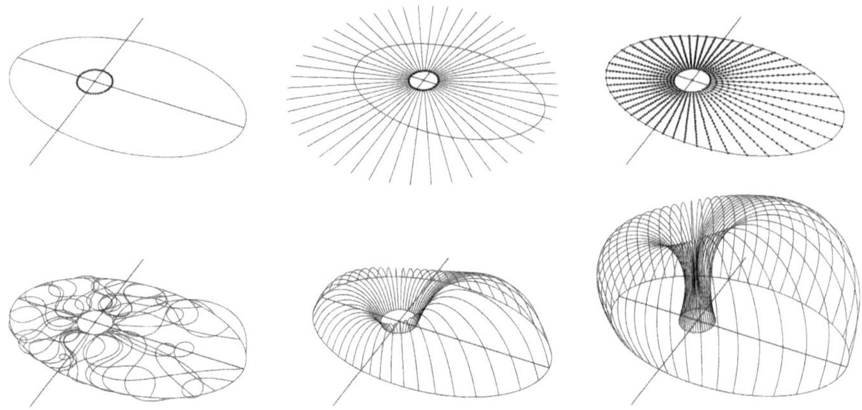

Fig. 2 Form-finding process for bending active shell

To simplify the process, these primary laths were initially assumed to be planar, vertically oriented and set out radially on plan. The setting out was based on a plan drawing by Ongreening Ltd with a central circular focus point and an elliptical boundary. A continuous shell was generated by lofting through these curves. The process (see Fig. 2) is described as follows:

1. Place node points on an inner circle
2. Create radial lines from each node point
3. Trim radial lines with outer ellipse and apply bending stiffness to the element
4. Increase natural length of elements and solve using dynamic relaxation
5. Constrain curves to a vertical plane by applying a small gravitational field
6. Adjust natural length of elements according to a trigonometric function

2.1 *Dynamic Relaxation*

Numerical form-finding techniques for large displacements were required to understand the equilibrium of bending-active shapes. A simple dynamic relaxation process using Euler integration was carried out using within Grasshopper with the Kangaroo physics plug-in (Piker 2013).

Following the initial form-finding it was found that with pinned connections the laths were clashing at the central area. This was resolved by replacing the inner pinned connections with fixed ones, constraining each lath at its end to a vertical direction and giving rise to a regular funnel structure at the centre of the pavilion. A total of 32 primary elements was found to give the required visual density for the pavilion whilst avoiding overlapping laths at the funnel (see Fig. 3).

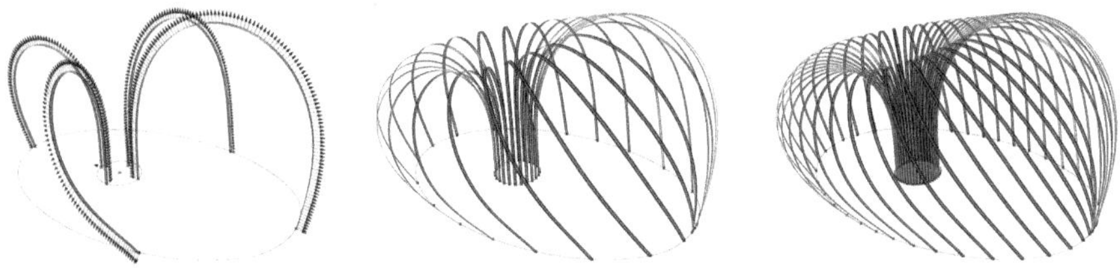

Fig. 3 Orientation of the primary laths showing the final density

2.2 Geodesic Replacement

A reference surface for the shell was created by performing a closed loft through the 32 primary curves. The original intention was that these curves would form the primary structural elements however due to the nature of the final surface when these curves were unrolled they were not straight and therefore harder to fabricate from standard timber sheets. In contrast, geodesic lines between two points on a continuous, doubly curved surface form straight lines when unrolled to a plane, making them very efficient for 3-axis CNC fabrication.

The original radial curves used to form the surface were therefore replaced by geodesics by seeding from the same points on the inner circle that were used to create the radial lines. These new curves were not restricted to the vertical plane and therefore introduced some torsion into the laths. It was assumed, however, that these resultant geodesics would still closely approximate the shape of a bent lath when supported at each end, assuming self-weight was negligible when compared to the pre-stress.

As well as generating a geodesic line between two given points on a surface, a geodesic or *plank line* may also be generated using an iterative approach (Kensek et al. 2002). By using a seed point, initial direction and step size, a new point is generated that is then projected back to the surface. This point becomes the next seed, with a new direction vector found by combining the previous direction with the current surface normal.

For the pavilion, this process continued until the projected point lay on the surface boundary. A smooth curve was then created by interpolating between the set of points. As with any integration approach, a relatively small step-size was required to produce an accurate result.

Each of the 32 primary laths were realised in 6.5 mm thick Finnish birch plywood and 100 mm wide. A second layer of laths interconnecting the primaries, was required to make the structure stable and act as a gridshell.

3 Placing of Secondary Laths

As previously mentioned, it is commonplace for gridshell structures to have a rectangular or diagrid topology. However, for the Ongreening Pavilion a more random aesthetic of laths was desired by the client, giving a woven-like appearance. Secondary laths were therefore wrapped around the primary elements again along geodesics in order to give the shell strength in all directions (see Fig. 4). This had the added benefit of making the shell membrane less directionally oriented (as with a diagrid) and therefore act more like a continuous shell or monocoque structure such as an egg shell.

Extending the method used for the primary elements, a number of secondary laths were created using a random distribution of seed-points and starting vectors. The secondary geodesics were generated in both directions from the their seed points, terminating at the shell boundary. As per the primary laths, the secondaries were realised again from 6.5 mm thick birch plywood, this time 75 mm wide.

3.1 Real-Time Structural Analysis

It was found that the initial random seeding of the laths led to significant bunching of the secondary geodesics. This was deemed unacceptable because: (a) an even distribution of laths was required by the client on aesthetic grounds, (b) it led to multiple laths weaving at nearby locations, thus negating the thinness of the shell, (c) the structural performance was poor due to large gaps with no secondary structure. Manual adjustment to the seed points was adopted in order to create a desirable arrangement of secondary elements.

This process was therefore driven by aesthetic, fabrication and structural grounds. For the latter, feedback was given due to the structural analysis being directly linked to the parametric geometry using the Grasshopper plug-in Karamba (Preisinger 2013) (see Fig. 5). This real-time linear analysis was conducted within

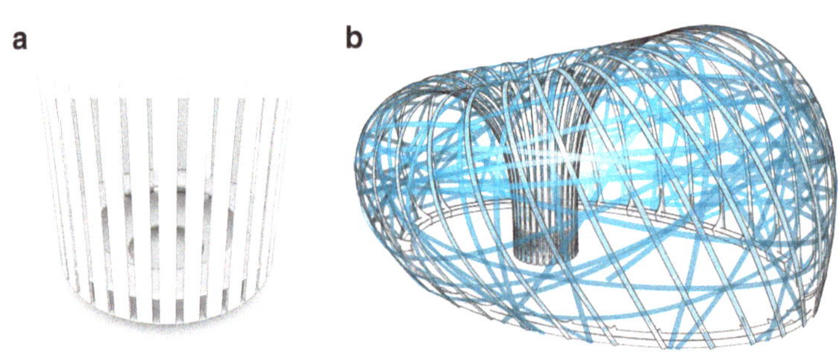

Fig. 4 (**a**) Central funnel detail. (**b**) Wrapping the primary laths with secondary elements in the computer model

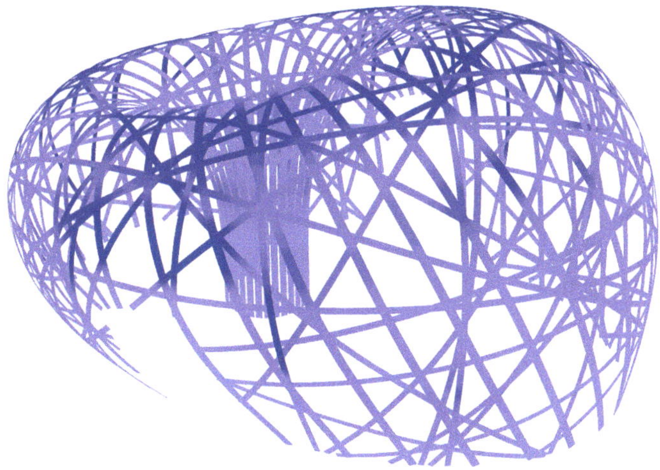

Fig. 5 Real-time structural analysis within the parametric model. *Darker shades* show larger deflections

the parametric model itself, allowing the performance of each configuration to be understood immediately.

The Karamba plug-in provided fast finite element analysis, perfectly suited for the conceptual design stage of small-scale, complex structures. Load combinations could be tested on the fly, along with redundancy checks, to better understand an appropriate density of secondary members for the shell. The ability to immediately interrogate a structure after each and every modification to the design was very useful when paired with the ability to interpret the results and make intelligent design decisions. Instead of finding an *optimal* result and excluding the human user, real-time analysis was used *in combination* with human intuition and qualitative judgement throughout the design process.

Once the secondary elements were finalised, the model was exported to Sofistik for further non-linear analysis and verification, which was found to give almost identical results. Both analysis models were set up to include self-weight and several imposed loads as prescribed by Ecobuild. The utilisation of each lath according to Eurocode 5 was determined using the sum of both the stresses due to loading and the stress induced by bending the lath to the required curvature. It was found that 6.5 mm thick Finnish Birch plywood was the best combination of high bending capability and high strength.

4 Details and Fabrication

The final topological layout of the timber laths is shown in Fig. 6. A numerical reference system was used for ease of assembly, with the pavilion being assembled as a kit of parts on site, with each primary element indicating which secondary was connected. These were simply pencilled onto the lath manually to avoid compromising the timber with engravings.

Although the random lacing of secondaries is visually interesting, when compared to the adoption of a diagrid there were several additional complexities that had to be addressed:

- The order of the overlapping laths is complex.
- Splice points have to be carefully set out so as not to clash with connections.
- A random topology means random connection locations along each lath.

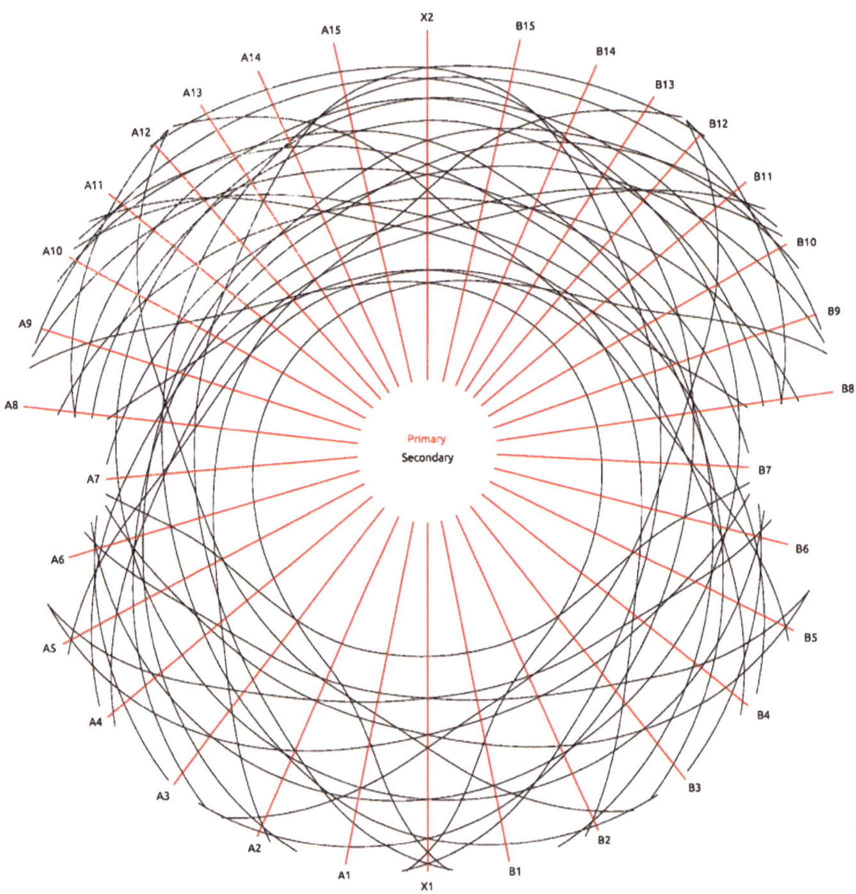

Fig. 6 Disk projection showing the topology of the radial primary elements (*red*) and secondary elements (*black*)

4.1 Connections

With the lath geometry fixed, a second parametric model was developed to produce fabrication information. With the laths represented by centreline curves, points of intersection could be easily identified. Pinned connections were to be made at each intersection between primary and secondary laths. These could be stored as curve parameters to allow the drilling locations to be mapped to the unrolled laths. Intersections between secondary laths (including self-intersections) were also identified and used when locating splice points.

The timber laths themselves were made up of multiple elements with temporary splice locations, thus making the structure demountable and able to be fabricated from standard sized plywood sheets. These splice locations were generated within the same parametric model as the connections. The length of each individual element was maximised, taking into account fabrication constraints, whilst simultaneously avoiding locating splice points near connections or overlapping laths (see Fig. 7). Secondary splice connections were slotted to allow for small adjustments during assembly and allow for the weaving of secondary laths over each other.

All connection and splice details (see Fig. 8) were designed to Eurocode 5, with physical testing of a single lath also conducted in order to verify the design when bent to tight radii, especially at the overlapping locations where stiffness is increased.

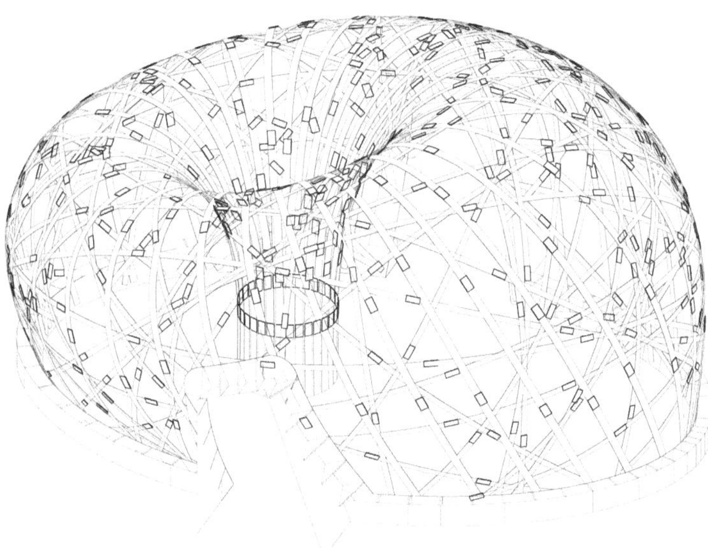

Fig. 7 Optimal splice locations avoided intersections between laths whilst minimising their number

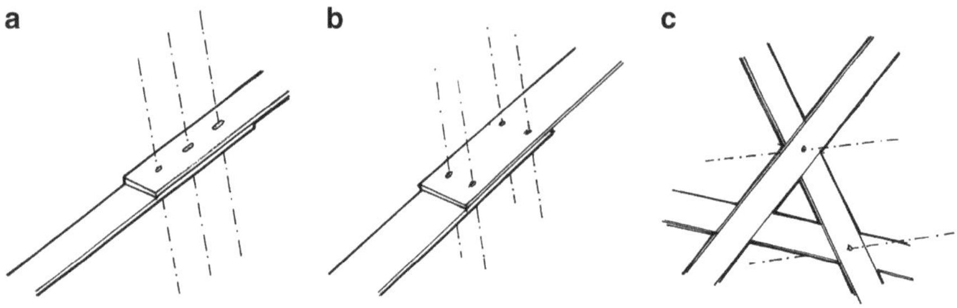

Fig. 8 Simple M6 bolted connection details used on the pavilion: (**a**) splice between two secondary elements with two slotted holes, (**b**) splice connection between two primary elements, (**c**) primary to secondary connections

4.2 Fabrication

The use of birch plywood was absolutely essential to achieve the bending radii whilst still retaining the required shell strength at such a thin size. The fact the shell could be made from such thin plywood (6.5 mm), also meant material use for this size of enclosure was minimised. Finishing materials were also made from birch plywood, for example the timber that wraps around the base of the structure.

The birch plywood used was FSC sourced and due to the pavilion being for internal use only, could be left with its original finish to express the timber. The use of timber also meant that the laths could be easily cut and connection holes drilled by a standard 3-axis CNC machine. Because of the geodesic geometry, when unrolled the laths were completely straight and therefore material wastage was almost zero when cutting the timber (see Fig. 9).

4.3 Perimeter Details

It was important to maintain a continuous boundary for attaching the primary and secondary laths and avoid free edges. Doorways were created along the line of the primary members at each side of the pavilion, with their edges a continuation of the base detail.

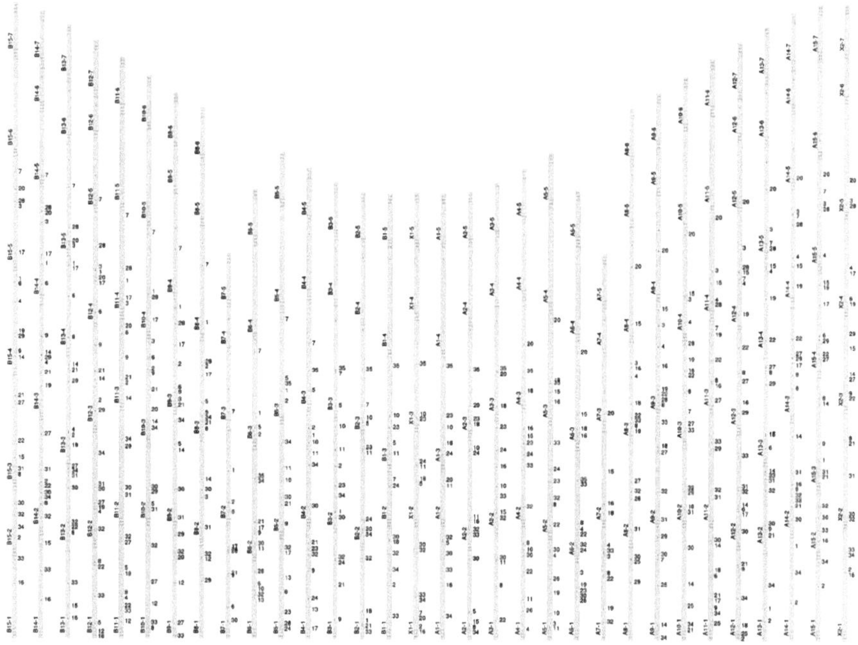

Fig. 9 As all the elements followed geodesic lines on the surface, when unrolled they were straight and simple and efficient to manufacture

5 Final Assembly

The gridshell took a small team 3 days to assemble, beginning with the perimeter and floor construction, moving onto the primary elements and then finally lacing with the secondaries.

The primary elements were prefabricated at their full length on the ground before bending and connecting to the perimeter. The secondary elements however were assembled in their actual position on the shell by wrapping around the primaries and connecting at splice points whilst moving around the structure.

Having real-time feedback from the analysis model again proved useful when determining the best sequence for constructing the pavilion. On their own each full length primary lath was unstable, relying heavily on the lateral support provided by the secondaries. Various combinations were explored until a secondary lath was identified that wrapped around the outside of the structure three times, at an approximate height of 2 m and therefore reachable from ground level. Each individual length of this particular secondary lath could be added incrementally, connecting the primary laths to each other and providing the necessary lateral restraint. In reality, during construction additional secondaries were also able to be added to provide additional stability.

Fig. 10 Internal view of realised pavilion at Ecobuild 2014

Fig. 11 External view of realised pavilion at Ecobuild 2014

The slotted connections at the secondary splice points proved to be essential in terms of allowing some tolerance to weave the elements, especially in areas of high connectivity. On one or two occasions, additional holes were drilled where several elements overlapped on site, however due to the nature of working with timber this was easily resolved.

The final structure is shown in Figs. 10 and 11. As the structure was open to the public is was important that it worked as predicted by the analysis model and this was indeed the case, with the entire structure feeling very robust.

Conclusion

The project was instigated and financed by Ongreening Ltd to mark the launch of a new web based platform for sustainable building. In this sense, the value of the pavilion was important not just financially, but also in showing the potential of timber as a versatile material.

The pavilion shows what can be achieved with computational design techniques when using traditional low-cost and low-carbon materials such as timber and relatively simple fabrication techniques as a constraint. The final result points the way to similar thin shell structures with bending active elements that do not require additional processes such as steam bending.

During the design process, real-time structural analysis provided crucial decision support, albeit only during the placement of the secondary laths. It would be interesting to examine how the initial form-finding could also be integrated in this process, as an alternative to the sequential ordering of design tasks.

Although initially assembled at the Ecobuild 2014 exhibition, due to its demountable and transportable design it is scheduled for future use by Ongreening Ltd. The structure follows a tradition of generating gridshells from timber, but in this case its geometry, demountability and assembly sequence suggests a unique approach to generating thin freeform shells.

Acknowledgements This work was completed while Harri Lewis was affiliated with Ramboll UK. The authors would like to thank Ongreening Ltd for the opportunity to be involved on the project.

References

Burkhardt, B., Otto, F., Arup, O.: Multihalle Mannheim. Institut für Leichte Flächentragwerke, Universität Stuttgart and Freunde und Förderer der Leichtbauforschung (1976)

Harris, R., Romer, J., Kelly, O., Johnson, S.: Design and construction of the downland gridshell. Build. Res. Inf. **31**(6), 427–454 (2003)

Harris, R., Dickson, M., Kelly, O.: The use of timber gridshells for long span structures. In: Proceedings of the 8th International Conference on Timber Engineering WCTE 2004, Lahti, pp. 1001–1006 (2004)

Harris, R., Haskins, S., Roynon, J.: The savill garden gridshell: design and construction. Struct. Eng. **86**(17), 27–34 (2008)

Kensek, K., Leuppi, J., Noble, D.: Plank lines of ribbed timber shell structures. In: Acadia 2000: Eternity, Infinity and Virtuality in Architecture, Washington, D.C., pp. 261–266 (2002)

Kudless, A.: The smartgeometry 2012 gridshell. Available from: http://matsysdesign.com/2012/04/13/sg2012-gridshell/ (2012). Accessed 16 July 2014

Levien, R.: The elastica: a mathematical history. Electrical Engineering and Computer Sciences University of California at Berkeley (2008)

Nicholas, P., Hernández, E.L., Gengnagel, C.: The faraday pavilion: activating bending in the design and analysis of an elastic gridshell. In: Proceedings of the Symposium on Simulation for

Architecture & Urban Design (SimAUD '13), San Diego, pp. 21:1–21:8. Society for Computer Simulation International (2013)
Piker, D.: Kangaroo: form finding with computational physics. Archit. Des. **83**(2), 136–137 (2013)
Pirazzi, C., Weinand, Y., et al.: Geodesic lines on free-form surfaces: optimized grids for timber rib shells. In: Proceedings of the World Conference on Timber Engineering, Portland, vol. 7 (2006)
Preisinger, C.: Linking structure and parametric geometry. Archit. Des. **83**(2), 110–113 (2013)
Van Mele, T., De Laet, L., Veenendaal, D., Mollaert, M., Block, P.: Shaping tension structures by actively bent linear elements. Int. J. Space Struct. **28**(3), 127–135 (2013)

The Caterpillar Gallery: Quadratic Surface Theorems, Parametric Design and Digital Fabrication

Roberto Narváez-Rodríguez, Andrés Martín-Pastor, and María Aguilar-Alejandre

Abstract The use of certain quadratic surface theorems has mainly been associated in architecture with the design of classical vaults, domes and piping. The work presented by the authors is intended to explore the potential of these theorems to be used in the generation law for more complex shapes in contemporary architecture. The paper shows the case study of a built full-scale prototype, The Caterpillar gallery, a project stemming from the combination of geometric research and teaching innovation.

Formal and structural experimentation take place in this project where, by starting from geometrical considerations, an efficient way of generating longitudinal spaces is proposed. One of the mentioned theorems applied to rotational cones provides the starting point for the generation of a set of concatenated surfaces that, once assembled, constitute a very stable self-supporting structure with a variety of possible applications.

1 Origin and Purpose

The use of parametric design and digital fabrication techniques is becoming widespread in architectural practice and provides the focus for many studies. The capacity of today's and tomorrow's software shifted conventional process into the realm of digital (Bentscheff and Gengnagel 2010), hence this should have direct repercussions within the academic sphere. Indeed, a number of courses on these topics are offered, both in undergraduate and graduate studies. However, these technologies tend to be implemented in the last years of the curriculum and are commonly unfastened from the beginnings of the training period (Fig. 1).

R. Narváez-Rodríguez (✉) • A. Martín-Pastor
University of Seville, Avda. Reina Mercedes 4A, 41012 Sevilla, Spain
e-mail: roberto@us.es; archiamp@us.es

M. Aguilar-Alejandre
University of Seville, C/Virgen de África 7, 41011 Sevilla, Spain
e-mail: maraguilar@us.es

P. Block et al. (eds.), *Advances in Architectural Geometry 2014*,
DOI 10.1007/978-3-319-11418-7_20

Fig. 1 Photographs of detail and general view of the installation

Fig. 2 Virtual model of the project. *Left*: East-front view. *Right*: South-side view

The Caterpillar is a project to materialise research carried out over several years to blend advances in geometry and technology with those of teaching, and to make them available to students at the earliest stage of their undergraduate training. The idea is that students learn the fundamentals of architectural geometry within a context where technological advances are commonplace Pottmann et al. (2007) and Weinstock (2010). This would allow students to acquire competences in the use of these technologies so that they can be developed and rendered during the rest of their training period in a more motivating and efficient way. On the other hand, attention must be paid in order not to miss essential competences and generate gaps of knowledge. Although these academic matters related to the development of the project were also revealing, they lie outside the scope of this paper, which focuses on the case study of the architectural process itself and the potential of its findings for application or use in other pieces of work or research (Fig. 2).

Within this context, the challenge was to investigate the creation of a motivating and attractive project to develop, fabricate and assemble a full-scale prototype. The main components were: simple generation law, covering of the fundamentals of architectural geometry throughout the whole process (with an emphasis on the use of developable surfaces), and the possibility of rendering the results through further research and development. Computational systems, basic parametric design, and digital fabrication provided the tools for the project to come to fruition.

The generation of a ten-meter-long gallery was proposed which symbolically connects the two schools (Architecture and Building Engineering of the University

Fig. 3 Photographs of the built full-scale prototype. *Left*: top view. *Right*: interior view

of Seville – Spain) in their adjoining gardens. The longitudinal path had predominance over the other dimensions and the generative law was articulated around that path to finally result in a set of conical surfaces. Parameterisation was also useful to respond to the location and environmental requirements during the design process (Fig. 3).

2 Quadratic Surface Theorems and Generation of the Form

Although simple uses of these theorems can be found in architecture in ancient times, no written evidence has been found about their systematisation until the modern day. The geometers who contributed towards the statement of theorems of the intersection of quadrics come from the French tradition in the 18th century; this statement was later developed in the nineteenth and twentieth centuries, and has even been continued by certain authors in recent times (Gentil-Baldrich 1997).

As a result of several studies on classic vaults, where theorems of the intersection of quadrics were involved, it was understood that their implicit geometric relationships were also applicable as a resource for the generation of compositions of a more complex nature through the use of parametric design. For the generative law proposed in this paper, the following statement by Gaspard Monge, has been chosen (Taibo 1983): *"If two quadratic surfaces C1 and C2 are circumscribed about a third C3 along their contact curves c1 and c2, their intersection curve decomposes into two planar curves i1 and i2, which pass through the points A and B, which are common to the contact curves c1 and c2."*

By *circumscribed* is meant that C1 and C2 osculate or touch C3 along curves on which the surfaces share the local tangent plane. In this work, quadrics or quadratic surfaces C1 and C2 are rotational cones and quadric C3 is a sphere. This ensures that the intersection curves of the cones are planar curves, particularly conic sections and, in this case, ellipses.

The interest of forcing the curves to be planar lies mainly in the analysis and construction simplicity. Cones are surfaces which offer an easy-to-model geometry and, since they are developable, are also easy to construct. Apart from this, from the design point of view, cones provide the possibility of defining various degrees of spaciousness since their geometry converges at a point.

2.1 Geometry

Due to the lineal nature of the required space, the first element of the generation law consists of a 3D polygonal line that travels through a determined path. This polygonal line is composed of vertices whose positions are alternately near the level of the floor or near the average height of a human. Each of the segments of this polygonal line constitutes the axis of the rotational cones that form the gallery. These cones should be built by circumscribing them about spheres so that their intersections result in planar curves (ellipses). From these curves, only those that connect the inner space of each pair of cones should be preserved. Finally, only the fragment of the conical surface between the planar curves that connect the cones and the plane defined by the floor is maintained (Figs. 4, 5, and 6).

At every point of the 3D polygonal line, there is the centre of a sphere which defines the spaciousness required by every point position along the path, thereby providing the possibility of generating a different space at the floor level from the space at the ceiling level. In the end, the parameter to control the volume of the interior space is the radius of the relevant sphere. For this purpose, six spheres at the lower level and five spheres at the upper level were created.

Figure 5 shows the polygonal, the spheres and the boundary resulting from the intersection between the set of cones and the plane defined by the floor. This boundary not only defines the limits of the gallery on a ground level, but it also

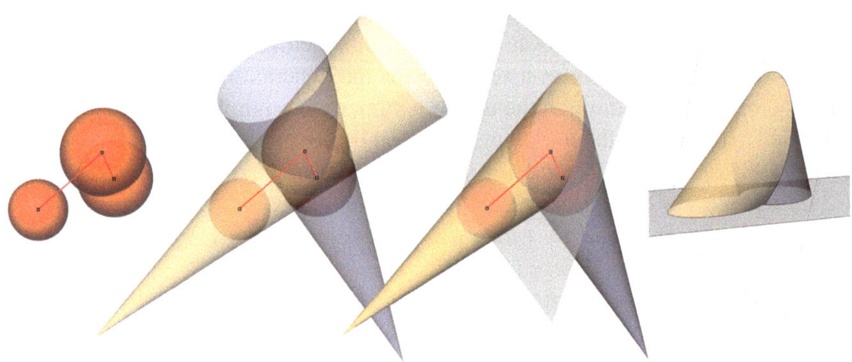

Fig. 4 Perspective view of the generation law for the first pair of conical fragments. From *left* to *right*: polygonal line and spheres at every vertex of the line, cones circumscribed about the spheres, cutting plane containing the common ellipse of connection for the two inner spaces, cutting plane defined by the floor

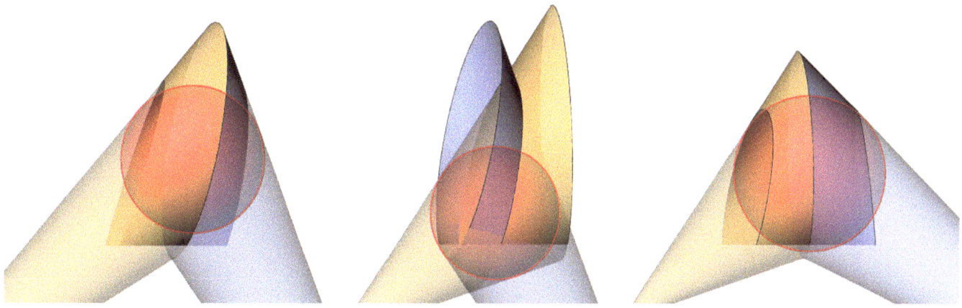

Fig. 5 Front view of the growth law for the first four conical fragments. *Left*: cones 1 and 2 circumscribed about the upper sphere 2. *Middle*: cones 2 and 3 circumscribed about the lower sphere 3. *Right*: cones 3 and 4 circumscribed about the upper sphere 4

plays a decisive role as an auxiliary element of support and acquisition of the specific curvature of each wooden surface.

2.2 *Parametric Definition*

The generation law was defined parametrically in Grasshopper, under Rhinoceros, to facilitate the creation and control of the form within the environment provided for the installation. The geometric principles defined in previous sections and the implicit relationship between the three-dimensional shapes and certain two-dimensional constructions were key concepts to define the parameterisation. The plane defined by the two axes of every pair of adjacent cones was the construction plane to carry out the relevant two-dimensional geometric constructions.

The definition was thought so that different solutions were easily created by modifying the two groups of input data: (i) the set of points defining the path of the 3D polygonal line and (ii) the spaciousness desired at every point (radius of the relevant sphere), where odd points refer to the lower level and even points refer to the upper level (Fig. 7).

These two groups of parameters control the position and the space created by the definition and are opened to be modified by the final user without modifying the geometric principles which govern the definition.

3 Structural Analysis and Experimentation

The use of developable surfaces considerably facilitated the experimentation with paper and cardboard as well as the relevant tests to transfer the scale model behaviours to the full-scale model. These are also surfaces which are simple to understand, generate and handle.

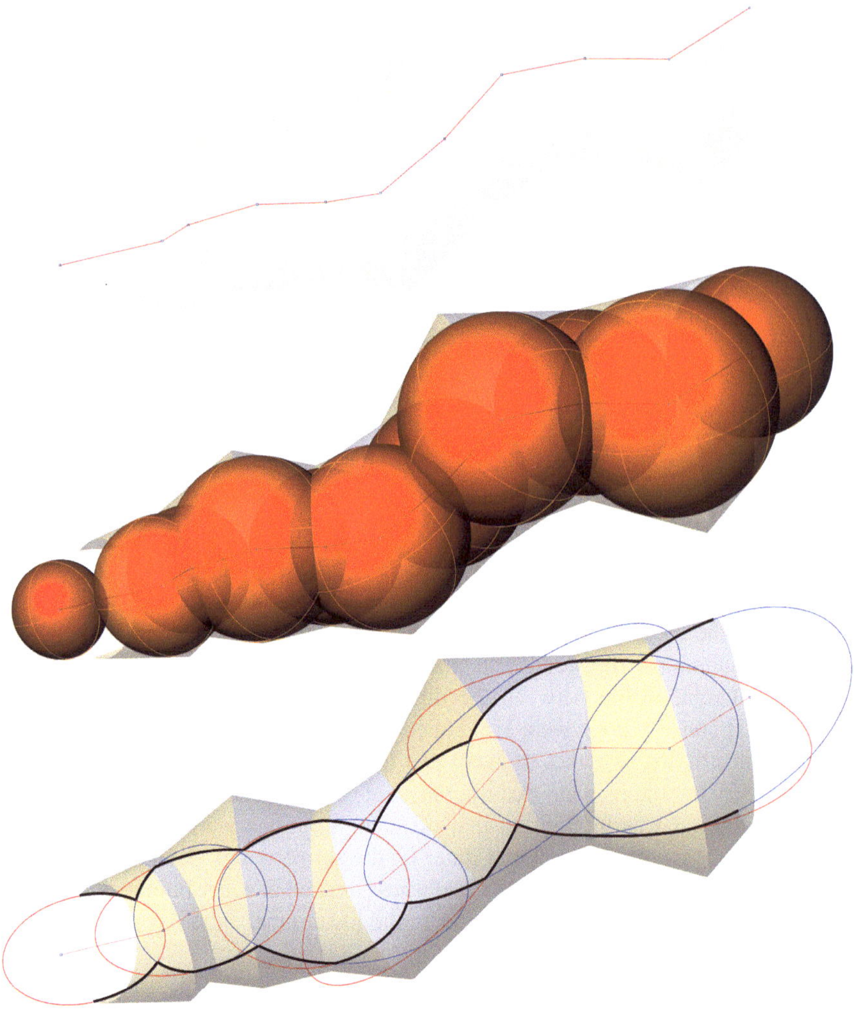

Fig. 6 Top view of the generation and growth processes. *Top*: axes defining the generation path. *Middle*: lower spheres (*dark*) and upper spheres (*light*) defining the spaciousness at the two levels. *Bottom*: intersection ellipses between the conical surfaces and floor defining the boundary at the floor level

The scale models turned out to be very useful to physically experiment with structural and constructive behaviour and with the assembly process. Furthermore, finite-element analysis was also implemented before the final step was taken. For this purpose, and before full-scale experimentation, two scale models were built; the first at 1:10 scale, and the second at 1:4 scale (Fig. 8).

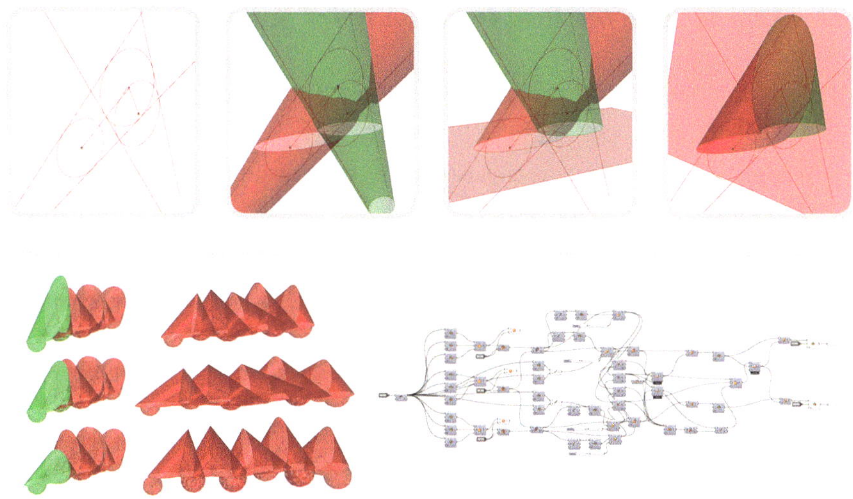

Fig. 7 Parametric definition of the model in Grasshopper for a pair of adjacent cones. *Top*: 2D construction to start the definition for a pair of cones. *Bottom left*: different variants of the resulting model after changing input data. *Bottom right*: Grasshopper screen definition

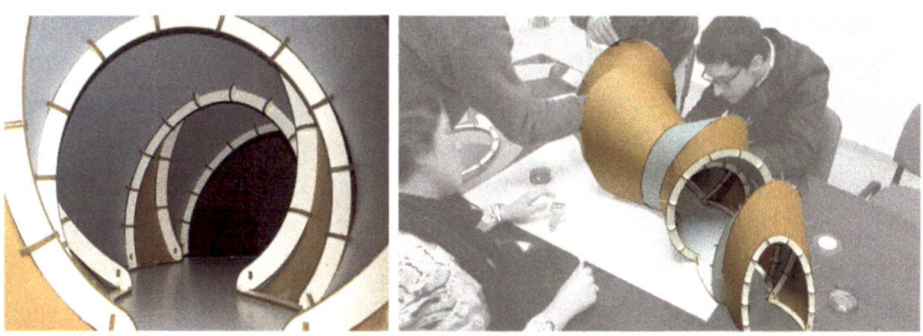

Fig. 8 Pictures of the students working on the initial model with an auxiliary structure. *Left*: interior view. *Right*: external view during the assembly process

3.1 Initial Model in Paper at 1:10 Scale

This model was mainly used as an introduction to the laser cutter machine within the academic tasks. It was useful to check certain spatial relationships as well as some perceptive matters. However, technical experimentation was very limited due to the small size and the difference in behaviour between the paper and the wood for the final prototype. An auxiliary structure made up of elliptic arches (common planar curves) was laid out as a guide to bend the paper and make it acquire the 3D form of the conical surfaces. It was later demonstrated that this structure was unnecessary. Nonetheless, this model was successful in generating sponsorship from companies.

3.2 Approaching the Real Behaviour; 1:4 Scale Model

After the first model, a new model was needed on a bigger scale (1:4) for the following purposes:

- To approach the real size, and consequently, the real structural behaviour of the set of surfaces. This allowed us to check that surfaces, once linked together, constituted a stable set with considerable rigidity.
- To define details and test their behaviour. Two major developments where achieved in this regard: (i) The use of joining strips for panel assembly, which had to satisfy the main target on this matter; to ensure curvature continuity between adjacent panels after bending the planar developments. For this purpose, the necessary width for the strips was established at 150 mm. Moreover, the joint had to be executed with four lines of screws; two along the longest boundaries of the strips and another two on both sides of the panel junction line. (ii) The solution of the junctions between the surfaces and the floor; thirty-millimetre-thick wooden wedges were used to hold the surface and cover the angle between the surface and the floor.
- To find a procedure to turn the installation into an easy-to-assemble construction. In this regard, and considering the ephemeral nature of the gallery, the use of nylon cable ties proved highly convenient. The key geometrical principle was the fact that joining points from the planar developments of different adjacent surfaces, whose positions in the space coincide, would make the surfaces acquire their 3D shape automatically. Thus the procedure consisted of using the nylon ties by way of threads to make the relevant points meet at their common position in the space. In order to ensure the correct operation of this principle, it was essential to first lay out the surface-floor edge along the rigid ellipses defined by the wedges. Otherwise, there would have been an infinite number of undesired 3D forms different from the original model.

3.3 Experimenting at 1:1 Scale

Finally, experimentation at 1:1 scale was also carried out, especially to test the behaviour differences as well as to check the feasibility of the assembly and of the set-up procedure. The main points and decisions were:

- Approve the material. After considering metal sheets and wooden panels, the final election was the use of only five-millimetre-thick black poplar wood laminate panels to materialise the project. The test carried out with an isolated conical surface revealed considerable deformations in all three dimensions as well as instability, with symptoms of near collapse. Nevertheless, the behaviour of the joining strips was very successful and curvature continuity after bending was no problem at all as can be appreciated in Fig. 9. Moreover, as expected,

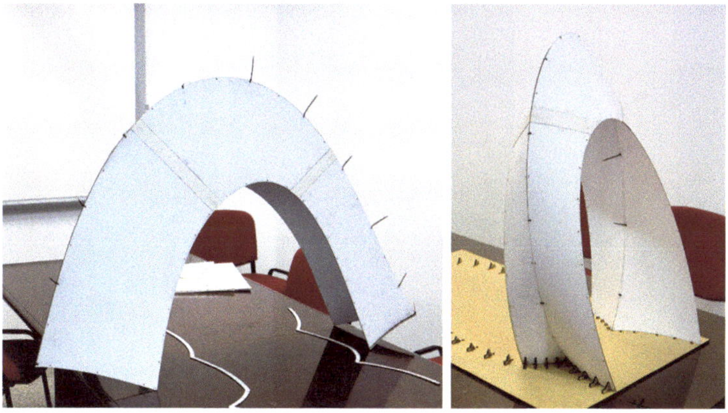

Fig. 9 *Left*: undesired form acquired by the linked surfaces without having been previously set to the rigid shape defined by the wedges on the floor. *Right*: in contrast, this picture shows the simple way to acquire the form by linking the surfaces first to the rigid shape of the floor. Joining strips between pieces to make up the planar unfolding of the surfaces can be appreciated, as well as the continuity of the surface after bending

the curvature radius was perfectly supported by the panels without cracking; the minimum existing radius in the project was 793 mm while the radius supported by the panels in the tests was under 600 mm before starting to crack.

- Test the structural behaviour of the linked surfaces. Although the test of an isolated surface was not successful in terms of structural behaviour, the gallery was expected to behave successfully after linking all the surfaces. The emergence of this new structural behaviour was proved with a finite-element structural analysis. Whereas the deformed shape of the isolated surface showed displacement values over 700 mm in the z axis, the analysis of the whole set of linked surfaces showed displacement values under 20 mm at the same point (Fig. 10).
- Check the feasibility of the assembly procedure. The trials carried out with the first two surfaces were enough to demonstrate that the procedure stated for the 1:4 scale model was perfectly valid at 1:1 scale. The details are described in the following section.

4 Fabrication and Assembly

The digital fabrication of the installation was carried out in Fab Lab Sevilla, with the collaboration of the Higher Technical School of Architecture and the Higher Technical School of Building Engineering of the University of Seville. A laser cutter machine was used for the fabrication of the scale models and a three-axis CNC milling machine for the full-scale prototype. The entire panel fabrication was done with 2D cutting profiles. That is not to say that fabricator was not working in 3D,

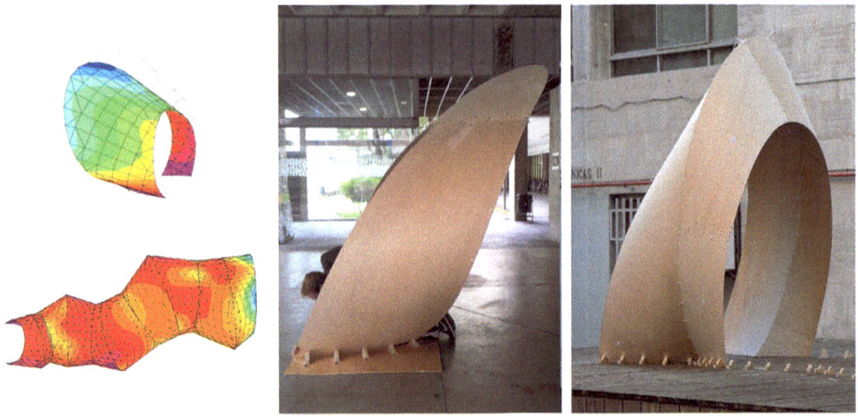

Fig. 10 *Left*: finite-element analysis. Maximum deformations are represented in *blue*: 700 mm for the isolated surface (*top*) and 40 mm for the set of surfaces (*bottom*). Photographs of the full-scale experiments. *Middle*: unacceptable deformation of the isolated conical surface. *Right*: successful behaviour of the joining strips to ensure curvature continuity after bending

but the understanding of 3D had more to do with assembly and fabrication of actual materials in space instead of digital models (Meredith and Kotronis 2012).

As stated in previous sections, the first step required for the correct wood bending and assembly of the different surfaces is the installation of the wedges. These were fabricated from 30-millimetre-thick black poplar plywood panels and would have to be laid out accurately in order to help wood acquire the correct shape. For this purpose, a laser total station combined with full-scale wooden templates were used to set them out on the definite location. Due to the different angle (measured on a vertical plane) that the surfaces form with the floor at every point, 100 hundred different wedges were fabricated and numbered (10 pieces per surface). The right position of these wedges facilitated considerably the later easy installation of the wooden conical fragments (Figs. 11 and 12).

Each conical surface needed a minimum of three standard (1,250 × 2,500 mm) plywood panels and a maximum of seven. For the development of the digital fabrication, the details were generated with the conventional use of CAD drawings and the auxiliary support of the parametric 3D model. At this level of undergraduate studies, students still lacked the necessary proficiency to generate automated 2D outputs from the parametric model. However, a full automated process is perfectly feasible for the system proposed.

The assembly of the panels to complete a whole fragment of unfolded conical surface had to be executed in two stages: (i) the first phase was the assembly of panels until they reach the transport limit size, (ii) the second phase, after transportation, was the completion of each unfolded surface before setting them up in the definite position. This work was carried out on the floor, without the need to operate on the two sides of the surface, thanks to the use of steel t-nuts hammered into the plywood panels.

Fig. 11 *Left*: auxiliary sectioning planes (in vertical position) to determine the shapes of the wedges. *Middle*: photograph of the definite position of the wedges after setting out with the total station and the full-scale templates. *Right*: installation of nylon cable ties to link the wooden surfaces to the wedges

Fig. 12 *Top*: detail of the different pieces/panels to make up the unfolded surface. *Bottom*: photograph of the second phase of the panels assembly

The final task was the installation and linkage of the conical surfaces. The steps taken for this purpose can be summarised as follows:

- Transportation of the unfolded surfaces from the assembly point to the installation place. Special attention had to be paid in order not to subject the surfaces to curvature radii under the limit supported by the wood. To ensure the integrity of the panels, they were transported trying to keep the original planarity until the final bending time.
- Bending of the surfaces. Once the surface was at the installation place, bending would be executed in the following way; first, the feet of the surfaces must be

Fig. 13 *Left*: transportation of the unfolded surface to the installation place. *Middle*: bending of the surfaces. *Right*: final bending of the surface and installation of ties

taken to the definite position at the same time as the upper part of the surface must be raised and held to avoid inappropriate deformations (Fig. 13).

- Installation of the nylon cable ties. To conclude the process, and once the surface has been bended and held by hand, cable ties must be applied by starting with the linkage of the surface and the wedges on the floor, and then, at the connection with the previously installed surface by starting with the keystone of the common arch.

Conclusion

The theorem of the intersection of quadrics studied, applied to rotational conical surfaces under the conditions stated in this work, together with their parameterisation, constitute an efficient generation law for the creation of longitudinal spaces.

The geometrical principles that govern the process provide the set of surfaces with very stable structural behaviour; a far cry from the weak properties of each of the isolated surfaces which make up the whole. This renders the use of any auxiliary structure unnecessary since the set operates as a single self-supporting surface. Hence, although the material needed to carry out this kind of installation is subject to only minor strength capacities, the main requirement is flexibility to support the minimum existing bending radius. In this work, a single layer of only five-millimetre-thick wooden laminate panels perfectly matched the needs of the project.

The simplicity of conception, fabrication and execution, in addition to high structural stability, draws interest not only to the solution achieved in this project, but also to the potential applicability of this concept to other uses; other applications on ephemeral architecture, non-longitudinal constructions by extension of the generation and growth laws to 2D and 3D laws for the design of spaces with a greater freedom of form, double-layer structures, as well as the generation of components that can be adapted for the materialization of other kinds of design surfaces (Fig. 14).

(continued)

Fig. 14 Progress of the installation and linkage of the ten wooden surfaces which make up the whole prototype

Acknowledgements We would like to thank Fab Lab Sevilla and the schools of Building Engineering and Architecture of the University of Seville, as well as the group of students and lectures from these schools for their support and collaboration throughout the development, fabrication and assembly of this project. Thanks to the sponsors which funded the fabrication of the gallery, especially to Luz Diaz for the photography and Visualmov for the production of the documentary short film about the fabrication and assembly (http://vimeo.com/88452240).

References

Bentscheff, I., Gengnagel, C.: Towards teaching generative design in architecture. In: Advances in Architectural Geometry, pp. 113–128. TU Wien, Vienna (2010)

Gentil-Baldrich, J. M.: Sobre la intersección de las cuádricas de revolución de ejes paralelos. Universidad de Sevilla – Departamento de Expresión Gráfica Arquitectónica, Spain (1997)

Meredith, N., Kotronis, J.: Self-detailing and self-documenting systems for wood fabrication: the Burj Khalifa. In: Advances in Architectural Geometry, pp. 185–198. TU Wien, Vienna (2012)

Pottmann, H., Asperl, A., Hofer, M., Kilian, A.: Architectural Geometry. Bentley Institute Press, Exton (2007)

Taibo-Fernández, A.: Geometría Descriptiva y sus aplicaciones. Tomo II. Tébar Flores, Madrid (1983)

Weinstock, M.: The Architecture of Emergence : The Evolution of Form in Nature and Civilisation. Wiley, Chichester (2010)

Constructing Complex Geometries: A Case Study on the Cité des Civilisations du Vin in Bordeaux, France

Benjamin Soquier, Raphael Mizzi, Daphné Dureisseix, and Jean-Baptiste Valette

Abstract A number of geometric optimization were applied to the Cité des Civilisations du Vin in Bordeaux, France. The façade is comprised of a curved panels attached to wood laminated, also called glulam, beams.

The purpose of this paper is to describe the research to construction process undertook by Vinci Construction France as part of the contracting consortium for the project when dealing with geometries involving concrete, laminated wood, glass panels and metallic panels.

1 Introduction

In 2012, the consortium mainly comprising of Vinci Construction France (VCF) began working on the production and construction phase of the *Cité des Civilisations du Vin* in Bordeaux designed by the French architecture firm X-TU. The complex geometry of the building, notably its free-form volume, poses a new set of challenges in digital and architectural geometry. Over the course of this past year, the Project Modeling team at VCF has led a number of form finding and fabrication optimization studies for the consortium from a simple initial mock-up of the concrete structure and foundation to the fabrication process of the wood laminated structure.

The main focus of this article will deal with the relationship between the digital 3D geometry of a project and its diverse possibilities of fabrication. It is an exploration of the thin line between the virtual and the reality, the analog and the digital whilst giving insight on the optimization methods in geometry and fabrication for the *Cité des Civilisations du Vin* (CCV) in Bordeaux. The in-house geometric research as well as the in-house fabrication create a unique process by streamlining the engineering and the industrial production within a single entity. Understanding materials, their fabrication methods and cost effectiveness in their

B. Soquier (✉) • R. Mizzi • D. Dureisseix • J.-B. Valette
Vinci Construction France, 61 avenue Jules Quentin, 92730 Nanterre, France
e-mail: benjamin.soquier@vinci-construction.fr; raphael.mizzi@vinci-construction.fr; daphne.dureisseix@vinci-construction.fr; jean-baptiste.valette@vinci-construction.fr

P. Block et al. (eds.), *Advances in Architectural Geometry 2014*,
DOI 10.1007/978-3-319-11418-7_21

geometrical transformations opens the way to truly express the architectural design. As the architectural and construction field continue to evolve in the rapid era of digital information, the CCV is an extraordinary example of the use of accessible technologies and the cohesion between traditional methods of fabrications and new and innovative methods.

The complexity of the project can be compared to a number of previous works. In the case of using glulam beams to build complex geometries, we can mention the Onepoto footbridge by Beca Architects in New Zealand which is composed of ribs of curved laminated wood or the main lobby of Credit Valley Hospital by Farrow Partnership in Mississauga, Canada. We can also mention the Centre Pompidou Metz or the Yeoju golf club, which deals with non-planar glulam beams. In the case of the CCV, the choice was made to keep all the glulam beams planar in order to reduce the fabrication complexity and to help the coordination between the wood structure and the skin.

2 Project Concept and Context

The *Cité des Civilisations du Vin* is located in the heart of Bordeaux, France in the old docks district of the city. As part of the urban renewal program, the project stands out as a key architectural expression of the city's desire to develop the region's attractiveness and assert the city's identity as the region's cultural and administrative capital (Fig. 1).

The architectural concept devised by the French firm X-TU is based on the movement of liquids. In reference to the waters of the Garonne River at the foot of the building as well as the flow of wine in a glass, the building is composed of two main levels allocated to exhibition areas along with a nine-story tower dedicated to administrative use and a restaurant.

Fig. 1 Architectural rendering of the project (Published with kind permission of © XTU architects 2013, Paris)

The unique form of the envelope spans over 8,000 m^2 of façade and is divided into two main parts: an exterior cladding comprised of glass and metal panels and a curved wood-laminated structure which is one of the main carrying load system for the panels. The fundamental architectural expression of the envelope must be soft, fluid and entangled and must evoke a movement rather than a fixed form.

3 Primary Structure: Concrete

3.1 *Purpose of a BIM Model*

The capabilities of a BIM model in a construction phase can be defined by the ways in which the model can translate and describe the consecutive construction phases and resources (i.e. money, workforce and access to equipment). In other words, the model is not only composed of geometric shapes in a three dimensional environment, but also information data that can evolve according to time and resource.

For the CCV, the concrete model was designed to output specific material quantities and methods of construction while considering a timeline of construction on-site. This meant creating an automated method of producing a 3D model according to different key phases. Sets of drawings are then automatically exported from the 3D model (Fig. 2).

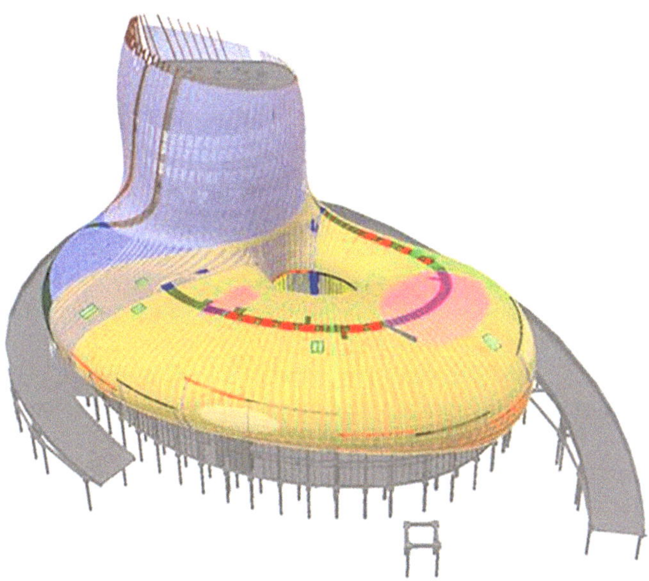

Fig. 2 Review model

4 Secondary Structure: Laminated Wood Beams

The properties of a laminated wood beam, also known as glulam beams, allows for a considerable span when compared to traditional the wood plank beam. This capability paired with the possibility of creating curved planar shapes can create intricate structures which best express the delicate and radical architectural expression of the free-form volume of the building. Given the knowledge in the history of the material, wood stands out as traditional material that never ceases to evolve with new methods of construction and architectural geometry.

The complex façade panels are held by a series of curved laminated wood beams spanning as much as 27 m with heights surpassing 8 m. Designed and constructed alongside VCF's subsidiary Arbonis – Caillaud-Lamellé-Collé, specialized in wood studies and fabrication, the laminated wood beams are a defining example of the possibilities of the material when combining innovative and traditional methods of fabrication. The main goal of the team was to develop a process between the production phase and the construction phase in order to retain detailing and information such as attributes from the 3D geometry to the shop fabrication.

Fabricated in two industrial workshops in France (Chemillé and Verosvres), the wood beams make up over 500 unique pieces. Each laminated wood beam is defined as a "planar section cut" of the envelope skin ensuring an aesthetic coherence inside and outside. As such, the uniqueness of each beam is defined by a unique planar curvature as well as a unique height and length.

It is then fundamental when addressing such imposing pieces and traditional time delays to develop a process of design to fabrication to ensure minimal errors and economically viable.

4.1 Shape and Form of the Wood Beams

4.1.1 Geometry Shape

One of the key challenges in the shaping of the wood beams in the production was the precise fidelity to the architectural envelope represented by a 3D geometric surface. The basis of the geometry for each beam is defined by a planar cut into the reference surface resulting in a single planar curve. This in turn gave way to the initial aspect of the beams' curvature (Fig. 3).

A minimal distance is set between the reference surface[1] and beams in order to make way for different façade components or elements such as insulation, fixtures

[1]This reference surface is defined by the Design Firm and represents the interior surface of all cladding panels.

Fig. 3 Interior rendering of design (Published with kind permission of © XTU architects 2013, Paris)

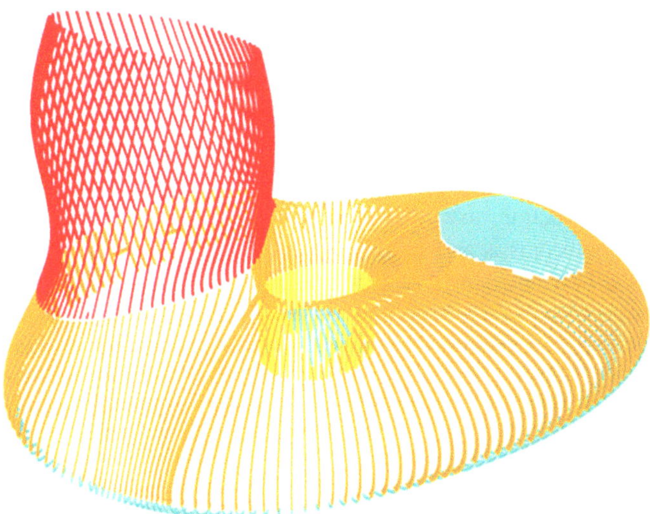

Fig. 4 Each of the four colored area indicates a different minimal distance between the exterior cladding and the edges of the laminated wood beams

and fastenings. Depending on the location of the beam, the variable distance is calculated perpendicularly from the reference surface to the linear offset of the beam on its section edge (Figs. 4 and 5).

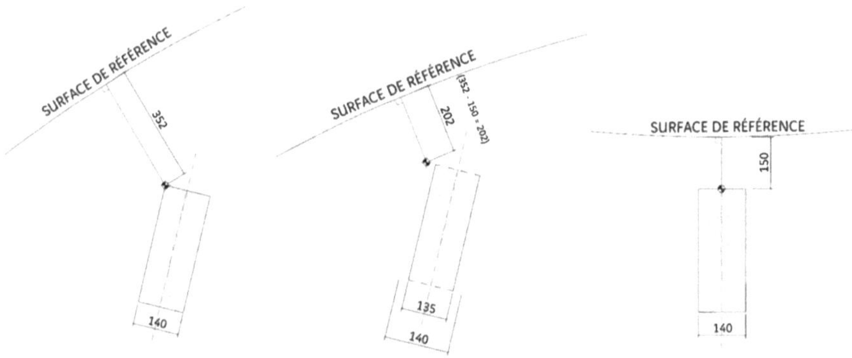

Fig. 5 Principle of minimal distance results. *Left*: section cut of wood beam and the exterior cladding in main area (*orange* in Fig. 1). *Middle*: tower area (*red* in Fig. 1). *Right*: patio area (*yellow* in Fig. 1)

4.2 Fabrication

4.2.1 Default to Factory Settings

The current post-modern era of architecture is characterized by the continuous research of form and aesthetics. Although ambitious and innovative in many ways, this has pushed certain fabrication processes into an exaggeration of digitalization with gauche applications to material transformations. It is clear and undoubtable that robotization and computer-aided tools have facilitated a large number of industry processes in production quality, delay and cost. The fabrication process for the laminated wood beams is an incredible example of the convergence of traditional methods which are instinctive and experienced based and modern technologies. Although we can go as far back as Philibert Delorme, architect in the sixteenth century who experimented on variable wood section assemblies, it is in 1890 when Otto Hetzer introduced adhesives to create laminated wood beams. Similarly, the laminated wood beams are made up of a number of smaller wood panels with the same section assembled together to form a larger beam.

The wooden beams are prefabricated in large hangars in the processing plant based on the geometry. A beam is composed of two or more pieces according to curvature, size, load calculation and transportation limits. Each pieces is formed using large parallel metal brackets shifted at different heights to recreate the beam curvature.

After the wooden panels receive their adhesive coats, workers align the panels onto the mold before applying heavy-duty fasteners to clasp the panels to each other (Fig. 6).

Fig. 6 Photos showing the bending process of the glulam timber

4.2.2 Adjusting Settings for Best Result

When using the metal brackets in material fabrication or transformation, the subject of reusability quickly arises to optimize time constraints and efficiency of production. In the case of a range of unique pieces, using the larger metal brackets to shape more than a single piece is a clear advantage when stacking the pieces as show in Fig. 6. It is then possible to assemble two or more elements using a sorting system which can pair and group a number of pieces together regardless of their position in the building. Using the traditional mold system in the factory's fabrication process and wedges inserted onto the brackets, the sorting system was devised in order to pair the inside surfaces of the beam (then represented by a planar curve) with a tolerance of 250 mm. This tolerance represents the maximum width of the wedges which slide on each bracket to accommodate the difference of shape between pieces. The sorting system takes into consideration for each piece the radii of curvature, the length and cord-length. Using a particle system solver (Daniel Piker's Kangaroo in GH), the pieces are then aligned in a best-fit scenario to adjust the beams' contour as close as possible.

Full scale fabrication began in spring of 2014 with the following process: fabrication groups of two to three pieces are calculated using a the sorting system; then the pieces are automatically exported as shop drawings and database tables which include the name, the number of fastening brackets, the radii of curvature, the number of panels forming the laminated beam as well as other relevant information for the workers; the fabrication team in the factory reviews the drawings and data before coordinating with the shop team; the molds are then created and the beams assembled; finally the prefabricated laminated wood pieces are delivered on-site for final assembly and mounting by crane.

As shown in Fig. 7, groups of one, two or three are composed to minimize the adjustment of the metal brackets which are tedious to manipulate. The drawing shows, from left to right, a reference drawing along with one or two consecutive pieces. Figure 8 follows the automated 2D drawing export which show the different values necessary for shop execution such as the spacing between the brackets (in this

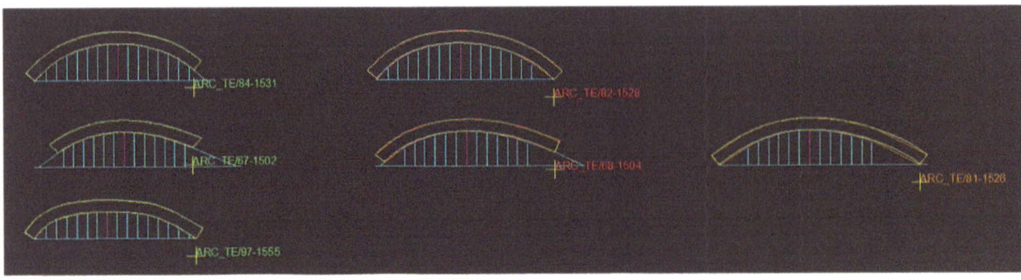

Fig. 7 Graphical output of optimization results for molds

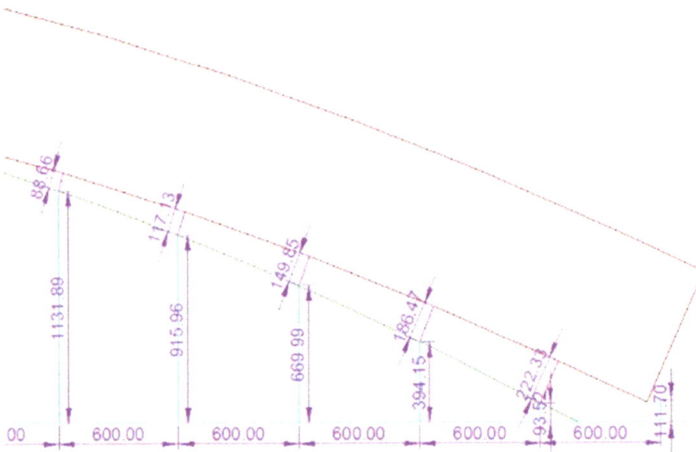

Fig. 8 The divergence is automatically measured after optimization

case, 600 mm), their position in length and the width of wedges to accommodate the following pieces of laminated wood.

This information is the result of the sorting and mapping system created which is then sent to the workshops.

4.2.3 Full Scale Tests and Prototypes

Several prototypes were fabricated in order to achieve proof-of-concepts and to validate aesthetics. The first full scale prototype was assembled at Chemillé in France outside the shops of Caillaud-Lamellé-Collé which is composed of three portions of beams, its linings and its wooden panels. Another full scale prototype is to be assembled on-site at Bordeaux in the summer of 2014 (Fig. 9).

Fig. 9 Full-scale prototype

5 Façade Paneling: Glass and Metal

The envelope of the façade is comprised of glass and metal cladding which are each defined by large scale panels. The unique shape of the façade calls for a research in the form-finding and fabrication processes for the two materials. A number of previous works such as the glass façade of the Eiffel Tower Pavilion have aided in the research of the definition of these panels.

5.1 Glass Panels

The glass portion of the façade are divided into large-scale panelized surfaces which represent approximately 900 panels (842 flat panels and 57 single curved panels). They are defined by approximate horizontal and vertical boundary curves, defined by the architect: In one direction (approximately vertically), the boundary curves were defined by the intersection of the reference surface and the plan of the glulam frame onto the attaches are fixed. In the other direction, the boundary curves were defined by the architect.

Depending on the geometry curvature, two different kinds of surface panels are used: 57 are cylindrical panels and 842 are flat (see Fig. 10). The flat panels are cold bent directly on site when the envelope is mounted onto the glulam timber structure. This flat skin is constrained by a limit of the divergence between adjacent panels, to respect the original freeform of the architect and to respect the possibilities of the attach system (Fig. 11).

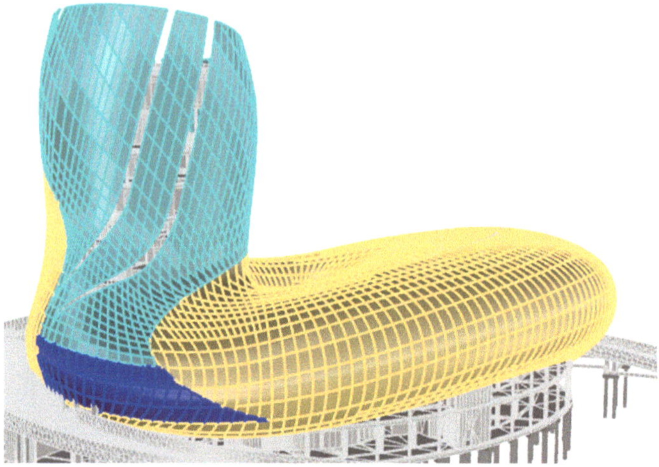

Fig. 10 Location of the glass panels (in *blue* the cylindrical panels)

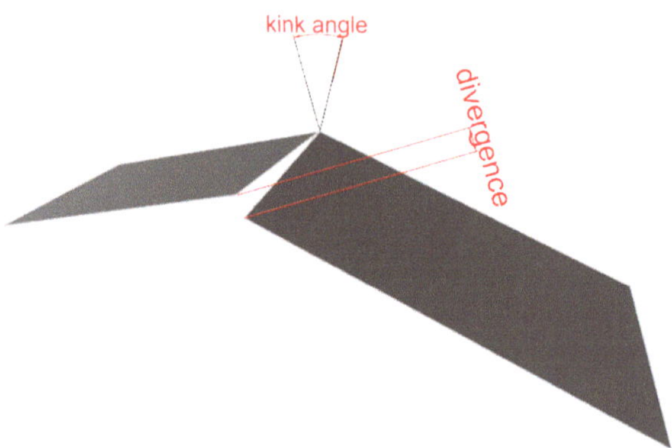

Fig. 11 Divergence constraints

5.1.1 Flat Panels

The majority of the glass panels (842) are defined by flat surfaces and are mostly located on the tower portion of the building.

Two attach systems are considered, one node based and rail based. Each one can manage a different limit of divergence. The rail based system allows a lower divergence than the node based system. Thus the divergence measured led to the possibility of choosing the fixation system for each panel.

As the panels can accept a deformation with cold bending, some structural analysis were made on different kinds of panels (sorted with the span dimension, the wind forces, and the divergence).

Fig. 12 Results mapping: *green*: 0–10 mm, *yellow*: 10–20 mm, *orange*: 20–30 mm, *red*: over 30 mm

From the free-form surface volume, the panels are defined by the boundary curves discussed previously. The planarity of each panel is calculated by taking the midpoints between the four corners. The panels have a maximum length of 5 m and a maximum span of 2 m. As initial first results showed areas of significant divergence, certain panels are then subdivided to limit this spacing. It should be noted that as the panels are reduced in size, the ease of deformation through cold bending is also decreasing.

The results are shown in the figure below (Fig. 12):

5.1.2 Cylindrical Panels

The single curved panels are in an area of significant curvature and count for 57 cylindrical panels. An iterative algorithm was used to minimize the divergence using the panel position, its orientation and its radius as variables. Through each iteration, the results are compiled and applied as new variables on the next iteration. Different techniques or methods of optimization could be realized by optimizing the fabrication techniques thus changing the tolerance values for cold bending, reconsidering the attachment systems or allowing for two radii for each panels thus creating bi-arc panels.

Divergence < 10 mm :	10mm < divergence < 20 mm	Divergence > 20 mm
32 panels	15 panels	11 panels

Fig. 13 Table of results for glass panels

Nonetheless, the table of results shown below summarize that given the tolerance of divergence set, this optimization was sufficient (Fig. 13).

5.2 Metal Panels

5.2.1 Ruled Developable Surfaces

The vast majority of the façade skin of the project is composed of perforated or uniform metal panels. With a maximum size of 1,650 × 3,000 mm, the panels are attached by metal rods which are fixated on the wooden beams. The edges are defined by boundary curves created by the design firm.

The boundary curves are generated on the surface of a freeform geometric surface. As such, the main challenge of the fabrication of the metal panels was to determine developable surfaces which were defined by the initial boundary curves. Ruled surfaces were created through a series of ruled developable surfaces or ribbons along the primary edges of the skin. The resulting 387 panels spanning over the building are then translated into its flattened form. Output data includes maximum radii of curvature, area, fabrication loss ratio between initial and finished material and divergence between panels.

5.2.2 Fabrication

The nature of the metallic panels given their resistance and pliability allowed for certain tolerances and on-site bending. Each panel is pre-bent during the fabrication process in a single direction with a value of the maximum radius of the surface's primary curvature.

Conclusions and Future Research

The presented methods for solving the construction and design of the CCV, mainly its façade structure and envelope are done with simple algorithms to best allow user review and modifications. Further development will continue,

(continued)

especially in view of the completion of the project in 2016 with a focus on fabrication methodology and construction material management given the thousands of unique pieces for the project's façade.

The thoroughness in the current ongoing development of the 3D model for geometry optimization has led to a level of technical achievement which ensures the quality of the project's construction and its aesthetic requirements.

Cité des Civilisations du Vin de Bordeaux Acknowledgements

Design Firm:
X-TU Architectes
Engineering Consultants:
RFR
Consortium Collaborators:
SMAC, Coveris, GTM Sud-Aquitaine and Caillaud-Lamellé-Collé.

The authors would like to thank Nicolas Leduc and the team at RFR for sharing their insights and expertise with regard to developable surfaces, Dr. Jörg Danzberg and Dr. Stefan Schmälzle of iCapp GmbH for their expertise on file conversion for CAD Works and François Josse-Tonnelier for his expertise on glulam timber. This project utilizes a number of software tools most notably Grasshopper developed by McNeel Associates through David Rutten.

References

Eigensatz, M., Schiftner, A., Kilian, A., Mitra, N. J., Pottmann, H., Pauly, M.: Freeform surfaces from single curved panels. ACM Trans. Graph. **45**, New York (2010)

Liu, Y., Wallner, J., Pottmann, H., Yang, Y.L., Wang, W.P.: Geometric modeling with conical meshes and developable surfaces. ACM Trans. Graph. **25**(3), 681–689 (2006). Proc. SIGGRAPH

Meredith, N., Kotronis, J.: Self-detailing and self-documenting systems for wood: the Burj Khalifa. In: Advances in Architectural Geometry, pp. 185–198. Springer, New York (2012)

Pottmann, H., Wallner, J.: Approximation algorithms for developable surfaces. Comput. Aided Geom. Des. **16**, 539–556 (1999)

Rose, K.: Modeling developable surfaces from arbitrary boundary curves. Master of Science Thesis, University of British Columbia (2007)

Schiftner, A., Raynaud, J., Baldassini, N., Bo, P., Pottmann, H.: Architectural freeform structures from single curved panels. In: Advances in Architectural Geometry, pp. 45–48. TU Wien, Vienna (2008)

Wallner, J., Pottmann, H.: Geometric computing for freeform architecture. In: Advances in Architectural Geometry. Springer, New York (2010)

The Geometry of the Error

Yota Adilenidou

Abstract This paper focuses on the importance of error in the evolution of form and the logic of matter distribution describing its relationship to randomness and repetitive behaviour. Using the logic of cellular automata systems on the origin curves of body formations, it displays a methodology for the creation of errors. Extracted cloud points are used as meshing guides for geometries that display a fine game balance between organization and disorder.

Through a series of experiments, a taxonomy of bodies' deviations and morphological errors is created resulting in a tooling system that can be applied in various scales and conditions according to the parameters specified, providing alterations to body form, optimization, and varied possibilities for interaction with the context, environment and other bodies.

1 Introduction

Variation in nature is actually a large assemblage of deviations. Errors are important in the creation of geometry and the emergence of form, since they exist in every instance as an origin of difference. Given this, an immediate question that arises is whether there is in fact a process for creating a geometry of the error, if it is controllable and how can we extract it and use it in order to achieve an "ideal" condition for form, structure and space. Error relates to the loss of symmetry and to mutations of eidetic forms but also to differentiated performance of repetition (Lynn 1998). Based on how form is created in the organic world via rules that guide the interactions between cells and similar parts of organic bodies, the goal is to extract a toolbox, a system that will allow us through taxonomies of morphological deviations (which in nature would be closely related to structural ones) to start building structures and spaces that will control the amount and reason for deviation. It is about formal explorations but actually connecting them to the procedure of creating

Y. Adilenidou (✉)
The Bartlett, UCL, 27 John Trundle Court, Barbican, EC2Y 8DJ London, UK

Arch-hives LTD, London, UK
e-mail: yota@arch-hives.net

P. Block et al. (eds.), *Advances in Architectural Geometry 2014*,
DOI 10.1007/978-3-319-11418-7_22

a topology while taking into account a necessity for the oblique in figure, in use, or in structural forces. This process will facilitate the emergence of structures which are able to evolve, calculate their limitations and deficiencies, and through their topology will provide strength, robustness, density variation, and spatial complexity being applied in various scales from an object to a city fragment. It is, also, about understanding how complexity relies on basic geometry and specific rules leading to detail concentration, disorder, intricacy and infinite variation.

Cellular automata (CA) patterns are found extensively in nature, so their logic is ideal to simulate the complexity of the build-up process of a body. The study is conducted using digital computation and CA as the main instrument. While cellular automata structures have been widely investigated in architecture, especially since the spread of programming and scripting tools, they focused mainly on self-organization and assemblage of components rather than investigating the cellular properties of an autonomous inefficient body, its geometry, its meshing subdivision, its hierarchy and symmetry breaking, its structural capacity, its necessity for extension, its immediate link to body patterning and the taxonomies of deviations that occur.

This research investigates the use of cellular structure formations and alterations through the notions of automatism and pseudo-randomness, repetition and cell communication, while introducing the notion of errors that produce noise into the paths of cell signals, loss of information and symmetry breaking. It explores errors in the development and evolution of form in bodies as local complexity and differentiation, and attempts to establish an analogy in design.

Through a number of tests, cellular automata rules are applied to the edit points of curves guiding the seemingly random process of cellular proliferation and formation. The local neighbourhoods' relations and cells' behaviour, are changing according to the specific error set conditions based on proximity, topology of monad and neighbourhood, resulting in a deviated local behaviour of the whole, creating body alterations like moulting or creating plugs/receptors for artificial extensions and connections with other bodies assemblages.

Errors consist of curves' (body longitudes and latitudes) mutations and sudden change of rules during the process, that drive the randomness of CA initial seeds and result in loss of information travelling between cells. Cell neighbours are not defined by mere distance and proximity but also, by interconnectivity strength and network hierarchies.

The outcome is a set of bodies that derive from the automated repetitive process of CA, with a degree of randomness and a very defined control mechanism. This mechanism is directed by rules that guide the random parameter while engaging errors in the evolving pattern following the controlled / random procedure. The difference of this cellular automata driven structures is found in the guiding system of coordinates that control the allocation of points in a geographical body map, via the topologically scanned longitudes and latitudes, as well as in the focus on the erroneous patterns and local formations. A meshing process is explored in order to optimize the visual and performance result affecting the way the maze is gathered in specific areas (Fig. 1).

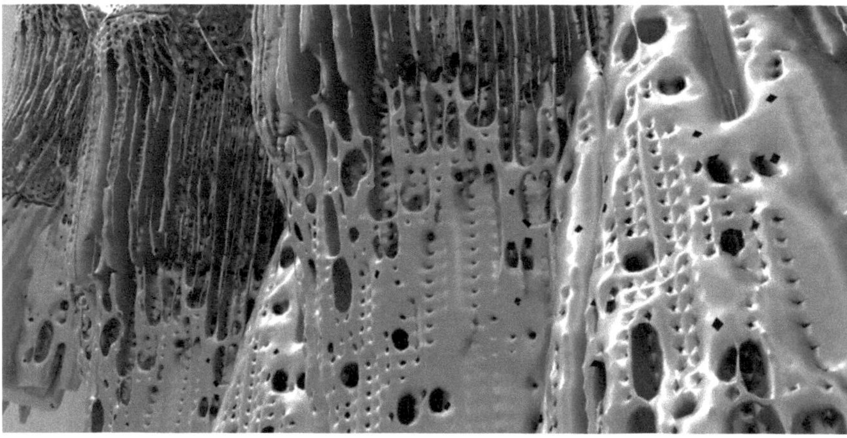

Fig. 1 Body patterns population

2 Error and Repetition || Error and Evolution

Repetitive behaviour is found in geometrical or formal patterns as well as in behavioural patterns, for example in the movement pattern of an autistic person while trying to create an envelope of protection from a disturbing environment. In the second case repetition relates to failure, describing a failing body with false interactivity and exchange of information with the context. In geometry, repetition relates to eidetic forms when their differentiated iterations are leading to mutations of perfect structures. In the behaviour of a machine on the other hand, accurate repetition following the rules, is what constitutes a robust engine (Fig. 2).

The evolution and development of a failing as well as a successful biological body happens through an automation procedure. The same very few genes result to a wide variety of animal species while switches are turned from on to off, providing operating instructions that guide the toolkit creating body pattern, layering and extensions as explained by Sean Carroll in his book *Endless Forms Most Beautiful* (Carroll 2005). This automated process unavoidably works with a structure of parallel events, −accepting precedent and creating new ones-, environmental changes or even chance. This collaboration of parallel procedures alters the inputs. Parts of the automated procedures are withheld, paths change and transmute leading to variation. The same process creates body failures, small little errors in symmetry, transformations of appendages or adjustments in uses of limbs, evolving to new species or other times staying as a single failure that would disappear.

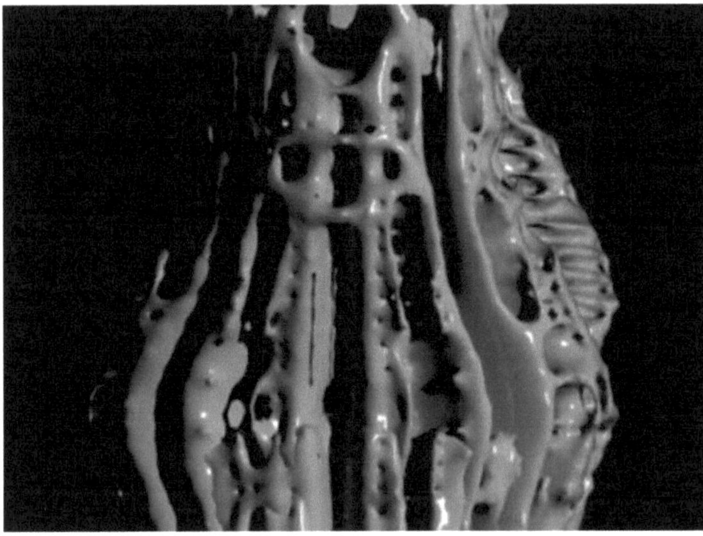

Fig. 2 Body patterns – loss of absolute symmetry, local differentiation

3 Error and Randomness

The amount of randomness, running parallel to a "*toolkit*" of defined rules, creating a new set of data, leads to a new combination, coined as Pseudo-Randomness. The same concept of complexity from simplicity is described by Stephen Wolfram (2002) in his studies on cellular automata structures. Same tools, same rules. Differently distributed initial "*switches*" assigned to the structure of initial seeds, changing the evolving patterns and boundaries of the structure, breaking the symmetry and creating local differentiated behaviours. Repetition through generations distributes behaviour by combining inputs and rules. The amount of randomness injected in the initial seeds provides a number of probable/uncertain results (Packard and Wolfram 1985).

Giuseppe Longo is describing randomness *"as part of the proposal for a structure of determination of physical processes"* (Longo 2009). Although he separates randomness from Pseudo-Randomness, he refers to random as not being the opposite of deterministic, despite the common knowledge and thought (Fig. 3).

Holland describes in his book *Hidden Order*, that *"diversity is neither accidental nor random"* and always depends on the context. "*If we remove one kind of agent from the system, creating a "hole", the system typically responds with a cascade of adaptations resulting in a new agent who "fills the hole". The new agent typically occupies the same niche as the deleted agent and provides most of the missing interactions*" (Holland 1995).

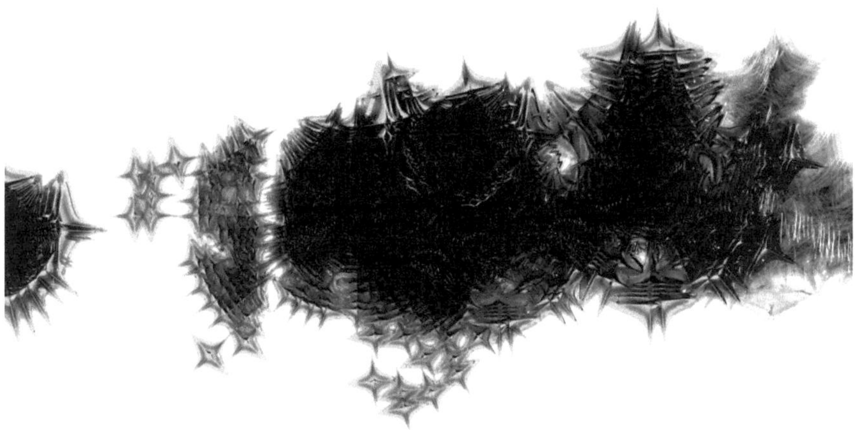

Fig. 3 CA Game of life patterns and distribution of magnitude and density to the assemblage units, based on mathematical relationships and graph functions

Bodies can be either natural or artificial, material or immaterial, humans or machines. A body system is composed by the body and its structure of development, growth and action. A context system is composed by other bodies and their corresponding structures. The Probability of a product/event resulting from the parallel activity and inter-extension of the body and the context system, is never clearly random. It is a combination of randomness with facts, a selection of possible outcomes derived from a certain set of initially defined possible outcomes. Locating a possible error, and inserting it into the automatic process of body growth and development is a mechanism of designing (programming) uncertainty.

4 Error and Geometry || Error and Symmetry

> This half-philosophical excursion was needed as a background for the discussion of the left-right symmetry in nature; we had to understand that the general organization of nature processes that symmetry. But one will not expect that any special object of nature shows it to perfection. Even so, it is surprising to what extent it prevails. (Weyl 1952)

Perfect bodies are described by the elimination of their imperfections. Through his work *Origin of Geometry,* Edmund Husserl was working with the progressive cancellation of variations in order to arrive to eidetic forms as in, for example, a sphere (Lynn 1998). But the most important invention of Husserl, according to Greg Lynn, is not the invention of the exact ideal geometries but the necessity of those that as being inexact provide the essential, measurable, and able to be eliminated, variations that can lead to the eidetic forms.

Bateson describes in his book *Materials on the Study of Variation* a series of abnormalities in nature. The examples of Bateson show a special interest to symmetry and secondary symmetries, supernumerary appendages, duplication of axis etc (Bateson 1894).

Using a cellular automata system, we can extract a typology of morphological errors that lead to a tooling system of formal deviations, optimizing a body while allowing it to interact and connect with other bodies in more varied ways. Bifurcation, duplication, secondary axis, extra or missing parts, extra digits, segmentation, forked limps, transformation of one part into another / change of use, exoskeleton condition, they can all introduce advanced behaviours that will incline from the ordinary system performance. All extracted through symmetry and its loss.

In the following examples, symmetry and, specifically, – rotational symmetry, is assigned via a polar array of curves, in order to be broken. Through the input of specific differentiated data, like resolution of curves, number of control points, number of curves and division possibilities, the symmetry is locally altered creating complexity at micro level evolving through the cellular automata generations (Fig. 4).

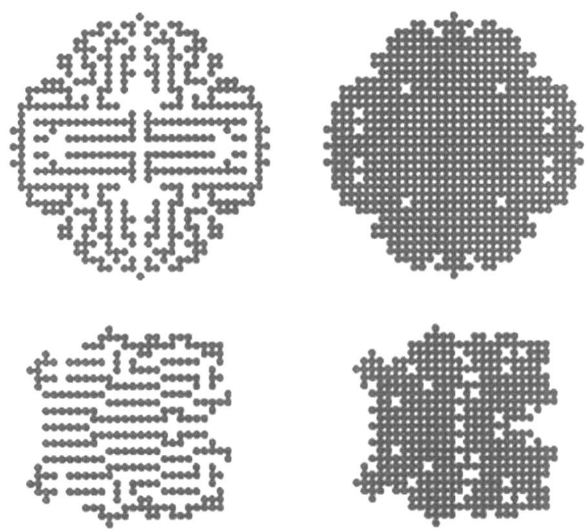

Fig. 4 2D cellular automata – diagrams using the Labyrinth Code 746 (Wolfram 2002) with seven initial seeds in ordered and disordered initial states

5 Error and Parameters of Body Growth: Methodology

5.1 Cellular Automata Properties

Controlled randomness is examined in a set of cellular automata based on curves' edit points (Fig. 5). Curves provide a more discrete context that can be mutated while organizing the CAs' main characteristics, orientation, intensity, seeds, patterns and boundary, producing a visible effect of control. Boundary is transformed according to the curves' deviation, while intensity and density are also controlled by the proximity of the curves that act independently from CA rules. The starting point / initial seeds' configuration is defined and whilst it is supposed to end up in a labyrinth though symmetrical, we end up, through levels of different information input, with a seemingly random result.

The experiments are using the rules of the labyrinth code 746 described by Wolfram (2002). The seven initial seeds of the code are re-arranged into a different configuration and the symmetrical labyrinth loses the clear boundaries and becomes disordered while keeping some of the labyrinth properties in the pattern (Fig. 4).

5.2 Solidification Process and Volume Formation

The proximity leads to solidification and to volume subdivision from which new patterns arise. Each body is subdivided according to the quantity and position of its neighbours. Larger distance between cells leads to dispersion, perforation, liquidity, and flexibility. Solidification happens in the same layer as well as during the stacking of cells of different layers – time-steps. The same layer cells are solidified when they come very close together. The resolution of the curve, that is the number of its divisions, will make the outcome more or less solid. The resulting point cloud is converted into a mesh with locally differentiated distributing matter.

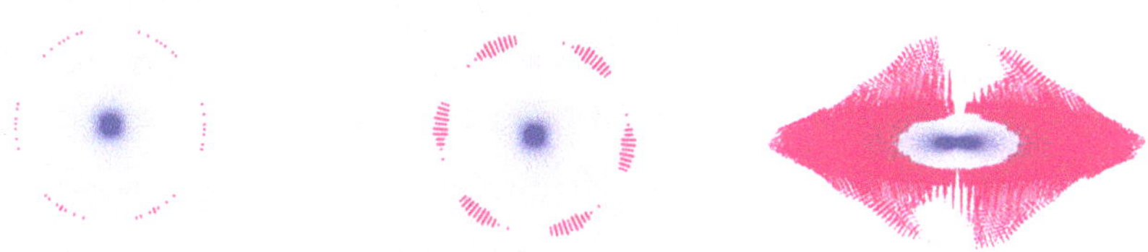

Fig. 5 Diagram showing a body formation by six initial seed groups in 1st, 10th and 50th generation

The density of the points' grid, the resolution / division of the curves (longitudes-latitudes of the body outcome) in control points, the number of these curves, the size of the cell, the shape of the cell, the orientation of curves, they are all parameters that cause the code to perform partially or in total. The code is also enriched with an extra information, the sub-grouping of the curves in order to create sub-volumes via the use of a fraction. If the number of curves is not divisible by the fraction number, parts of the rotationally symmetrical body are led to collapse.

Different steps of cellular automata organization based on polar curves are stacked one on top of the other and produce time-based 3d formations and volumes. The time – steps are also distributed in multiple centre points around the main volume. Different polarities and different centres are introducing at each step a new seed of CA. The timeline – steps provide information for landscape extruding the cells in a third dimension. Multiplication of symmetry and recursion is combined with distortion and horizontal flips of the erosion volumes that result from the time steps. Erosion in nature is a cell defined and time – based procedure where automation based on specific rules along with nature dynamics and forces like fluids, air and animal interaction, change the process of the "*game*" by differentiating each time the automated rules applied to the cells of matter that is being eroded. Randomness and disorder appear in the outcomes of these procedures. The repetition of cells in several continuous steps creates the volume effect.

The way the cellular automata are distributed according to xy and yz planes with the insertion of time intervals, contributes to the build-up of a grid topology and resembles the way the embryo is transformed to body, applying the information found on the u, v, w locations of its longitudes and latitudes.

Repetition, annihilation, initialization, iteration, redefinition of parameters, change of cell type from 0 to 1 (dead to alive) and reset of values are the basic actions needed for the cells' organization. Errors are introduced to the process by the deletion of the objects of the previous time-step, in order to repeat the procedure. Resetting the objects for the new layers gives possibilities for disorder of the instances (copies on the grid) at a next time interval in order to insert new data to the system while the process has already started. Another way of inserting an error (a controlled deviation) into the initialized procedure is to alter the curves while the formation of CA has started. The information of cells is distributed through the edit points of curves that constitute the grid. A change of the curve immediately affects the cell location but not the initial rules. The tests are examining the relationship of control authorship over the action. The automatic process is integrated with the curves' set of data that can be altered during the stacking.

The experiments provide information on structural behaviour through explorations of cell communication, solidification, interlocking, and attaching to neighbouring structures. Material deposit possibilities are investigated focusing on density, directionality, boundaries and structure initializers.

The network of nodes, points, is meshed through different software describing different techniques and different resolutions. The technique that is finally used is the import of a point cloud via Mel scripting into Autodesk Maya as n-particles (Fig. 6). After import, the n-particles system is converted into a mesh

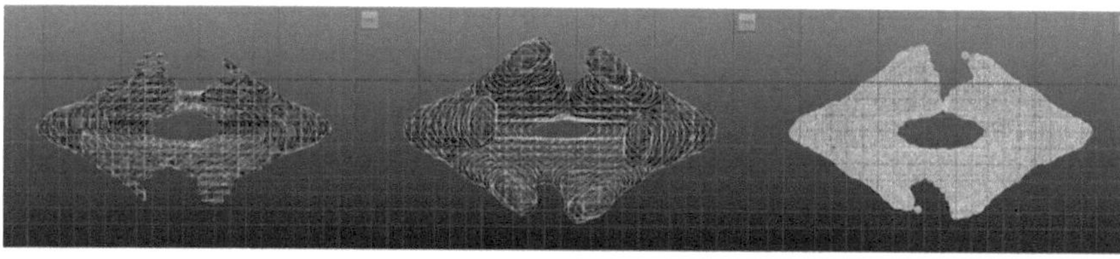

Fig. 6 Mesh conversion and optimization in Autodesk Maya

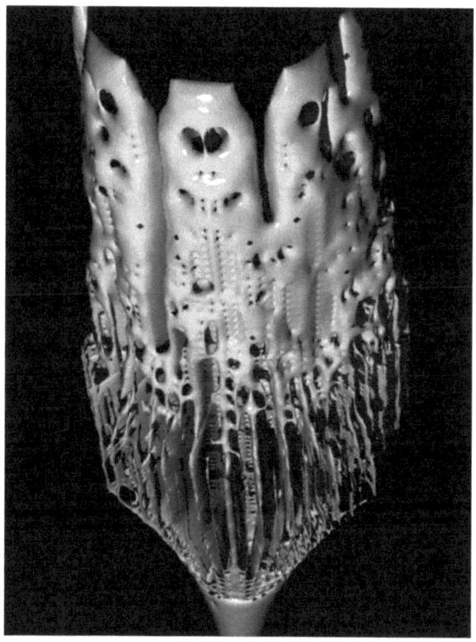

Fig. 7 Body patterns – resolution

where we can adjust certain parameters. The meshing quality and final form is affected by the radius scale of the mesh sub-components located at the points of the point cloud, the mesh triangle size, and the max triangle resolution. Altering these values we can adjust the solidification, the detail and the complexity of the mesh-body.

A perfect geometrical pattern being a network of points and coordinates that obeys into certain rules, is described and explored at a great extend by the mutated repetitions and the disordered iterations. These changes elect the whole set of perfection by showcasing the characteristics that should be eliminated (Fig. 7).

6 Design Experiments

6.1 Symmetry and Local Differentiation

The first set of design experiments (Fig. 8) is consisting of vertical bodies. The origin curves are a more clear polar array that is reshaped while copied through the generations. The width of the hull formed by the curves, is changing while the

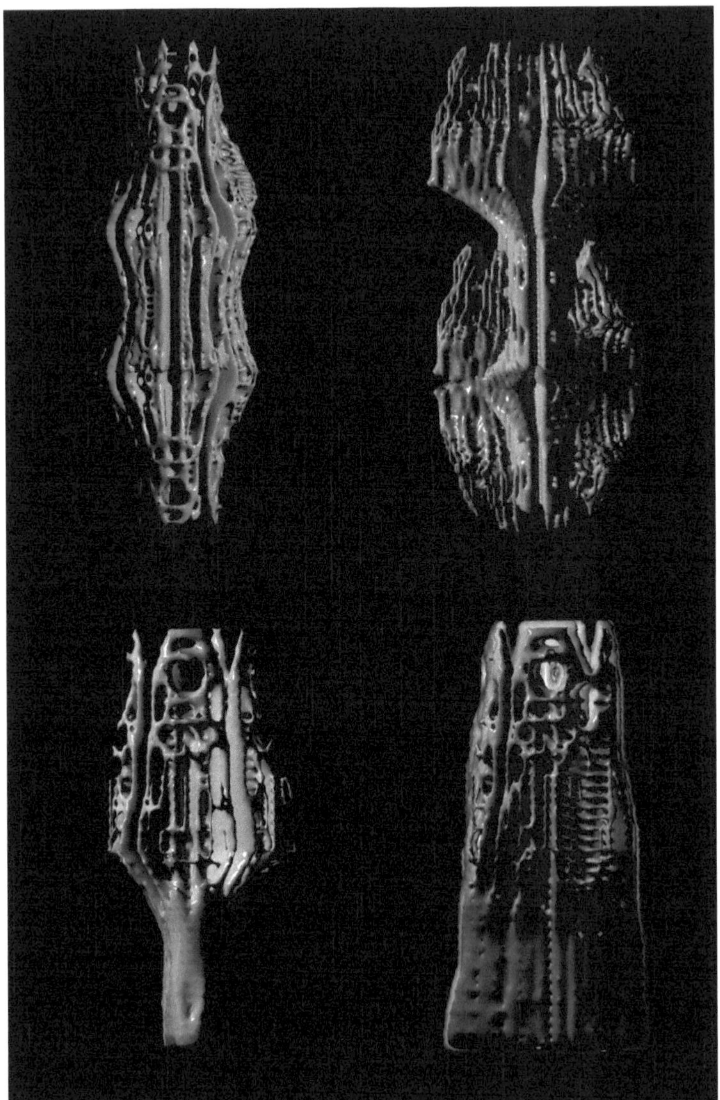

Fig. 8 Body patterns

points are constructing the multiple layers. They are more symmetrical structures with local differentiations from right to left. Four initial seed groups have been selected in disordered states. The origin curves are revealed in the body formations in the type of detail and sculpting. The initial seeds give their place to new points positioned in the 3-dimensional grid, building a solid/porous form. The number of points increases while the seeds are moving from the first generations to the next and the structure is growing.

According to different settings on the hull curves transformation and on the order and number of initial seed groups, the resulting figures are resembling the digit formation of an animal body, revealing structure merged with skin, animal exoskeleton, geological formations, sand deposition, a material between liquid and solid.

In the next set of experiments (Figs. 9 and 10), the loss of symmetry was applied to the whole structure instead of a local positioning of errors. This was

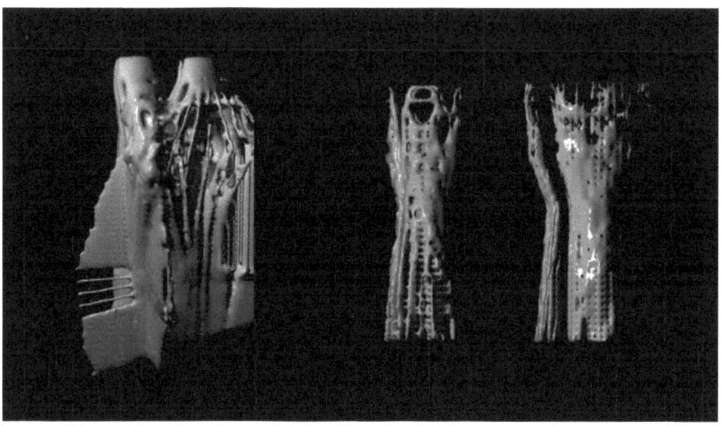

Fig. 9 Body patterns – symmetry breaking

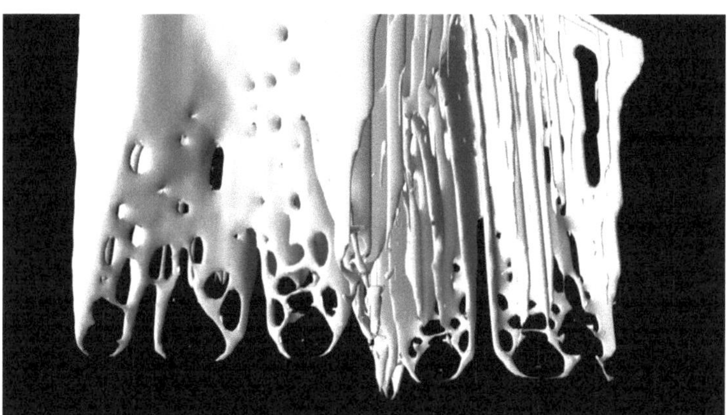

Fig. 10 Body patterns – liquidity

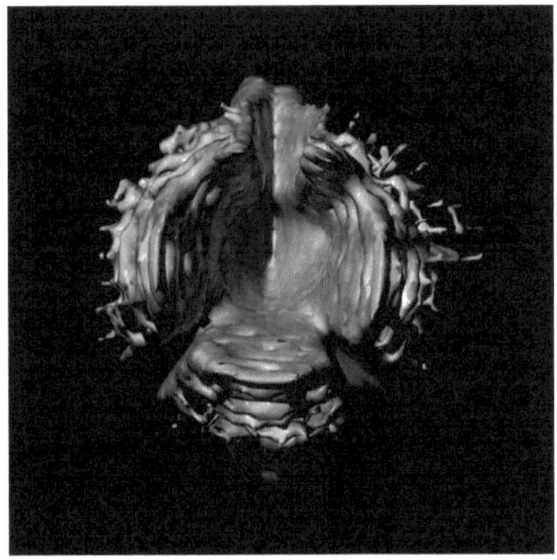

Fig. 11 Body pattern in plan

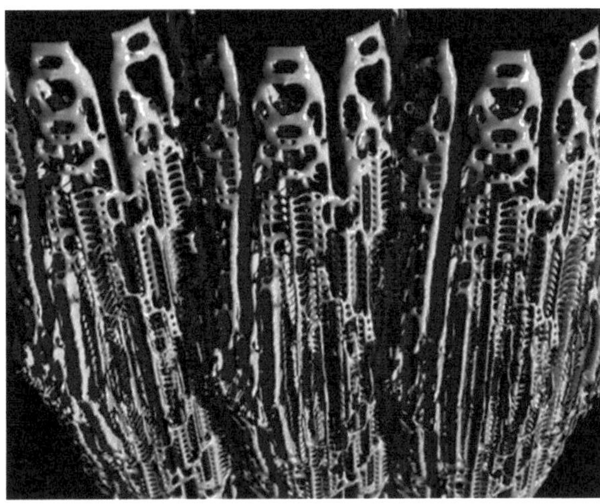

Fig. 12 Body patterns – introducing larger populations

succeeded through non-symmetrical arrangement of the initial seed groups and the non- proportional deformation of the hull curves.

In the final set of experiments (Figs. 1, 7, 11, 12, 13, 14), the focus was given to detail, resolution and larger population of elements. The vertical structural components were now transformed into wider structures, multiplications, using more initial seeds and polar curves assemblages. The resolution of the curves and their number in each assemblage was increased leading to bigger point clouds. The aim is a number of structural elements that will give spatial quality through the complexity that will act both as decorative detail locally, and as a space distributor in larger scale.

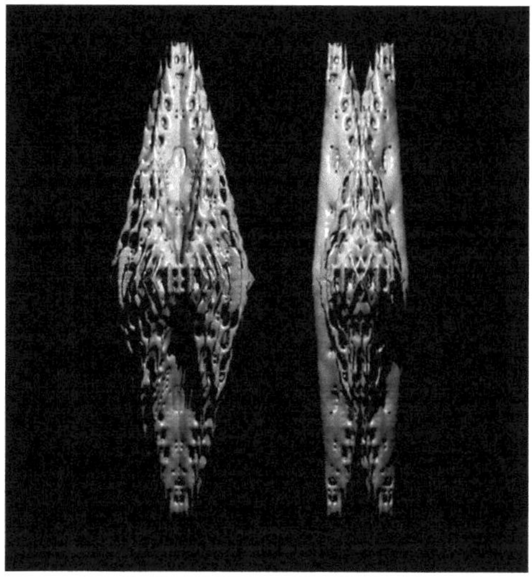

Fig. 13 Body patterns – symmetry

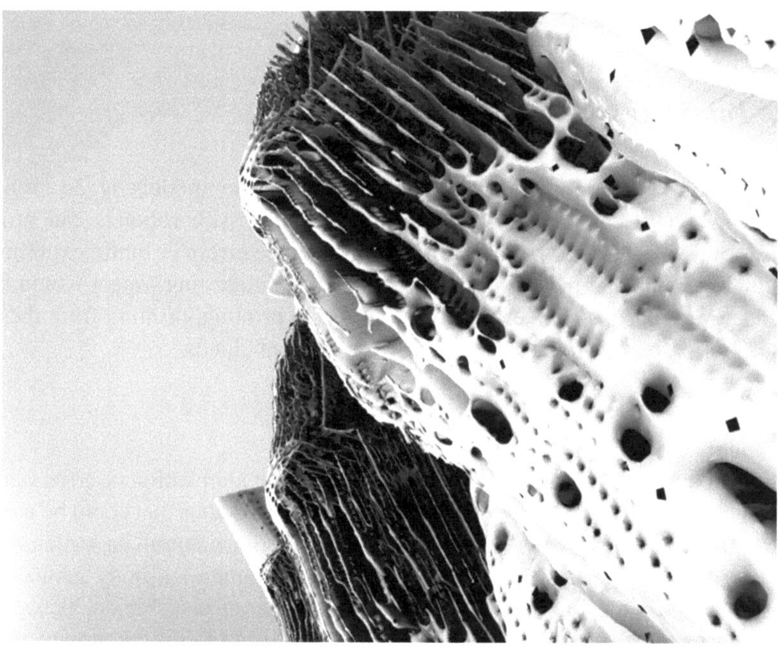

Fig. 14 Larger populations

7 Application and Materiality

7.1 Application and Scale

Although the intension of the research was mainly to produce a taxonomy of body errors, deviations and a toolbox entailing all the tools that would lead various parameters affect different hulls in versatile ways, it is very important to name a number of ideas on application as well as on materialization of the resulting bodies.

The methodological process can be applied to various scales, from the object to an urban scale, setting criteria as porosity, permeability of the structure, material arrangement, load transfer etc. Due to the solidification process explained earlier in the paper, the results could be seen as structural components.

The first experiments were based on the idea of vertical elements and specifically, the column. The examples move forward to surfaces and structural components as load bearing walls. The scope is to produce a taxonomy of components that in additive mode will expand to the creation of a complex spatial configuration being applied to specific plans, sections, vaults and bridging.

The body-components are indeed answering a question on distribution of matter through paths extracted from body frames.

7.2 Fabrication

The fabrication technique used to create larger scale models is 3d printing. Interacting phase changing materials in combination with robotics can provide room for more controlled randomness and predictable errors in future experiments working on paths and points that interact. Another interesting aspect would be to introduce varied axis of material deposition for 3d printing, printing from the core to the exterior, based on the polar distribution of point clouds.

Conclusion

In this paper, the logic of cellular automata and the implication of error were applied through a series of experiments, resulting to bodies that could be used as structural components. The application of cellular automata structures, with semi-random initialization, on the outline frame of a body geometry has displayed the link between pattern and material distribution, between geometry and resolution. It revealed the importance of cell communication for the topology of structures and the configuration of matter through patterns of sub-components. Symmetrical structures with local disorders and disordered structures with local organization resembling configurations in

(continued)

nature provide comparisons with artificial architectural elements that through the acknowledgement of their own body language find rules to self-transform and expand to the immediate and distant environment.

Next step of this research will be the generation of larger populations of assemblages using the same methodology, allowing for growth patterns to emerge. Using supercomputing, the body elements will be combined in series of spatial and structural load bearing elements, from column to surfaces, ceilings, floors, vaults using topology optimization embedding structural intelligence into the system or, more accurately, discovering and using the existing one. Error geometry will be used to answer the questions of how a structural component can become space, how solidity can transform to porosity, lightness and openness, while providing gravity control and room for living.

Acknowledgements This paper presents part of the Ph.D. by Architectural Design Research "Scripting Errors" pursued by Yota Adilenidou at the Bartlett School of Architecture at UCL, under the supervision of Dr. Marjan Colletti and Prof. Stephen Gage.

References

Bateson, W.: Materials for the Study of Variation. Macmillan and Co, New York/London (1894)

Carroll, S.: Endless Forms Most Beautiful: The New Science of Evo Devo and the Making of the Animal Kingdom. Phoenix, London (2005)

Holland, J.H.: Hidden Order: How Adaptation Builds Complexity. Addison-Wesley, Reading (1995)

Longo, G.: Randomness and determination, from physics and computing towards biology. In: Springer, Series: Lecture Notes in Computer Science, vol. 5404, Subseries: Theoretical Computer Science and General Issues, 2009, XVII, p. 670 (2009)

Lynn, G.: Folds, Bodies & Blobs Collected Essays. La Lettre Volée, Bruxelles (1998)

Packard, N., Wolfram, S.: Two-dimensional cellular automata. J. Stat. Phys. **38**, 901–946 (1985)

Weyl, H.: Symmetry. Princeton University Press/Oxford University Press, Princeton/London (1952)

Wolfram, S.: A New Kind of Science. Wolfram Media, Champaign (2002)

LAR-ABC, a Representation of Architectural Geometry from Concept of Spaces, to Design of Building Fabric, to Construction Simulation

Alberto Paoluzzi, Enrico Marino, and Federico Spini

Abstract This paper discusses the application of LAR (Linear Algebraic Representation) scheme to the architectural design process. LAR is a novel representation scheme for geometric design of curves, surfaces and solids, using simple, general and well founded concepts from algebraic topology (Dicarlo et al., Comput Aided Des 46:269–274, 2014). LAR supports all topological incidence structures, including enumerative (images), decompositive (meshes) and boundary (CAD) representations. It is dimension-independent, and not restricted to regular complexes. Furthermore, LAR enjoys a neat mathematical format, being based on chains, the domains of discrete integration, and cochains, the discrete prototype of differential forms, so naturally integrating the geometric shape with the supported physical properties. The LAR representation find his roots in the design language PLaSM (Paoluzzi et al., ACM Trans. Graph 14(3):266–306, 1995; Paoluzzi, Geometric programming for computer aided design. Wiley, Chichester 2003), and is being embedded in Python and Javascript, providing the designer with powerful and simple tools for a geometric calculus of shapes. In this paper we introduce the motivation of this approach, discussing how it compares to other mixed-dimensionality representations of geometry and is supported by open-source software projects. We also discuss simple examples of use.

1 Introduction

Geometric models, called mesh in computational science and engineering, are of primary importance in old and novel problem areas, such as architectural geometry, material science or biomedicine. In particular, we see that novel applications require the convergence of shape synthesis and analysis, from computer graphics, computer imaging, and computer-aided geometric design, with discrete meshing of domains used for physical simulations.

In this paper we discuss a novel linear algebraic representation, supporting topological methods to represent and process mesh connectivity information

A. Paoluzzi (✉) • E. Marino • F. Spini
Roma Tre University, Roma, Italy
e-mail: paoluzzi@dia.uniroma3.it

P. Block et al. (eds.), *Advances in Architectural Geometry 2014*,
DOI 10.1007/978-3-319-11418-7_23

for dimension-independent cellular complexes, including simplicial, cuboidal, and polytopal cells. This representation works even with the awkward domain partitions—with non-convex and/or non-manifold cells, and cells non homeomorphic to balls—that might arise in Building Information Modeling (BIM) and Geographic Information Systems (GIS) applications. Simplicial and cuboidal cell complexes provide the standard mesh representation used in most science and engineering simulations, whereas complexes of possibly non-convex or non-contractible cells may be needed to represent the built environment in software applications for the Architecture, Engineering, and Construction (AEC) sector.

Our approach provides a unified representation where concepts and techniques from computer graphics (graphics primitives), geometric modelling (curves and surfaces) and solid modelling (solids and higher dim manifolds) converge with those from computational science (structured and unstructured meshes) in a common computational structure. This representation may be used in novel and highly demanding application areas,[1] and may be supported by modern silicon-based APIs[2] on last and next generation hardware.

The descriptive power of a representation scheme is measured by the extent of its domain. In this sense, the LAR scheme—based on *chain complexes*, where chains are sets of cells—is defined on a quite large mathematical domain, and may represent a wide class of finite cell decompositions of a space. The LAR scheme may denote symbolically a large class of geometric shapes in e^d, and makes use of basic concepts from algebraic topology (mostly from mod two homology), and of standard representations from basic computational algebra. The aim is to provide a representation that can support (at least in principle) all topological and geometric queries and constructions that may be asked to the corresponding model. LAR can be used from initial concept of spaces, to the additive manufacturing of design models, to the meshing for CAE analysis, to the detailed design of components of building fabric, to the BIM processing of quantities and costs.

2 Definitions

A central notion in solid modelling is the concept of representation scheme, a mapping from mathematical solid models to actual computer representations.

A compact topological subspace is a *convex cell* if it is the set of solutions of affine equalities and inequalities. A *face* of a cell is the convex cell obtained by replacing some of the inequalities by equalities. A *facet* of a cell is a face defined by just one equality. The *dimension* n of a n-cell is that of its *affine hull*, the smallest affine subspace that contains it.

[1]It is being tested within the IEEEP3333.2 – Standard for Three-Dimensional Model Creation Using Unprocessed 3D Medical Data.

[2]A prototype implementation with OpenCL and WebCL is on the way.

A *convex-cell complex* or *polytopal complex* P is a finite union of convex cells in e^n such that: (i) if A is a cell of P, so are the faces of A; (ii) the intersection of two cells of P is a common face of each of them. A simplicial (respectively, cuboidal) complex is a polytopal complex where all cells are simplices (respectively, cuboids).

A *filtration* is an indexed set S_i of subobjects of an algebraic structure S, such that $i < j$ implies $S_i \subset S_j$. A *CW-structure* on the space X is a filtration $\emptyset = X^{-1} \subset X_0 \subset X_1 \subset \cdots \subset X = \bigcup_{n \in N} X_n$, such that, for each n, the space X_n is homeomorphic to a space obtained from X_{n-1} by attachment of n-cells of X in $\Lambda_n = \Lambda_n(X)$. A space endowed with a CW-structure is a *CW-complex*, also said a *cellular complex*. The space X_n is called the *n-skeleton* of X. Each subcomplex of X is a closed subset of X. A cellular complex is *finite* when it contains a finite number of cells.

The dimension of a complex is the maximal cell dimension. The r-skeleton X_r is the subcomplex formed by the cells of dimension $\leq r$. The 0-skeleton coincides with the set $V(X)$ of *vertices*.

Let be given a Hausdorff space X, and a finite cellular complex $\Lambda(X)$ such that $X = \Lambda_0 \cup \cdots \cup \Lambda_d$, and let $M_p := M_n^m(\mathbb{Z}_2)$ be binary matrices, with $m = \sharp\Lambda_p$ rows, and $n = \sharp\Lambda_0$ columns.

Definition 1 (Math models) The LAR domain is defined on the set M of *chain complexes* supported by a finite cellular complex $\Lambda(X)$.

Definition 2 (Computer representations) The codomain of the LAR scheme is the set R of d-tuples of CSR sparse matrix representation[3]

$$\mathrm{CSR}(M_n^m(\mathbb{Z}_2)), \quad n = \sharp\Lambda_0, m = \sharp\Lambda_p$$

where d is the dimension of the space $X = \bigcup_p \Lambda_p$, and $1 \leq p \leq d$.

Every matrix $M_p = M_p(\Lambda(X))$ is an *incidence matrix*, i.e. it represents a *characteristic function* $\chi : \Lambda_p \times \Lambda_0 \to \mathbb{Z}_2$, so that $M_p(\mu, \lambda) = 1$ if and only if the cell $\mu \in \Lambda_p$ contains the cell $\lambda \in \Lambda_0$.

The interesting point is that, for a given cellular d-complex $\Lambda = \Lambda(X)$, all the M_p matrices $(1 \leq p \leq d)$ contain the same number of columns, and can be operated by standard matrix multiplication or transposition, providing a simple and convenient tool for computing boundary and coboundary operators and topological relations between cells.

The LAR scheme is a paradigm change in shape representation: whereas topological relations cannot be easily combined, linear operators are readily composed via matrix multiplication.

For *regular polytopal* d-complexes, where each cell is contained in a d-cell, only a single M_d matrix is needed to compute the homology of a space and

[3]Compressed Sparse Row (CSR) format, for which efficient implementations on high-performance hardware exist. See Bell and Garland (2008), Buluç and Gilbert (2012), and Lokhmotov (2012).

its (co)boundary operators. Efficient methods exist to compute a vertex-based representation of $(p - 1)$-cells from that of p-cells, for instance by using the *qhull* algorithm by Barber et al. (1996) for polytopal cells, or a $O(1)$ formula for simplices.

Therefore, giving only CSR(M_d) is sufficient to fully denote the chain complex induced by $\Lambda(X)$. For a given decomposition, no ambiguity may arise. Using the language of representation schemes: *on the domain of polytopal complexes, including simplicial and cuboidal ones, all LAR representations are valid*. It is worth noting that CSR(M_d) exactly coincides with the bulk of common representations in Computer Graphics—for instance, the OBJ or PLY file formats.

3 Models, Structures, Assemblies

In LAR-ABC we make a distinction between geometric *models*, *structures*, and *assemblies*. Some terminology and definitions are given below.

Geometric models A geometric *model* is a pair (*geometry*, *topology*) in a given coordinate system, where *topology* is the LAR specification of highest dimensional cells of a cellular decomposition of the model space, and *geometry* is specified by the coordinates of *vertices*, the spatial embedding of 0-cells of the cellular decomposition of space. From a coding viewpoint, a model is either an instance of the `Model` class, or simply a pair (`vertices`, `cells`), where `vertices` is a two-dimensional array of floats arranged by rows, and where the number of columns (i.e. of *coordinates*) equals the dimension n of the embedding Euclidean space e^n. Similarly, `cells` is a list of lists of vertex indices, where every list of indices corresponds to one of d-dimensional *cells* of the space partition, with $d \leq n$.

Structures A *structure* is the LAR representation of a hierarchical organisation of spaces into substructures, that may be organised into lower-level substructures, and so on, where each part *may* be specified in a *local coordinate system*. Therefore, a structure is given as an *(ordered) list of substructures and transformations* of coordinates, that apply to all the substructures following in the same list. A structure actually represents a *graph of the scene*, since a substructure may be given a name, and referenced more than one time within one or more other structures. The *structure network*, including references, can be seen as an acyclic directed multigraph. In coding term, a structure is an instance of the *Struct* class, whose parameter is a list of either other structures, or models, or transformations of coordinates, or references to structures or models.

Assemblies An assembly is an *(unordered) list of models* all *embedded in the same coordinate space*, i.e. all using the same coordinate system (the *world coordinate system*). An assembly may be either defined by the user as a list of models, or automatically generated by the *traversal* of a structure network. At traversal time, all the traversed structures and models are transformed from their local coordinate system to the world coordinates, that correspond to the coordinate frame of the root of the traversed network, i.e. to the first model of the structure passed as argument to the `evalStruct` function, that implements the traversal algorithm. In few words, we can say that an assembly is the linearised version of the traversed structure network, where all the models are using the world coordinate system.

3.1 *The Design of LAR-ABC*

LAR-ABC is a Python library for geometric design of building objects with the Linear Algebraic Representation (LAR) specialised for Architecture, Building and Construction (ABC). In the present first prototype implementation of the library, we concentrate on the first two letters of this specification, namely the organisation of spaces (architecture) and the specification of physical, concrete components (building).

3.1.1 Concept Design and Project Plan

The client needs and wishes are initially specified by some initial set of requirements, traduced by the architect firm into a design concept, better specified by an initial project plan. When the project plan is accepted by the client, giving a definite shape to the first architectural concept, the model of construction is usually a 2.5D model, made by opaque or transparent 2D surfaces embedded in 3D space. In this stage, LAR-ABC allows for the computation of every topological or geometrical property of interest, including the evaluation of the surface of the building envelope and its partitioning into subsets with different thermal requirements, as well as the computation of the internal volume, and its partitioning into any classes of internal space, and will grant any other geometric computation or simulation (for example of the thermal behaviour) of possible interest for the architect or the client.

3.1.2 Building Objects: Components and Assemblies

The LAR description of the topology and its geometric embedding, defined by the position vectors of vertices or control points of the surfaces, makes possible to (mostly) automatically generate a first 3D model of the physical construction, i.e. of the concrete instances of building components. This (semi-)automatic transformation from a 2.5D model formed by surfaces to a 3D model formed by assemblies of solid objects, is obtained using the boundary operator, that allow to discriminate between the various subsystems of the building fabric, i.e. between the horizontal and vertical enclosures, the horizontal and vertical partitions of the interior, the elements of horizontal and vertical communications, and so on, as we show in Sect. 4.

3.1.3 Operators for Assemblies and Derived Subclasses

Remember that an assembly is an unordered *list* of geometric models, i.e. a list of pairs made by vertices and cells, and hence with named components, implementable as a dictionary in Python or as an object in Javascript. Therefore, few specialised higher-level functions are needed to apply the typical operations for models both to basic assemblies and to all derived specialised subclasses, associating specialised semantics to the basic geometric information. Some higher-order general utilities for handling basic geometric assemblies are given for this purpose (Figs. 1–3).

3.2 Boundary Operator

The LAR representation of the cellular complex $\Lambda(X)$ in Fig. 1d, with $\Lambda_1 = \{e_0, e_1, \ldots, e_9\}$, and with non-convex faces $\Lambda_2 = \{f_0, f_1\}$, is:

```
FV = [[0,1,3,5,6,7],[0,2,3,4,5,6]]
EV = [[0,1],[0,2],[0,6],[1,3],[2,3],[3,5],[4,5],[4,6],
     [5,7],[6,7]]
```

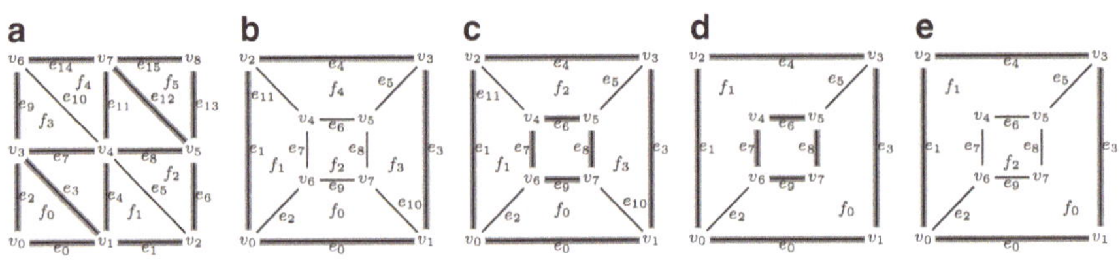

Fig. 1 Cellular 2-complexes and their (*blue*) boundaries: (**a**) simplicial complex; (**b**, **c**) cuboidal complexes; (**d**, **e**) cellular complexes with non-convex 2-cells

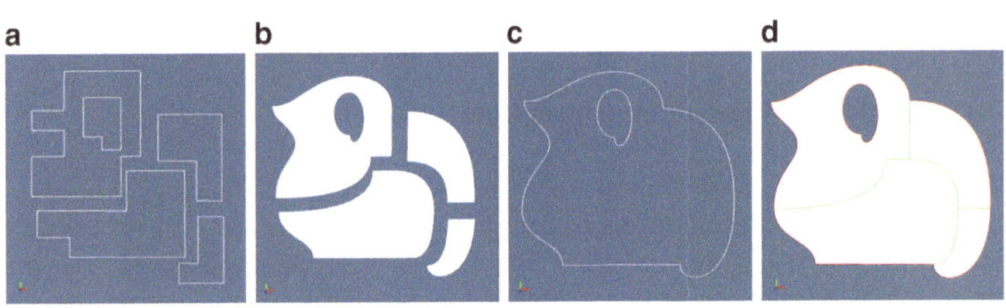

Fig. 2 A cellular complex $B = \Lambda(X)$, with curved 2-cells non necessarily homeomorphic to the 2-ball: (**a**) input Bézier polygons, defined by cycles of control points of Bézier curves of various degrees; (**b**) exploded 2-cells; (**c**) boundary 1-chain; (**d**) boundary 1-chain (*red*) and interior 1-chain (*green*). See "Appendix: Implementation Example" for a detailed presentation of the model, including the actual Python code

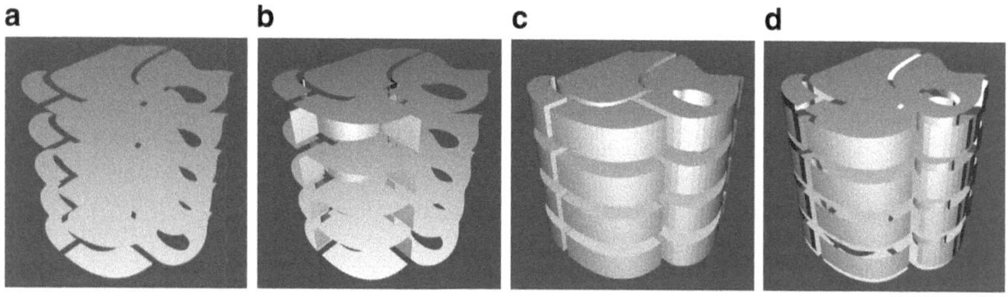

Fig. 3 A cellular complex $B = \Lambda(X)$, with curved 2-cells non necessarily homeomorphic to the 2-ball: (**a**) input Bézier polygons, defined by cycles of control points of Bézier curves of various degrees; (**b**) exploded 2-cells; (**c**) boundary 1-chain; (**d**) boundary 1-chain (*red*) and interior 1-chain (*green*). See "Appendix: Implementation Example" for a detailed presentation of the model, including the actual Python code

The corresponding binary matrices M_2 and M_1, indexed on the columns by 0-cells, and indexed on the rows by either by 2- or 1-cells are, respectively:

$$M_1 = \begin{pmatrix} 1&1&0&0&0&0&0&0 \\ 1&0&1&0&0&0&0&0 \\ 1&0&0&0&0&0&1&0 \\ 0&1&0&1&0&0&0&0 \\ 0&0&1&1&0&0&0&0 \\ 0&0&0&1&0&1&0&0 \\ 0&0&0&0&1&1&0&0 \\ 0&0&0&0&1&0&1&0 \\ 0&0&0&0&0&1&0&1 \\ 0&0&0&0&0&0&1&1 \end{pmatrix}, \quad M_2 = \begin{pmatrix} 1&1&0&1&0&1&1&1 \\ 1&0&1&1&1&1&1&0 \end{pmatrix}$$

In order to compute the matrix representation of the boundary operator $\partial_2 : C_2 \to C_1$ with respect to the standard bases (see Dicarlo et al. 2014 for detail), the edge-face incidence matrix is computed first:

$$[I_{1,2}] = M_1 M_2^t = \begin{pmatrix} 2&1&2&2&1&2&1&1&2&2 \\ 1&2&2&1&2&2&2&2&1&1 \end{pmatrix}^t.$$

where the (i, j) entry, by definition equal to $\langle \mu_i, \lambda_j \rangle = \mu_i(\lambda_j)$, denotes the number of vertices shared by the (elementary) chains $\mu_i \in C_1$ and $\lambda_j \in C_2$.

Let us recall, from Dicarlo et al. (2014), that

$$[\partial_p]_{ij} = \begin{matrix} 1 & \text{if } A_{p-1,p}(i,j) = \sharp\mu_i = 2 \\ 0 & \text{otherwise.} \end{matrix}$$

Hence we get:

$$[\partial_2] = \begin{pmatrix} 1&0&1&1&0&1&0&0&1&1 \\ 0&1&1&0&1&1&1&1&0&0 \end{pmatrix}^t$$

Finally, the boundary 1-chain is computed as the (mod 2) product of $[\,\partial_2\,]$ times the coordinate representation of the *total 2-chain* Λ_2 as element of C_2:

$$[\,\partial_2\,][\Lambda_2] = [\,\partial_2\,]\mathbf{1} \bmod 2 = [1,1,0,1,1,0,1,1,1,1] = \{e_0, e_1, e_3, e_4, e_6, e_7, e_8, e_9\}$$

The result is shown in blue colour in Fig. 1d.

4 Example: Apartment Block Design

A simple example of housing design is discussed in this section. The concept design of the dwelling is produced by using a vectorial drawing program, and shown in Fig. 4. The input file is parsed, producing the LAR `model` given in the script below, as a pair `V, FV` of vertices `V` and 2-cells `FV`.

4.1 LAR Model Input

Input of the dwelling concept The vertices list `V` contains pairs of coordinates within a local reference frame. The 2-cell list `FV` of references to vertices is given counterclockwise, in order to automatically get a complete LAR representation of topology, i.e. the pair `FV, EV` of 2-cells and 1-cells.

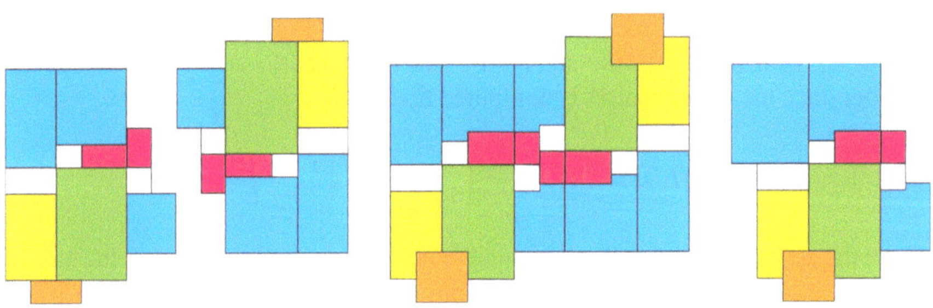

Fig. 4 Concept design: living/eating area (*green/yellow*); bedrooms (*cyan*), lavatories (*magenta*); entrance (*white*)

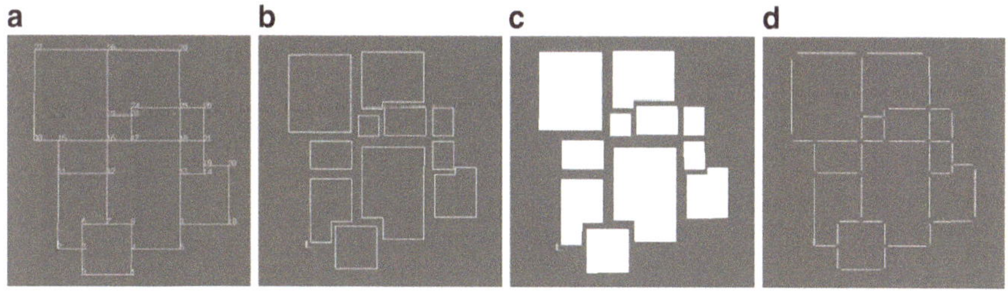

Fig. 5 (**a**) LAR drawing as close polylines; (**b**) exploded polylines; (**c**) exploded 2-cells; (**d**) exploded 1-cells

```
V = [[3,-3],
[9,-3],[0,0],[3,0],[9,0],[15,0],[3,3],[6,3],[9,3],[15,3],[21,3],
[0,9],[6,9],[15,9],[18,9],[0,13],[6,13],[9,13],[15,13],[18,10],
[21,10], [18,13],[6,16],[9,16],[9,17],[15,17],[18,17],[-3,24],[6,
24],[15,24],[-3,13]]
```

```
FV = [
[22,23,24,25,29,28], [15,16,22,28,27,30], [18,21,26,25],
[13,14,19,21,18], [16,17,23,22], [11,12,16,15],
[9,10,20,19,14,13], [2,3,6,7,12,11], [0,1,4,8,7,6,3],
[4,5,9,13,18,17,16,12,7,8],[17,18,25,24,23]]
```

Simple data transformations The LAR model of the `dwelling` is simply given by the pair `V,FV`. Some simple transformations of the input data are given below, and displayed in Fig. 5. The FL operator `AA` stands for *Apply-to-All* (a function to a list of arguments). The `pyplasm` primitive `TEXT` is used to show the enumeration of vertices in Fig. 5a.

```
from myfont import *
dwelling = V,FV
poly = AA(POLYLINE)(lar2polylines(dwelling))
bU = AA(SOLIDIFY)(poly)
EV = face2edge(FV)
VIEW(STRUCT(poly + [T([1,2])(v)(S([1,2])([.1,.1])(TEXT(str(k))))
    for k,v in enumerate(V)]))
VIEW(EXPLODE(1.2,1.2,1)(poly))
VIEW(EXPLODE(1.2,1.2,1)(bU))
VIEW(EXPLODE(1.2,1.2,1)(MKPOLS((V,EV))))
```

4.2 Partitioning the 1-Cells

The subdivision of the 1-cells of the complex, between boundary cells and interior cells, is executed by computing the boundary operator ∂_2, and multiplying it by the coordinate representation **1** of the 2D basis of cells.

Subdivide the 1-cells of the concept plan The input to `bUnit_to_eEiP`, to compute the 1D external envelope and interior partitions starting from 2D from building units, is given below and shown in Figs. 6–8.

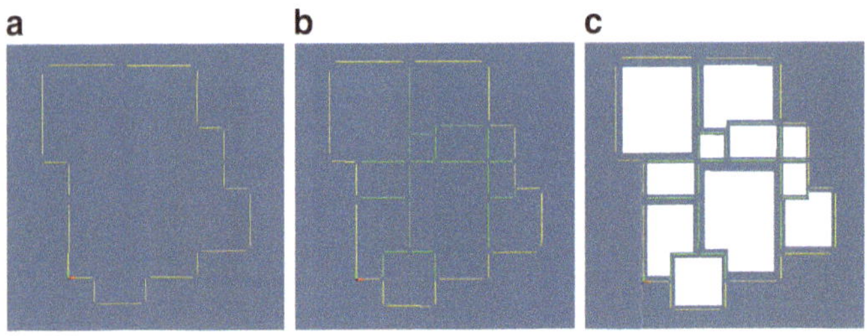

Fig. 6 (**a**) 2-cells, interior 1-chain (*green*), boundary 1-chain (*red*); (**b**) boundary 2-chain; (**c**) interior 2-chain

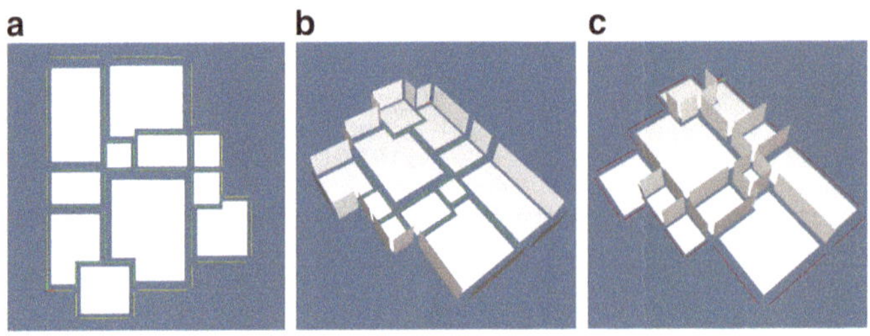

Fig. 7 Partitioning of the 2-chains: (**a**) 2-cells: interior 1-chain (*green*), boundary 1-chain (*yellow*); (**b**) boundary 2-chain; (**c**) interior 2-chain

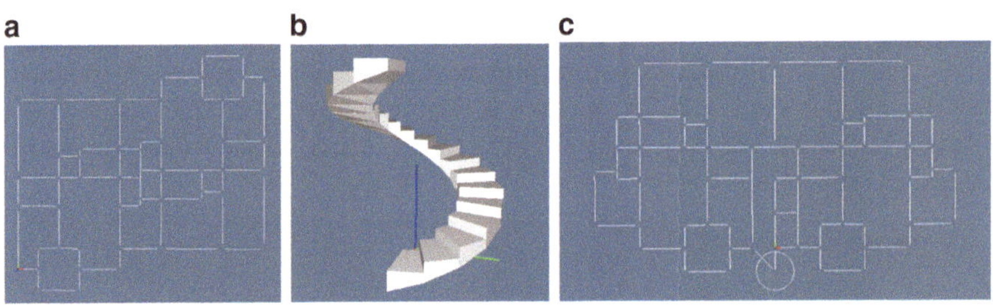

Fig. 8 Concept design: (**a**) aggregation of two building units; (**b**) fully parametric spiral stair; (**c**) the plan of a building flat

```
eE,iP = bUnit_to_eEiP(FV,EV)
eE1D = AA(COLOR(YELLOW))(MKPOLS([V,[EV[e] for e in eE]]))
iP1D = AA(COLOR(GREEN))(MKPOLS([V,[EV[e] for e in iP]]))
VIEW(EXPLODE(1.2,1.2,1)(bU + iP1D + eE1D))
```

Spiral stair A fully parameterised LAR model of a spiral stair is given below. When instancing the `spiralStair` function, the user may specify the `thickness` of the slab, the major and minor radiuses `R` and `r`, the `riser` height of the step, the stair `pitch`, i.e. the vertical distance after a 2π turn, and the number of steps, corresponding to the angular subdivisions of the stair. Notice that

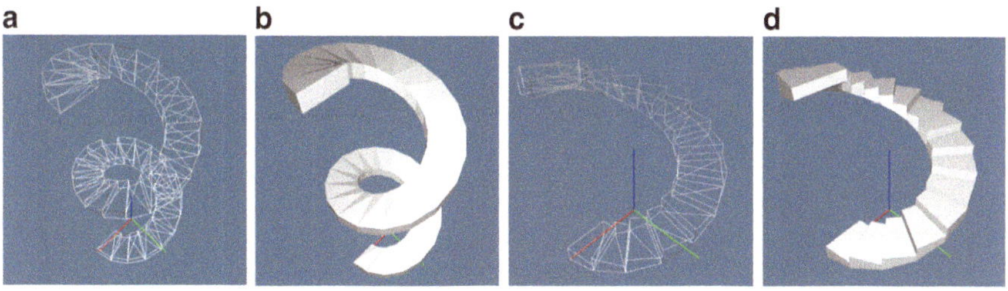

Fig. 9 The construction of the spiral stair; (**a**) wire-frame of two turns of the solid spiraloid; (**b**) image of solid spiraloid, made of polytopal cells; (**c**, **d**) one turn of the spiral stair, obtained by suitable translation of the vertices of the spiralled cells

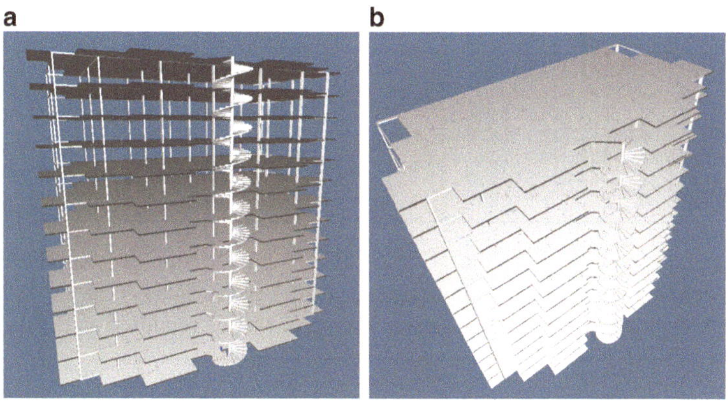

Fig. 10 3D frame of apartment block: (**a**) view from above; (**b**) view from top.

the returned value is a LAR model, i.e. a pair (*vertices, cells*), given by `W,CW` (Figs. 9 and 10).

```
def spiralStair(thickness=0.2,R=1.,r=0.5,riser=0.1,pitch=2.,
                nturns=2.,steps=18):
    V,CV = larSolidHelicoid(thickness,R,r,pitch,nturns,steps)()
    W = CAT([[V[k],V[k+1],V[k+2],V[k+3]]+
        [SUM([V[k+1],[0,0,-riser]]),SUM([V[k+3],[0,0,-riser]])]
        for k,v in enumerate(V[:-4]) if k%4==0])
    for k,w in enumerate(W[:-12]):
        if k%6==0: W[k+1][2]=W[k+10][2]; W[k+3][2]=W[k+11][2]
    nsteps = len(W)/12
    CW =[SUM([[0,1,2,3,6,8,10,11],[6*k]*8])
                for k in range(nsteps)]
    return W,CW
```

Complexes with non-convex and curved cells Since the LAR representation is purely topological, the actual shape of the cells does not matter. Examples of non-linear and non-convex cells are given in Fig. 11b–d.

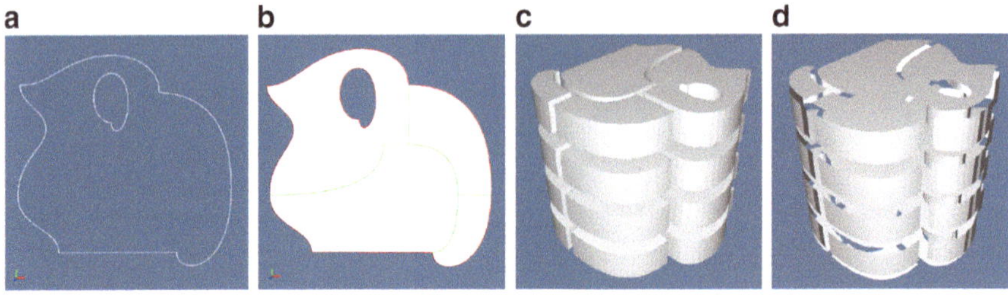

Fig. 11 A cellular complex $B = \Lambda(X) \subset \mathrm{e}^2$, with curved 2-cells non homeomorphic to the 2-ball: (**a**) boundary 1-chain; (**b**) boundary 1-chain (*red*) and interior 1-chain (*green*); (**c**) solid complex $B \times C$, with $C \subset \mathrm{e}$; (**d**) 2D complex $(B \times \partial C) \cup (\partial B \times C)$

```
# input of vertex locations (numbered from zero)
V = [[5.,29.],[17.,29.],[8.,25.],[11.,25.],[14.,25.],[0.,
23.],[5.,23.],[17.,23.],[27.,23.],[0.,20.],[5.,20.],[8.,
19.],[11.,19.],[11.,17.],[14.,17.],[0.,16.],[5.,16.],[14.
,16.],[17.,16.],[23.,16.],[0.,10.],[14.,10.],[23.,10.],[
27.,10.],[0.,6.],[5.,6.],[5.,3.],[20.,3.],[23.,3.],[20.,
0.],[27.,0.]]
```

Edges and faces describe the 2D and 1D cells partitioning the geometry, named here according to their meaning, i.e. FV (faces-by-vertices) and EV (edges-by-vertices), respectively. Edge lists are ordered and will be rendered as Bézier curves.

```
# input of faces as lists of control points indices
FV = [[0,1,2,3,4,5,6,7,9,10,11,12,13,14,15,16,17,18,20,
21],[7,8,18,19,22,23],[17,18,19,20,21,22,24,25,26,27,28],
[22,23,27,28,29,30]]
# input of edges as lists of control points indices
EV = [[5,6,0,1,7],[7,18],[18,17],[17,21,20],[20,15,16,10,
9,5], [12,11,2,3],[3,4,14,13,12],[7,8,23],[22,23],[22,19,
18],[22,28,27],[26,27],[26,25,24,20],[23,30,29,27]]
```

A *canonical* (numerically ordered) description of 2-cells is used for FV, without any 'topological' ordering. Conversely, the control points of Bézier curves must be ordered according to the geometric shape of edges (point-valued polynomials of various degrees).

4.3 Contributions

The use of sparse matrices for representing the topology in Python was independently developed by Bell and Hirani (2012), by implementing the DEC (Discrete Exterior Calculus) introduced by the Hirani (2003) thesis. At the authors' knowledge, LAR is the single approach allowing for a *complete*—in the sense of Requicha (1980)—representation of all curves, surfaces, solids, and higher-dimensional manifolds and non-manifolds, storing only a family of (index) subsets of vertices or control points. The common representations used in BIM, including the parametric representation used by *Revit*, are a derivative of *very* intricate non-manifold data structures used by MCAD and PLM, embedding an explicit graph of relationships and constraints between geometric entities (holes, surfaces, loops, etc.). Conversely, LAR may *compute* every topological relationship and query by a single SpMV (sparse matrix-vector) multiplication implemented in GPGPU.

Of course, in its present state, the LAR system discussed here is only a research environment, and would need both an advanced user-friendly interface, and some software re-engineering and optimisation. With some more work, we believe that it would be possible to show to some architect—reluctant to use computational tools—that just by using a suitable interface, the LAR-ABC calculus of shapes could docilely follow the flow of architect ideas as expressed through the more advanced UIs, providing them a rigorous mathematical expression, and allowing the designer to exploit the full power of modern computational environments.

Results and Conclusions

The main concept developed in this paper is the description of cells of a CW-complex decomposition of a space by the subsets of 0-cells they contain. This choice allowed us to represent a large class of solid models, including spaces formed by assembling curves, surfaces, solids and higher-dimensional cells, under rather weak assumptions on the cell topology and geometry. When connectedness of cells is assumed, the LAR domain coincides with the domain of the SGC representation introduced by Rossignac and O'Connor (1990), allowing cells non contractible to a point. If cells are convex pointsets, then every LAR representation is valid, and requires as input no more than the description of the highest-dimensional cells. Using algebraic methods and mod two (co)homology, we were able to fully compute the model topology, i.e. all the relations of incidence/adjacency between cells, as well as the chain-complex supported by the space partition, through the algebraic product of sparse binary matrices. Finally, our research was also driven by the search for techniques allowing the convergence of methods from graphics, geometric design, solid modeling and computational science, as well as by the search for a fast implementation with advanced tools for parallel computation over distributed and/or accelerated geometric processors. Representing geometry with sparse binary matrices was a major progress on this direction. Both `lar-cc` and `pyplasm` projects are being implemented as opensource software, and can be downloaded from GitHub.

Appendix: Implementation Example

In this appendix we discuss a straightforward implementation of the LAR scheme, and discuss step by step the generation of geometric models in Fig. 11. They should be considered as simple BIM (Building Information Modeling) applications in the

AEC (Architecture, Engineering, Construction) domain. The coding is in Python, which exhibits its full power in rapid prototyping developments. The *Pyplasm* package for *geometric programming* (see Paoluzzi 2003) is used, with higher-level functionals from FL (Backus 1978; Backus et al. 1989, 1990). *Pyplasm* operators are written all-caps.

Shape Input

The model data were entered through an editor for SVG (simple vector graphics), the W3C standard for vector graphics on the web. Several tools exist for interactively editing a SVG shape. We used the free web application *svg-edit* within a browser. Once parsed the input file, the points were scaled by a 1/10 factor and snapped to a grid of integers, to compensate for low precision of the free input tool, that did not provides any snap, finally generating a list of points, given as a list of pairs of floats, named `V`, and two lists of lists of point indices.

```
# input of vertex locations (numbered from zero)
V = [[5.,29.],[17.,29.],[8.,25.],[11.,25.],[14.,25.],
[ 0.,23.], [5.,23.],[17.,23.],[27.,23.],[0.,20.],[5.,
20.],[8.,19.],[11.,19.],[11.,17.],[14.,17.],[0.,16.],
[5.,16.],[14.,16.],[17.,16.],[23.,16.],[0.,10.],[14.,
10.],[23.,10.],[27.,10.],[0.,6.],[5.,6.],[5.,3.],
[20.,3.],[23.,3.],[20.,0.],[27.,0.]]
```

Edges and faces describe the 2D and 1D cells partitioning the geometry, and are named here according to their meaning, i.e. **`FV`** (faces-by-vertices) and **`EV`** (edges-by-vertices), respectively. Edges and faces are given as lists of control points indices. Edge lists are ordered and will be rendered as Bézier curves.

```
# input of edges as lists of control points indices
EV = [[5,6,0,1,7],[7,18],[18,17],[17,21,20],[20,15,16,
10,9,5], [12,11,2,3],[3,4,14,13,12],[7,8,23],[22,23],
[22,19,18],[22,28,27],[26,27],[26,25,24,20],[23,30,29,
27]]

# input of faces as lists of control points indices
FV = [[0,1,2,3,4,5,6,7,9,10,11,12,13,14,15,16,17,18,
20,21],[7,8,18,19,22,23],[17,18,19,20,21,22,24,25,26,
27,28],[22,23,27,28,29,30]]
```

A *canonical* (numerically ordered) description of 2-cells is used here, without any 'topological' ordering. Conversely, the control points of Bézier curves must be ordered according to the geometric shape of edges (point-valued polynomials of various degrees). Of course, this ordering does not affect the topological computations, but determines the geometric embedding of the shape. Notice, from Fig. 11, that the 2-cells are *non convex*, and that one of them is even non contractible to a point.

Kernel Functions

Only a few simple functionalities are required to implement the LAR scheme. As discussed in the paper, **FV** and **EV** are a CSR representation of binary matrices. In particular, kernel functions are needed to: (a) set-up the internal representation of CSR matrix format; (b) display the matrices in a readable way; (c) execute matrix transposition and matrix product operations; (d) update the value of non-zero matrix elements.

Conversion to Standard CSR Format

We use the **sparse** subpackage, provided by the *Scipy* package for scientific computing in Python, to handle sparse matrices efficiently. Faster implementations in OpenCL and WebCL for use from C^{++} and JavaScript, respectively, will be available soon. Presently, the matrix conversion to the internal format is executed by the **csr** function.

Display of Binary CSR Matrices

The display of a binary matrix starting from the CSR format is provided by the **csr2mat** function.

```
# characteristic matrices
print "FV =\n", csr2mat(csr(FV))
>>> FV =
[[11111111011111111111011000000000]
 [00000001100000000001100110000000]
 [00000000000000000011111101111100]
 [00000000000000000000000110001111]]

print "EV =\n", csr2mat(csr(EV))
>>> EV =
[[11000111000000000000000000000000]
 [00000001000000000001000000000000]
 [00000000000000000011000000000000]
 [00000000000000000010011000000000]
 [00000100011000001100010000000000]
 [00110000000011000000000000000000]
 [00011000000001110000000000000000]
 [00000001100000000000000010000000]
 [00000000000000000000000110000000]
 [00000000000000000001100100000000]
 [00000000000000000000000100001100]
 [00000000000000000000000000011000]
 [00000000000000000000010001110000]
 [00000000000000000000000010001011]]
```

CSR Matrix Transposition and Product

Only two basic computational linear algebra operators are used, namely (a) to *transpose* a CSR matrix and (b) to *multiply* two CSR matrices. The computation of the **EF** : $C_2 \to C_1$ operator is just implemented as follows. It is transposed below to **FE** for the sake of space. Notice that the product of binary matrices is an integer matrix, whose entry (i, j) denotes the number of vertices connoting the incidence of the edge j upon the face i. For instance, the first column below tell us that the (curved) edge 0 has 5 control vertices belonging to the face 0, whereas only 1 vertex belongs to face 1.

```
# product and transposition
EF = csrProd(csr(EV), csrTrans(csr(FV)))
FE = csrTrans(EF)
print "FE =\n", csr2mat(FE)
>>> FE =
[[5 2 2 3 6 4 5 1 0 1 0 0 1 0]
 [1 2 1 0 0 0 0 3 2 3 1 0 0 1]
 [0 1 2 3 1 0 0 0 1 3 3 2 4 1]
 [0 0 0 0 0 0 0 1 2 1 3 1 0 4]]
```

Computation of (Co)boundary Operators

The function **maxFilter** is used to get either the matrix $[\partial_2]$ from **EF** or the matrix $[\delta_1]$ from **FE**. Of course, a boundary edge belongs to a face iff all edge vertices belong to (or generate a curved edge of) the face. Hence we search each row of the boundary matrix (column of the coboundary matrix) for its relative maxima element. The reader should compare columnwise the **FE** matrix above with the **coboundary** matrix below.

```
# boundary and coboundary operators
boundary = maxFilter(EF)
coboundary = csrTrans(boundary)
print "coboundary =\n", csr2mat(coboundary)
>>> coboundary =
[[1 1 1 1 1 1 1 0 0 0 0 0 0 0]
 [0 1 0 0 0 0 0 1 1 1 0 0 0 0]
 [0 0 1 1 0 0 0 0 0 1 1 1 1 0]
 [0 0 0 0 0 0 0 0 1 0 1 0 0 1]]
```

Making Solid Cells from b-Reps

The coordinate representation of boundary 1-chains of 2-cells is given by the rows of $[\delta_1]$, i.e. by the rows of the **coboundary** matrix. Let first compute the boundary 1-chain of **_2cells** (from rows of the coboundary matrix). Here numbers denote *edge numerals*.

```
# boundary 1-chains of unit 2-chains
_1chains = format(coboundary,shape="csr")
_2cells = [_1chains[k].tocoo().col.tolist()
    for k in range(len(FV))]
print "_2cells =\n", _2cells
>>> _2cells =
[[0,1,2,3,4,5,6],[1,7,8,9],[2,3,9,10,11,12],[8,10,13]]
```

Notice that **_2cells** is a list representation of boundary 1-chains, whereas **cells2D** below is a list of geometric values, whose "exploded" value is shown in Fig. 2a.

```
# 2D cell complex as a list of polyline structures
cells2D = [ STRUCT([ POLYLINE([V[v]
    for v in EV[edge]]) for edge in cell ])
    for cell in _2cells ]
VIEW(EXPLODE(1.2,1.2,1)(cells2D))
```

A *solid* representation of 2-cells is computed by applying the **SOLIDIFY** algorithm[1] to the piecewise-linear approximation of 1D edges (generated by the **bezier** function) of the face boundary. An exploded view of the 'solid 2-cells' is shown in Fig. 2b.

```
# approximation with 20 segments of a Bezier curve
def bezier(points):
    return MAP(BEZIERCURVE(points))(INTERVALS(1)(20))

# 2D cell complex as a list of solid cells
cells2D = [ SOLIDIFY(STRUCT([ bezier([V[v]
  for v in EV[edge]]) for edge in cell]))
    for cell in _2cells ]
VIEW(EXPLODE(1.2,1.2,1)(cells2D))
```

Boundary and Interior Computation

The *boundary* and *interior* 1-cells of the complex are computed in this section. The CSR rep of the *total 2-chain* (the set of all 2-cells) is put in **_2chain** as a column matrix.

```
# coordinate rep of the largest 2-chain
_2chain = csrTrans(csr([range(len(FV))]))
```

Then the boundary 1-chain, as a list of triples (row, column, 1), is computed by the **csrFilter(1, 0)(csrProd(boundary, _2chain))** expression, and the 1-chain **_1boundarycells** is constructed. A **boundary1D** piecewise-linear approximation of the boundary is generated and displayed in Fig. 2c.

[1]The interested reader may find it on page 615 of Paoluzzi (2003) book.

```
# boundary 1-chain computation
_1boundarycells = [triple[0] for triple
    in csrFilter(1,0)(csrProd(boundary,_2chain))]
print "1-boundary =\n", _1boundarycells
>>>1-boundary =
[1,2,3,4,6,7,8,9,10,11,13,14,15,16,18,19,20]

boundary1D= AA(bezier)([[V[v] for v in EV[e]]
    for e in _1boundarycells ])
VIEW(STRUCT(boundary1D))
```

Finally, the chain **_1interiorcells** is computed as the complement to **_1boundarycells** and displayed in color in Fig. 2d.

```
# computation of interior 1-cells
_1interiorcells = [e for e in range(len(EV))
    if e not in _1boundarycells]
print "1-interior =", _1interiorcells
>>> 1-interior = [0,5,12,17]

interior1D = AA(POLYLINE)([[V[v] for v in EV[e]]
    for e in _1interiorcells])
VIEW(STRUCT(AA(COLOR(RED))(boundary1D) +
    cells2D + AA(COLOR(GREEN))(interior1D)))
```

Assembling the 3D Model

The extruded 3D models of Fig. 3 are computed in this section. A 0-complex and an 1-complex, both embedded in e^1, are needed in order to perform the extrusion of the 2D model as a Cartesian product of pointsets of dimension 2, 1, and 0.

Computation of the 0-Complex

A list **V_0** of five 1D points gives the z-elevation heights. The CSR binary matrix **VV_0** provides the description of the highest dimensional cells (vertices), by vertices itself. Of course, it coincides with the identity matrix 5×5. The complex **floors2D** generated by Cartesian product $\Lambda_2(X) \times \Lambda_0(Y)$ cell pairs is shown in Fig. 3a.

```
# computation of a 0-complex
V_0 = [[0.],[50.],[100.],[150.],[200.]]
VV_0 = AA(LIST)(range(len(V_0)))
print "VV_0 =", VV_0
>>> VV_0 = [[0],[1],[2],[3],[4]]

cells0D = AA(MK)([V_0[v[0]] for v in VV_0])
floors2D = AA(PROD)(CART([cells2D,cells0D]))
VIEW(EXPLODE(1.2,1.2,1.5)(floors2D))
```

Computation of the 1-Complex

Analogously, **EV_1** describes the edges by means of vertices. The matrix of boundary operator **boundary1** := $[\partial_1]$ equates by definition the matrix [**VE_1**], since the max() value of each [**VE_1**]'s row is 1.

```
# computation of a 1-complex
EV_1 = [[0,1],[1,2],[2,3],[3,4]]
VE_1 = csrTrans(csr(EV_1))
print "VE_1 =\n", csr2mat(VE_1)
>>> VE_1 =
[[1 0 0 0]
 [1 1 0 0]
 [0 1 1 0]
 [0 0 1 1]
 [0 0 0 1]]
boundary1 = VE_1  # bydef: max(VE_1[v]) == 1, forall v
```

The function **format** is from *Scipy*'s **sparse** package, and transforms a set of (row,column,value) triples into its internal format (**"csr"**). The **.tocoo().col.tolist()** method composition extracts the columns (edge indices) of each unit 1-chain.

```
# 1D cell complex
chains_1 = csrTrans(boundary1)
chains_1 = format(chains_1,shape="csr")
cells_1 = [chains_1[k].tocoo().col.tolist()
    for k in range(len(EV_1))]
print "cells_1 =", cells_1
>>> cells_1 = [[0,1],[1,2],[2,3],[3,4]]

cells1D = [ POLYLINE([ V_0[v] for v in edge ])
    for edge in cells_1 ]
VIEW(EXPLODE(2,2,2)(cells1D))
```

Cartesian Products

Finally, the sets of cells of various dimensions are generated and variously assembled by the below script, using the *Pyplasm*'s Cartesian product of pointsets. The various complexes (implemented as lists of *Pyplasm*'s geometric values) are displayed exploded in Fig. 10c, d, respectively.

```
# Cartesian product complexes
boundary2D=AA(PROD)(CART([boundary1D,cells1D]))
interior2D=AA(PROD)(CART([interior1D,cells1D]))
VIEW(EXPLODE(1.1,1.1,1.5)(interior2D+floors2D))
VIEW(EXPLODE(1.1,1.1,1.5)(boundary2D+floors2D))
solid3D = AA(PROD)(CART([cells2D,cells1D]))
VIEW(EXPLODE(1.1,1.1,1.5)(solid3D))
```

References

Backus, J.: Can programming be liberated from the von Neumann style?: a functional style and its algebra of programs. Commun. ACM **21**(8), 613–641 (1978)

Backus, J., Williams, J., Wimmers, E., Lucas, P., Aiken, A.: FL language manual, parts 1 and 2. Technical report, IBM Research Report (1989)

Backus, J., Williams, J.H., Wimmers, E.L.: An introduction to the programming language FL. In: Research Topics in Functional Programming, pp. 219–247. Addison-Wesley/Longman Publishing, Boston (1990)

Barber, C.B., Dobkin, D.P., Huhdanpaa, H.: The quickhull algorithm for convex hulls. ACM Trans. Math. Softw. **22**(4), 469–483 (1996)

Bell, N., Garland, M.: Efficient sparse matrix-vector multiplication on CUDA. Technical report NVR-2008-004, NVIDIA Corp (2008)

Bell, N., Hirani, A.N.: PyDEC: software and algorithms for discretization of exterior calculus. ACM Trans. Math. Softw. **39**(1), 3:1–3:41 (2012)

Buluç, A., Gilbert, J.R.: Parallel sparse matrix-matrix multiplication and indexing: implementation and experiments. SIAM J. Sci. Comput. (SISC) **34**(4), 170–191 (2012)

Dicarlo, A., Paoluzzi, A., Shapiro, V.: Linear algebraic representation for topological structures. Comput. Aided Des. **46**, 269–274 (2014)

Hirani, A.N.: Discrete exterior calculus. PhD thesis, Pasadena (2003). AAI3086864

Lokhmotov, A.: Implementing sparse matrix-vector product in OpenCL. In: OpenCL Tutorial, 6th International Conference on High Performance and Embedded Architectures and Compilers (HiPEAC'11), Crete (2012)

Paoluzzi, A.: Geometric Programming for Computer Aided Design. Wiley, Chichester (2003)

Paoluzzi, A., Pascucci, V., Vicentino, M.: Geometric programming: a programming approach to geometric design. ACM Trans. Graph. **14**(3), 266–306 (1995)

Requicha, A.G.: Representations for rigid solids: theory, methods, and systems. ACM Comput. Surv. **12**(4), 437–464 (1980)

Rossignac, J.R., O'Connor, M.A.: SGC: a dimension-independent model for pointsets with internal structures and incomplete boundaries. In: Geometric Modeling for Product Engineering. North-Holland, Amsterdam/New York (1990)

Offset Folding

Alexander Stahr and Hannes Löschke

Abstract Facing a growing world population and a general economic model which focuses on growth on the one hand and on respectively decreasing limited natural resources on the other, the issue of resource efficient construction becomes more and more important in architecture and building industry. It is well known that double curved structures represent one of the most efficient principles of natural design, as shell structures are able to fulfil different requirements with a minimal consumption of material. Implementing large scale spatial curvature in a world full of the linear and flat products inherent to mass-production is one of the most discussed structural topics of the last two decades in architecture and civil engineering. This paper presents a new, structural approach to implementing resource efficient double curved structures using flat panels. The basic idea is to utilize a quasi-double-layer system by folding rhombic quadrangular panels and connecting them with a simple plug-connection which is stiffened by the different spatial orientation of neighboring elements. As a result it is possible to create a wide range of semi-permeable structures which are adaptable double curved surfaces.

1 Introduction

Typically mass produced, semifinished products are linear or flat. They mirror an efficient, linear workflow which is often based on rotating production techniques. The application of those elements directly led to a more or less orthogonal building system.

Of course a lot of products around us show complex geometries. Depending of the production process and the material used they are made of one piece (e.g. a teapot) or of a relatively clear number of complex formed elements (e.g. a car body) made out of formed metal sheet. One thing is uniting all these products: they are produced on mass scale.

A. Stahr (✉) • H. Löschke
Hochschule für Wirtschaft, Technik und Kultur Leipzig, University of Applied Sciences, Karl-Liebknecht-Straße 132, D 04277 Leipzig, Germany
e-mail: stahr@htwk-leipzig.de; hannes.loeschke@hlae.de

P. Block et al. (eds.), *Advances in Architectural Geometry 2014*,
DOI 10.1007/978-3-319-11418-7_24

Fig. 1 Bubble pavilion (*left*), Vela Roof (*middle*), Forst pavilion (*right*)

The design potential of double curved surfaces as well as their efficiency in material usage has been known for a long time; just as the problem of implementing these structures in large scale as individual pieces is – this field is a current topic of scientific discussion and practice-oriented research. In the following three different approaches will be described. Common to all is the basic approach of assembling the structure out of prefabricated elements (Bechthold 2008; Beorkrem 2012; Hensel and Menges 2008).

The Bubble pavilion (Fig. 1, left) designed by Bernhard Franken for BMW at the IAA 1999 in Frankfurt shows at an architectural small scale a strong, curved form. The structural approach bases on the decoupling of cladding and load-bearing system. The overall surface was discretized by projecting a quadrangular grid (in y-z-plane). The framing elements were made out of flat curved cut aluminum elements which were connected by additional angle profiles and screws. The transparent envelope elements are double curved acrylic glass (Schmal and Flagge 2001).

The Vela project (Fig. 1, middle) designed by Massimiliano Fuksas for the new Milan fair in 2004 shows at an architectural large scale a very inhomogeneous, curved form. The structural approach is based on decoupling of cladding and load-bearing system. The overall surface was discretized by mixed approach of projecting a quadrangular grid in less-curved areas and a radial grid in the funnel parts. The framing elements were made out of straight t-shaped steel elements which were connected by cylindrical nodes and screws. The transparent envelope elements are flat safety glass (Schober et al. 2004; Klimke 2004).

The *Forst pavilion* (Fig. 1, right) designed by a team of the Universität Stuttgart led by *Achim Menges* and *Jan Knippers* for the national horticultural show of Baden -Württemberg in Schwäbisch Gmünd in 2004 shows at an architectural small scale a strong, curved form. The project represents an integral construction system. Flat, non-transparent wooden elements forming a shell structure which is the load bearing structure and envelope rolled into one. The overall surface was discretized by a special algorithm which focuses on hexagonal elements which describe a polyhedral structure in final state (Sigmund 2014).

2 The Fold and Plug Approach

All approaches described in the previous chapter aim to achieve completely closed, weather-proof constructions according to different principles of the relation of structure to envelope.

The approach described in the following is based on the idea of a semi-permeable structure which results from the intersection of folded rhombic elements (Fig. 2).

2.1 *Basic Element*

The basic element of the structure is a single folded flat element which has a rhombic outline (Fig. 3). Connecting two opposite corner points of the quadrangle one can identify two diagonals. Reading one of them as an edge it is possible to transform the element from a flat geometry to a spatial one by a single folding process. By notching the outline in a minimum of locations it becomes possible to join adjacent elements in a very simple way (plug-and-stay-connection).

Fig. 2 Basic model showing the principle of the offset-fold-approach

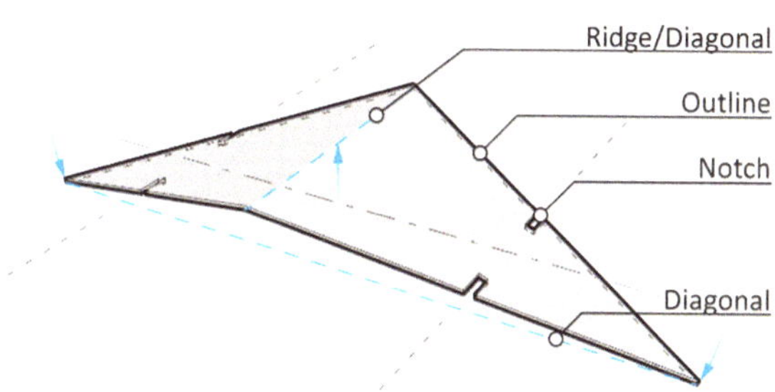

Fig. 3 Basic rhombic element (folded)

Arranging several rhombic elements leads to a surface pattern system which is semi-permeable, much deeper than that of a patterns system which is solely connected along its edges, and much easier to adapt to various geometries. Furthermore it is easy to assemble because there is basically only one point per edge to connect to the adjacent element. Last but not least in case of a reversal folding mechanism it is possible to transport a large amount of elements in a flat collected situation – a big advantage for applications in the field of mobile architecture.

2.2 Arranging the Basic Elements to Create a Areal Structure

The above mentioned simple plug joint leads more or less directly to the question how the basic elements (called *panels* in the following) have to be arranged to determine a structure which is stiff enough to transfer loads on a more basic level to be kinematically determined.

As shown in Fig. 4 the adjustment of panels along two straight directrix arises a structure which is buildable out of identical panels. From the statical viewpoint it represents an undecided equilibrium status which is detectable at the parallel vertical joint-connection lines. The top view shows a rhombic space in between the panels describing a classical "4-hinge-system" which acts kinematically like a worm fence. Such structures are really fragile and not usable for the application in the field of architecture and engineering.

Adjusting the panels along a single curved horizontal directrix and a straight vertical one (Fig. 5) can be interpreted as a deflected flat system which uses the kinematical behavior of a 4-hinge-system. That means that it's also realizable connecting equal panels at each position of the structure. In case of reacting to horizontal loads the system becomes more stable than the flat configuration described before due to the bigger second moment of area and moment of resistance of the footprint figure.

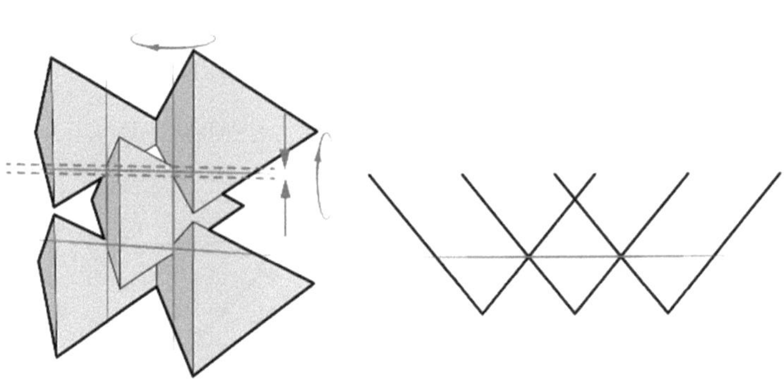

Fig. 4 Increasing structural height by folding out of plane

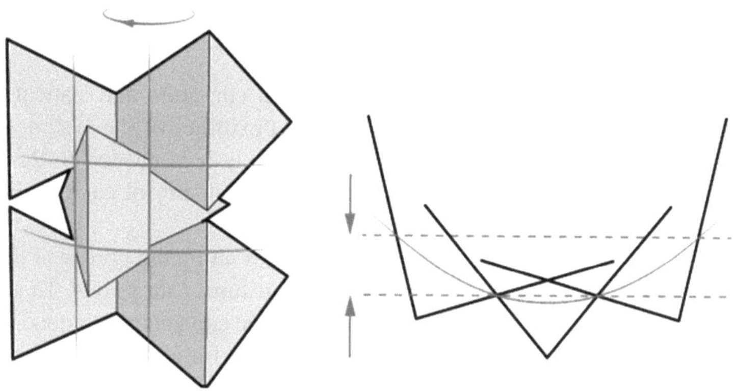

Fig. 5 Increasing structural depth by curving in horizontal plane

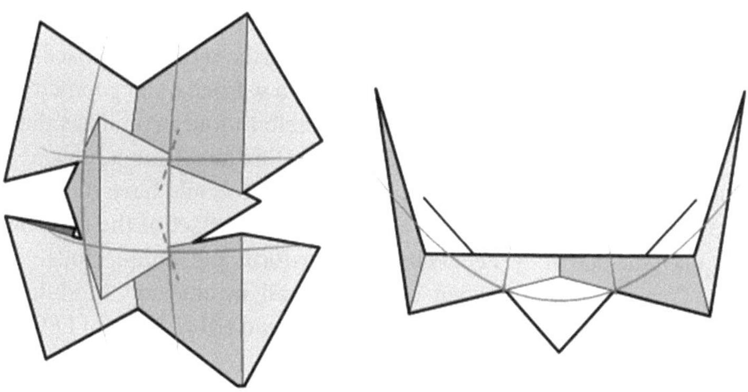

Fig. 6 Intersections of all four notches of a single panel do not line up

Obviously the consistent vertical paths prove to be a problem for the overall structural integrity because they define a system of *linear hinges* between the panels respectively a rotational degree of freedom (cf. Fig. 4; marked orange in the left part of Fig. 5). To Curve that vertical *line of hinges*, eliminates that singularity and transforms the arrangement from a line to a system of skew lines. This greatly improves the performance as the intersections can easily take forces perpendicular to cut direction. The system of curved directrix in horizontal and vertical direction shows Fig. 6.

In case of a variable curvature in both areal directions, the structure isn't buildable out of equal elements. There is no common *hinge axis* or *rotational axis* in the system (orange marked broken line in Fig. 6). Therefore the kinematic mechanism of the planar 4-hing-system is eliminated and the structure becomes much more stiff respectively the character of a real structural system.

2.3 Generating the Folding Line

In front of the above mentioned relations between curvature and statical as well as kinematical behavior on the one hand and the flexibility of the system to adapt it to arbitrary curved surfaces on the other hand side it is needed to develop an "backwards algorithm" which describes the unique geometry of each folded panel based on the local geometry of the surface.

Each basic element with the exception of the ones along the border of the entire structure is connected to four adjacent panels at (in sum) four points. To adapt the structural principle on an arbitrary curved surface, one can generate a quad mesh on top of that and use the mesh nodes as the starting point for generating the geometry of the elements.

Figure 7 shows the situation in case of a planar situation. By defining a middle axis between the two vertical directed *hinge axes* and moving them vertically it is easy to create a folded panel form from a planar situation.

But usually arbitrary four points would not describe a plane in space. Thus it is necessary to describe an alternative way based on a warped quad geometry. For any configuration of opposite linear edges, it is possible to find two planes that include each of the edges and intersect in a third line. This line may form a ridge as in Fig. 7 but this line will be rotated with respect to the line, we would have found by simply moving the middle line (see Fig. 8). To determine the offset of the ridge regarding the basic quad surface it is necessary to transform the unique local curvature parameters. This topic is described in more detail in chapter "Modular Fibrous Morphologies: Computational Design, Simulation and Fabrication of Differentiated Fibre Composite Building Components ".

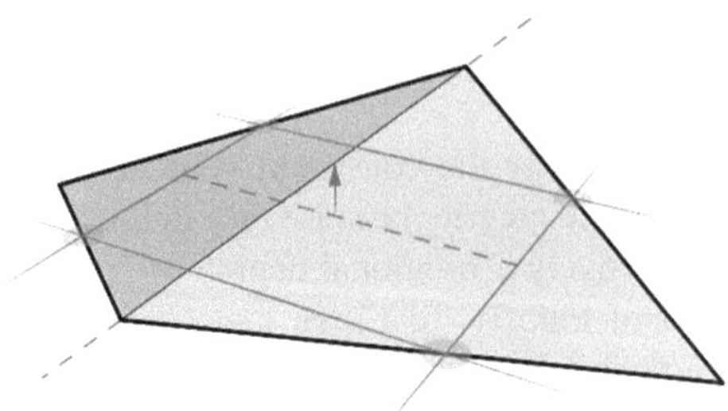

Fig. 7 Regular panel on a planar, rectangular grid

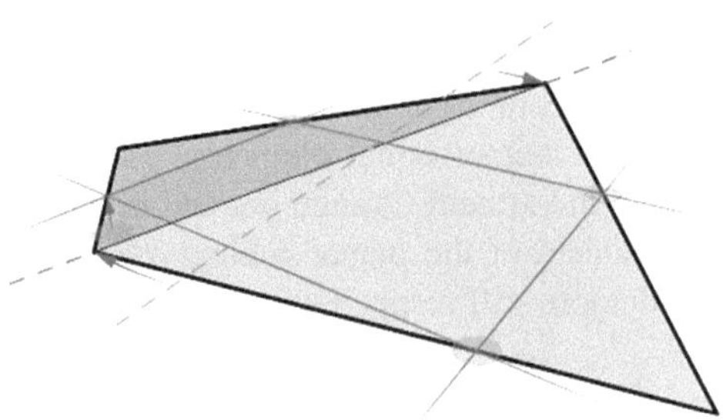

Fig. 8 Panel deformation by moving one corner of the grid out of plane

3 Some Geometrical Tasks in Front of Structural Idea

3.1 Mesh Size and Panel Geometry

Because form and size of panels directly depend on the configuration of the mesh nodes it is necessary to create a homogeneous quad mesh most favorable with a rhythm and density directly connected to the local curvature of the surface.

Figure 9 shows a ribbon structure of variable height which is lifted up partly. Within the arc-like subarea the ribbon is considerably narrower than the rest. The left part is distinctly wider than the right hand one. The overall geometry is characterized by a very inhomogeneous curvature. Red colored parts describe a positive Gaussian curvature, while light blue areas depict negative curvature. Areas shown in green are less curved.

The unique form and size of the folded panels depends on the local curvature of the underlying surface geometry and on the geometry of the individual mesh. Figure 10 shows the connection between these. The left picture describes the location of two "example panels" on the overall grid of a test structure. The middle

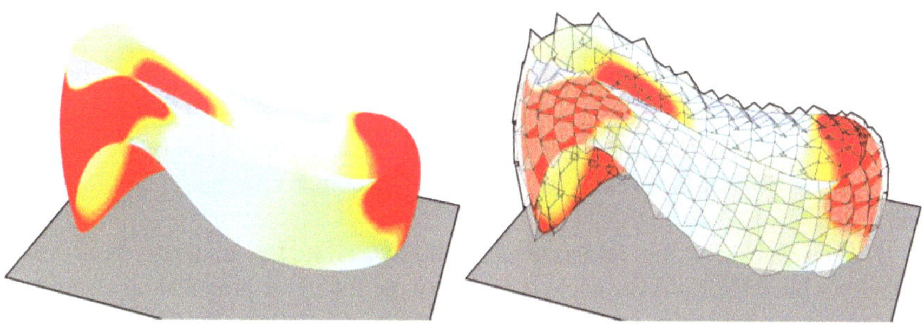

Fig. 9 Surface curvature (*left*) and folded panel structure (*right*)

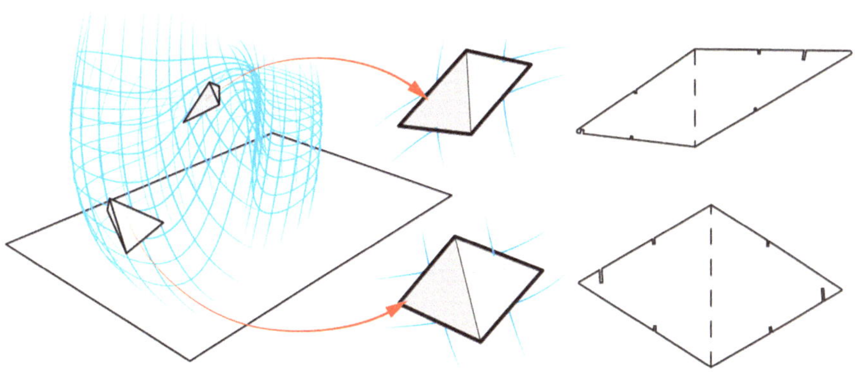

Fig. 10 Panel deformation by mesh proportion

one contains the representation of both in a position parallel to the image plane and the right picture shows the unfolded panel outline.

The basic panel as described in chapter "Post-tensioned Discrete Concrete Elements Developed for Free-form Construction " is generated by lifting a ridge line from a quad mesh described by four corner points. The position of the ridge is determined by a constant offset related to the original surface. This ensures a less degree of folding within small curved sections as well as an increasing offset in areas of stronger curvature.

The second aspect defining the final shape is intersection depth. In order to make the panels intersect after creating them from the corner points of a single mesh, they need to be extended beyond that. The extension and thus the depth of the intersection of two adjacent panels sharing the same grid point is set to a percentage of the cell height. This decision was made mostly for aesthetical reasons and controllability.

The final parameter controlling the shape is the length of the ridge. This length is defined as a percentage of the single mesh height. Using those three parameters, the individual shape of the panel is bound to the local mesh geometry. That makes it possible to realize a wide range of designer's idea using the structural idea of offset folding.

3.2 Folding Angle

Folding means to bend material along a (mostly) straight line. In early design stages paper or cardboard is often used for studying geometry or building small scale mock-ups (cf. Fig. 14).

The thickness of the folded panels and therefore the material thickness is directly related to the constructive detailing of the fold. One practical opportunity to implement the fold is to weaken the material by a partial cut. The depth of the cut defines the "rest-height" of the material. This reduction of the material cross-section decreases the rigidity of the material and reduces the bending stresses. The width of the cut controls the possible folding angle of the panel, whereby the wider the cut is, the greater the angle is that can be folded (Fig. 11).

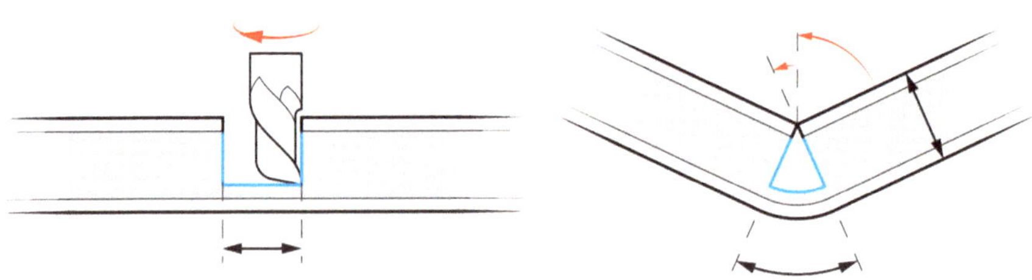

Fig. 11 Defining maximum folding angle by cut width on the inside

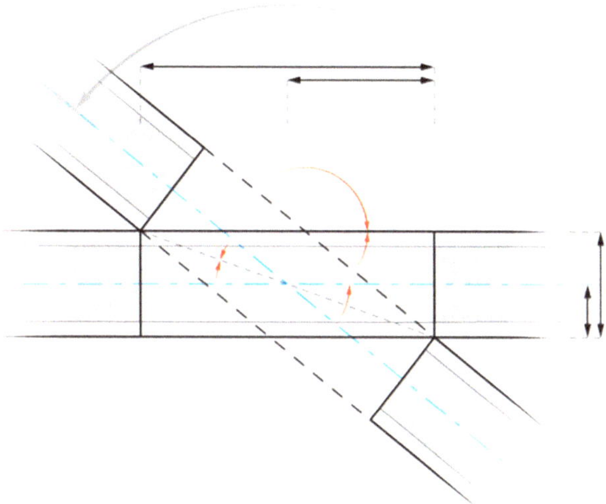

Fig. 12 Stopped hinge function and geometry of the notch connection

3.3 *Double Plug Connection*

At each connection point two flat panels have to be connected. The varying curvature of the master surface with respect to the unique size of each mesh results in differing crossing angles. The constructive challenge is to always connect two panels in a very straightforward yet variable way.

Figure 12 shows a cross-section detail of the intersection of two panels. The rotational axis, which is the line of intersection, is directed perpendicular to the image plane. The width of the cut defines the possible rotating angle (marked by the orange curved arrows) between two panels of a specified thickness. The grey arrow marks the possible rotational movement as well as the final resting position.

The behavior described in the preceding paragraph is of great significance due to the mechanical behavior of the system. Basically each hinged connection of two elements in space has a degree of freedom of 6. That means the connection must be able to transfer forces in all three independent spatial directions with a simultaneous possibility of twisting of elements.

The simple *plug and stay connection* showed in Fig. 3 (and others) is able to restrain translations (and transfer loads) in 2½ directions (with the exception of tension forces along the hinge axes). Moreover the above mentioned *stroke mechanism* limits the twisting of adjacent elements and brings a lot of additional stiffness to the system.

While using rhombic elements like described before another "phenomenon" emerges. Due to the spatial configuration of the panels a double intersection of adjacent elements takes place. This double interlock increases the stiffness

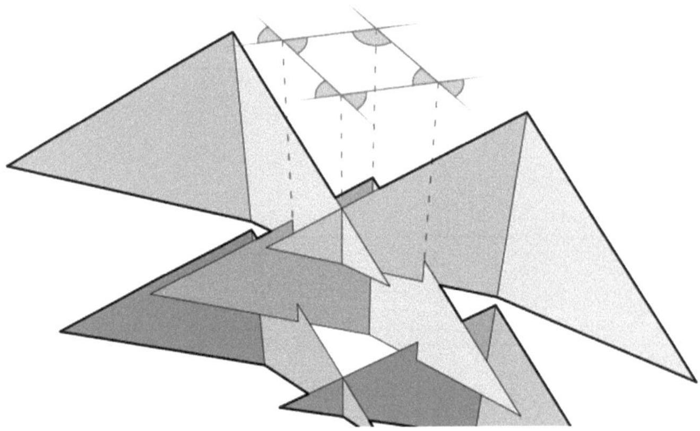

Fig. 13 Double interlock of rhombic panels

of the structure additionally. Figure 13 shows the arising quadrangle and the crossing- angles between the panels and folding-angle at the Backfold of each rhombic panel. Orange sections show the limits for each hinge. The elements planes (shown as blue lines) cannot rotate into those areas while every other angle would be acceptable. As shown, each intersection in itself is free to rotate to a certain degree, yet in combination with all the other edges, they form a stiff quad.

4 The Lizard Cloud Project

To verify the results of the above mentioned research approach a pavilion project was initiated at the HTWK Leipzig. A research team consisting of seven master students of architecture from the HTWK Leipzig started to work on this project in March 2013. Their first task was to design a "medium-scale-pavilion" which would be big enough to walk around inside and would show all the elements of an offset folding structure. Using the basic principle of creating semi-permeable surface structures out of flat folded elements they could immediately begin to design with a hands-on approach using cardboard and scissors (Fig. 14).

Only a short time after finishing the early design phase it quickly became clear, that three main issues have to be solved before the planning process could continue

Find a material which is light and stiff to simplify the handling of the elements during the assembling process.

Find a material which is foldable – and unfoldable! – by hand, also with regard to assembly and disassembly of the pavilion structure.

Fig. 14 Cardboard design model

Fig. 15 Cutting up aluminum top layer

Find a way of weakening a stiff material along a line to avoid expensive additional hinge elements (like e.g. piano hinges) for achieving the back fold.

As the solution to all three issues a 3-layered composite material (Alucobond ®) often used in façade construction was selected for application. The sandwich consists of a layer of aluminum on both sides and a core of polyethylene. It is available in different thicknesses and offers high quality stiffness and strength parameters. By cutting up one of the aluminum top layers it was possible to locally weaken the material and to then precisely cut a semi-rigid linear hinge (See Fig. 15).

Another very important subject of the project was the educational goal of modeling the whole digital workflow from early design phases to fabrication of the unique panels inside the university labs and workshops. This means using available software, scripting transformation of geometric data and production of the panels by means of numerical controlled machines. After a comparatively easy assembling process which took about 6 h (including the assembly of the wooden base plate) the pavilion was presented at a local design fair in Leipzig (Fig. 16).

Fig. 16 The lizard cloud pavilion at the Designers' Open Fair in Leipzig

Conclusion and Further Research

The fold and plug approach, for first time described in this paper shows a geometrically very flexible principle to cover double curved surfaces with a semi-permeable structure. Applications in façade construction, usage for shading structures or in the field of acoustics are conceivable. Double curvature and the in-plane stiffness of the panels is combined to stiffen the fragile structure irrespective simple "plug and stay" hinged connection between the panels.

The implementation of the lizard cloud pavilion shows the potential of the structural approach. To develop the principle and its solution further, the connection between curvature of design surface, size of elements and stiffness of structures needs to be analyzed in much more detail. For shading applications and acoustical demands it would be necessary to know much more about controlling the density of the structure by maintaining the structural cohesion. Furthermore of course the question remains of how to achieve a tension-proof joint which is still pluggable (and possibly also unpluggable). It would be interesting to develop this concept to practical application.

Acknowledgements The authors would like to thank the students Anne Wörfel, Christian Bobsin, Max Schuster, Marius Zwigart, Nathan Ashton and Martin Böttcher for their ideas and diligence as well as their endurance to finish the pavilion in time for the fair. Moreover we would like to thank *3A Composites* (Alucobond) and *Egger* (OSB-boards) for their material support. Special thanks to *lichtraum3* (Torsten Müller) for consulting in the field of lightning design.

References

Bechthold, M.: Innovative Surface Structures – Technologies and Applications. Taylor& Francis, Abingdon/New York (2008)
Beorkrem, C.: Material Strategies in Digital Fabrication, pp. 46–51. Routledge, New York (2012)
Hensel, M., Menges, A. (eds.): Versatility and vicissitude: performance in morpho-ecological design. Archit. Des. **78**(2), 6–11. Wiley Academy, London (2008)
Klimke, H.P.: Structural layout of reticulated free form envelopes. Nouv Mèm Ac Sc de Belgique **2**, 172 (2004)
Schmal, P., Flagge, I.: Digital Real, Blobmeister: First Build Projects. Birkhäuser, Basel/Boston (2001)
Schober, H., Kürschner, K., Jungjohann, H.: Neue Messe Mailand – Netzstruktur und Tragverhalten einer Freiformfläche. Stahlbau **73**, 541–551 (2004)
Sigmund, B.: Robotik im Holzbau: Pavillon aus digital vorgefertigten Einzelteilen. Detail. http://www.detail.de/research/forschung-ent-wicklung/robotik-im-holzbau-pavillon-aus-digital-vorgefertigten-einzelteilen-023047.html (2014)

MIX
Papier aus verantwortungsvollen Quellen
Paper from responsible sources
FSC® C105338

If you have any concerns about our products,
you can contact us on
ProductSafety@springernature.com

In case Publisher is established outside the EU,
the EU authorized representative is:
Springer Nature Customer Service Center GmbH
Europaplatz 3, 69115 Heidelberg, Germany

Printed by Libri Plureos GmbH
in Hamburg, Germany